A Comprehensive Checklist Approach To Selecting and Using

Personal Protective Equipment

Jeffrey O. Stull

Government Institutes
Rockville, MD

Government Institutes, Inc., 4 Research Place, Rockville, Maryland 20850, USA.
Phone: (301) 921-2300
Fax: (301) 921-0373
Email: giinfo@govinst.com
Internet address: http://www.govinst.com

03 5 4 3

The reader should not rely on this publication to address specific questions that apply to a particular set of facts. The author and publisher make no representation or warranty, express or implied, as to the completeness, correctness or utility of the information in this publication. In addition, the author and publisher assume no liability of any kind whatsoever resulting from the use of or reliance upon the contents of this book.

Library of Congress Cataloging-in-Publication Data

Stull, Jeffrey O.
PPE made easy : a comprehensive checklist approach to selecting and using personal protective equipment / by Jeffrey O. Stull.
p. cm.
Includes index.
ISBN: 0-86587-558-8
1. Protective clothing. I. Title.
T55.3.P75S78 1998
620.8'6'0284--dc21
97-52720
CIP

Printed in the United States of America

To Grace and Christina, whose loving support made writing this book worthwhile.

Summary of Contents

Table of Contents

Foreword

People have used clothing and equipment for protection against their environment almost since almost the dawn of civilization. Personal protective equipment (PPE) used in the work environment can be traced back to at least the first century A.D. when workers used cloths as breathing filters and animal bladders as crude respirators for protection against lead and other noxious materials in mining. With the establishment of the industrial age and the First World War came recognition of the need for protective equipment. Nevertheless, the greatest advancements have come more recently. In fact, there has been an impressive technological evolution in the design, testing, and effectiveness of personal protective equipment spanning the last 50 years with the most rapid and significant advances taking place over the last two decades.

One result of this has been an explosion of individual product and brand offerings within existing and new categories of protective equipment. While the equipment options for protecting workers have greatly increased, so have the knowledge and technical sophistication required of the purchaser. In fact, this dilemma extends beyond the issue of simply purchasing the safest, most cost effective and efficient equipment. Current OSHA rules and precedent setting liability cases suggest that all personal protective equipment be purchased and used based upon a thorough evaluation of the hazard to the user and be maintained under a managed program approach. It is in response to this need that Jeff Stull has developed a guide to help in the selection and use of equipment that has been called the "last line of defense" against occupational injury and illness.

Personal protective equipment should not be considered until other methods of control that do not rely on the individual worker (such as design and engineering controls) are proven to be infeasible or impractical. Nevertheless, it is clear that a great number of work assignments will still rely on personal protective equipment to control hazardous work conditions. In fact, due to the establishment of OSHA and EPA requirements for protective equipment in many of their recent standards (e.g., substance standards such as lead and performance standards such as those for emergency response to chemical spills) use of personal protective equipment has increased dramatically. The situation today is that it is rare to find a work environment that does not require the use of personal protective equipment for some segment of the work performed. It is quite likely that this trend will continue.

One of the difficulties facing those who select and purchase protective equipment is the general lack of regulatory or advisory specifications as to what constitutes the best or proper equipment for selection. In other words, the regulations may specify that certain types of protective equipment be used (e.g., eye protection) but not specifically what or which protect. The area best defined is repiratory protection; however, even this area requires a great deal of "professional" judement. This is because each work situation will likely be a little different from the next, hence requirements for personal protective equipment will also be different. In spite of this, the ultimate responsibility for proper selection, use, and maintenance of PPE lies with employers—irrepective of their level of technical knowledge and experience in making the proper selection. This book should be very useful to the reader in helping to understand what protection is needed, the differences in types of equipment, equipment performance, and the development of specifications for equipment purchase.

This book spans the broad range of protective clothing, headwear, eyewear, hearing protection, and respiratory protection. It includes checklists for the completion of required hazard assessments including the identification of physical, chemical, environmental, radiological, and biological hazards. This is followed by a section on person-position and person-equipment hazards and sections on relative risk issues, types and classifications of PPE design, and PPE features and sizing issues. In the sections on the various equipment categories, there is a wealth of information on difficult to find standards and the interpretation of performance testing results. This information is presented in a simple, easy-to-follow, outline format with warnings and helpful hints to purchasers.

Jeff has spent most of his career successfully helping others in the design, testing, selection, use, and maintenance of personal protective equipment. In this book, he shares that wealth of knowledge with the rest of us in an informative and easy-to-use manner. Those of us responsible for the selection, purchasing, or use of PPE and those having to use what will be selected will both be the better for it.

Zack Mansdorf, Ph.D., CIH, CSP, QEP
Beverly, Massachusetts
December 1997

About the Author

Jeffrey O. Stull is President of International Personnel Protection, Inc. which provides expertise on personnel protection to end users and manufacturers in the United States and around the world. Mr. Stull gained much of his experience in the PPE filed when he was President of TRI/Environmental, Inc. in Austin, Texas. In that capacity he was responsible for a company which engaged in industrial hygiene and personal protective equipment testing services. Mr. Stull obtained a Bachelor of Science Degree from the U.S. Coast Guard Academy and Master of Science Degrees in Chemical Engineering from the Georgia Institute of Technology and in Engineering Management from the Catholic University of America. He is a member of the American Society for Testing and Materials F23 Committee on Protective Clothing and D13 Committee on Textiles, the National Fire Protection Association Technical Correlating Committee on Fire and Emergency Medical Services Protective Clothing and Equipment, American Industrial Hygiene Association, and American Institute of Chemical Engineers. Mr. Stull represents the United States in the development of International standards on protective clothing with Technical Committee 94. He has written over 100 articles, chapters, and guides in the area of protective clothing and equipment.

WARNING

This book is an attempt to address several types of personal protective equipment (PPE) for protection against a variety of different workplace hazards. While many areas of protection are covered in detail, not all types of PPE or hazards are described. Organizations or individuals who are responsible for the selection of PPE must follow all applicable federal, state, and local regulations regarding workplace safety, including the consideration of all possible hazards for selecting PPE, be trained and capable of property using selected PPE, and understanding all warnings, instructions, and user information provided by the manufacturer of the selected PPE. Failure to properly select and use PPE may result in serious injury or death.

A NOTE ON STANDARDS

Standards provide a very important tool for the specification of PPE. This book references standards from several organizations. While an attempt has been made to cite the relevant and most current editions of these standards, some standards may have been ovelooked, modified, or withdrawn. Most standards are subject to periodic revision and can significantly change. Users of this information are cautioned to check with the respective responsible standards organizations to determine the current edition and availability of other applicable standards.

Section 1

Introduction to PPE

The purposes of this section are to describe the general roles, uses, and limitations of personal protective equipment (PPE) and to describe the scope and purposes of this book.

1.1 PPE and Its Role in Providing Worker Protection

- ☐ Personal Protective Equipment (PPE) comprises items of clothing and equipment which are used by themselves or in combination with other protective clothing and equipment to isolate the individual wearer from a particular hazard or a number of hazards.
- ☐ PPE can also be used to protect the environment from the individual such as in the case of cleanroom apparel and medical devices for infection control.
- ☐ PPE is considered the "last line of defense" against particular hazards when it is not possible to prevent worker exposure by using engineering or administrative controls.
  - Engineering controls should be first used to eliminate a hazard from the work place by modifying the work environment or process to prevent any contact of workers with the hazard. An example of an engineering control is the replacement of a manual task involving potential hazard exposure with an automated process.
  - In the absence of engineering controls, administrative controls should then be used to prevent worker contact with the hazard. An example of an administrative control is to establish a procedure that dictates that workers are out of an area when the hazards are present.
  - Finally, when neither engineering nor administrative controls are possible, PPE should be used.

☐ Workers should not rely on PPE exclusively for protection against hazards, but should use PPE in conjunction with guards, engineering controls, and sound manufacturing practices in the workplace setting.

☐ While PPE is designed for protection of personnel against various hazards, **PPE cannot provide protection to the wearer against all hazards under all conditions.**

☐ Use of PPE itself may create additional hazards for the wearer including heat stress, reduced mobility, dexterity, and tactility; and impaired vision or hearing.

1.2 Scope and Purpose of This Reference

☐ This reference covers a broad range of PPE, including primary PPE items such as:

- Full and partial body protective garments
- Protective gloves and other handwear
- Protective footwear
- Protective headwear
- Protective face and eyewear
- Respirators
- Hearing protectors or hearing protection devices

☐ Other types of clothing and equipment, often considered as PPE, are described in this reference but not discussed in as much detail as the primary PPE above; these items include:

- Fall protection devices
- Communications equipment
- Personal flotation devices
- Cooling devices

☐ This reference also covers selection and use of PPE with information relevant to several different industries, including:

- Physical protection
- Environmental protection
- Chemical protection

- Biological protection
- Thermal protection
- Electrical protection
- Radiation protection

☐ Table 1-1 (see page 4) provides examples of specific industries in each of the protection areas covered in this reference.

☐ The purposes of this book are:

- To provide a comprehensive source describing a process which can be used to assist end users in selecting appropriate PPE
- To describe different types of PPE in terms of design configurations, features, and sizing
- To discuss different performance properties that can be applied to specifying appropriate PPE
- To indicate other factors which end users should consider in selecting PPE
- To list relevant regulations, standards, and guidelines that can be used in selecting PPE

Table 1-1. Examples of Relevant Industries by Protection Area

Type of Protection	Examples of Relevant Industries
Physical	Construction Meat cutting Forestry/logging operations Mining Machining Material handling Law enforcement
Environmental	Construction Outdoor work Agriculture Emergency response Road maintenance
Chemical	Chemical process industry Oil refining General manufacturing using chemicals Agriculture (pesticide/insecticide handling) Semiconductor manufacturing Pharmaceutical manufacturing Emergency response
Biological	Healthcare Emergency medical operations Agriculture Animal handling and processing Textiles Outdoor work
Thermal	Metal refining and smelting General manufacturing with hot operations Welding Fire fighting
Electrical	Utility work Power generation Construction (electrical) Disaster relief
Radiation	Nuclear power generation Radioisotope work Operations involving lasers Telecommunications

Section 2

Industry Regulations Governing PPE Selection, Use, and Training

The purpose of this section is to list and describe many of the OSHA regulations which apply to personal protective equipment (PPE) selection, use, and training. Canadian and European regulations are also discussed.

2.1 Sources of Regulations and General Regulations on PPE

2.1.1 United States PPE Regulations

- ☐ In the United States, the majority of PPE-related regulations are promulgated by the Occupational Safety and Health Administration (OSHA).
- ☐ Other countries have different responsible
- ☐ OSHA provides regulations that address selection and use of PPE in Title 29, Code of Federal Regulations (CFR) Subpart I, Sections 1910.132 through 1910.140.
    - Section 1910.132 provides general requirements for PPE
    - Section 1910.133 pertains to eye and face protection
    - Section 1910.134 pertains to respiratory protection
    - Section 1910.135 pertains to head protection
    - Section 1910.136 pertains to foot protection
    - Section 1910.137 pertains to electrical protective equipment
    - Section 1910.138 pertains to hand protection.

- Section 1910.139 provides sources of standards on which Sections 1910.132 through 1910.137 are based
- Section 1910.140 provides a list of standards organizations relative to PPE.

☐ Examples of other OSHA requirements for PPE are contained in the following:

- Section 1910.95, paragraphs (i) and (j) pertain to occupational noise exposure and hearing protectors
- Section 1910.120, paragraphs (g)(3) through (g)(5) pertain to use of PPE in hazardous waste operations and emergency response
- Section 1910.156, paragraphs (e) and (f) pertain to specific requirements for PPE worn in fire fighting
- Sections 1910.1000 through 1910.1050 pertain to specific occupational toxic and hazardous substances and address PPE use relative to these substances
- Section 1910.1030 (c)(3) pertains to PPE used for protection against bloodborne pathogens
- Various sections of Part 1926 address use of PPE for the construction industry
- Various sections of Part 1915, 1917, and 1918 address use of PPE for the maritime industry.

☐ Table 2-1 summarizes relevant OSHA standards for PPE selection and use.

☐ States with their own OSHA plans must meet or exceed these requirements. Appendix A lists those states with their own Occupational Safety and Health Plans.

2.1.2 PPE Regulations in Canada and Europe

☐ Canadian regulations requiring the use of PPE exist through the regulations of the individual provinces (Appendix A lists sources for these regulations).

☐ In Europe, there are two specific directives that apply to protective clothing:

- EEC/89/656—Minimum Health and Safety Requirements for the Use by Workers of PPE at the Workplace

Table 2-1. OSHA Personal Protective Equipment Requirements

Section and Title	Personal Protective Equipment (PPE) Referenced in Regulation						
	Protective Clothing	Gloves or Handwear	Footwear	Headwear	Eye or Facewear	Respirators	Hearing Protectors
General Industry Standards							
1910.94 Ventilation (abrasive blasting, spray finishing, and open surface tanks)	X	X	X		X	X	
1910.95 Occupational noise exposure							X
1910.111 Storage and handling of anhydrous ammonia						X	
1910.120 Hazardous waste operations and emergency response	X	X	X	X	X	X	X
1910.132 General PPE requirements	X	X	X	X	X	X	
1910.133 Eye and face protection					X		
1910.134 Respiratory protection						X	
1910.135 Head protection				X			
1910.136 Foot protection			X				
1910.137 Electrical protective equipment	X	X	X				
1910.138 Hand protection		X					
1910.141 Sanitation (wet processes)			X				
1910.156 Fire brigades	X	X	X	X		X	
1910.183 Helicopters				X	X		

Table 2-1. (continued)

Section and Title	Personal Protective Equipment (PPE) Referenced in Regulation						
	Protective Clothing	Gloves or Handwear	Footwear	Headwear	Eye or Facewear	Respirators	Hearing Protectors
General Industry Standards (continued)							
1910.218 Forging machines	X	X			X		
1910.219 Mechanical power transmission apparatus	X						
1910.243 Guarding of portable powered tools				X	X		
1910.252 Welding, cutting and brazing	X	X		X	X	X	
1910.261 Pulp, paper and paperboard mills	X	X	X	X	X	X	X
1910.262 Textiles	X	X		X	X	X	
1910.265 Sawmills	X	X	X		X	X	
1910.266 Pulpwood logging		X	X	X	X	X	
1910.268 Telecommunications	X	X	X	X			
1910.422 Procedures during diving		X					
1910.430 Equipment for diving	X					X	
1910.1001 Asbestos	X	X			X	X	
1910.1003 4-Nitrobiphenyl	X	X	X			X	
1910.1004 alpha-Napthylamine	X	X	X			X	
1910.1006 Methyl chloromethyl ether	X	X	X			X	
1910.1007 3,3'-Dichlorobenzidine	X	X	X			X	
1910.1008 bis-Chloromethyl ether	X	X	X			X	

Table 2-1. (continued)

Section and Title	Personal Protective Equipment (PPE) Referenced in Regulation						
	Protective Clothing	Gloves or Handwear	Footwear	Headwear	Eye or Facewear	Respirators	Hearing Protectors
General Industry Standards (continued)							
1910.1011 4-Aminodiphenyl	X	X	X			X	
1910.1012 Ethyleneimine	X	X	X			X	
1910.1013 beta-Propiolactone	X	X	X			X	
1910.1014 2-Acetylaminofluorene	X	X	X			X	
1910.1015 4-Dimethylaminoazobenzene	X	X	X			X	
1910.1016 N-Nitrosodimethylamine	X	X	X			X	
1910.1017 Vinyl chloride	X					X	
1910.1018 Inorganic arsenic	X	X	X		X	X	
1910.1025 Lead	X	X	X		X	X	
1910.1027 Cadmium	X	X	X		X	X	
1910.1028 Benzene	X	X	X		X	X	
1910.1029 Coke oven emissions	X	X	X		X	X	
1910.1030 Bloodborne pathogens	X	X	X		X		
1910.1043 Cotton dust						X	
1910.1044 1,2-dibromo-3-chloropropane	X	X	X		X	X	
1910.1045 Acrylonitrile	X	X	X		X	X	
1910.1047 Ethylene oxide	X	X			X	X	
1910.1048 Formaldehyde	X	X			X	X	
1910.1050 Methylenedianiline	X	X			X	X	

Table 2-1. (continued)

Section and Title	Personal Protective Equipment (PPE) Referenced in Regulation						
	Protective Clothing	Gloves or Handwear	Footwear	Headwear	Eye or Facewear	Respirators	Hearing Protectors
Construction Standards							
1926.52 Occupational noise exposure							X
1926.55 Gases, vapors, fumes and mists	X						
1926.106 Working over or near water	X						
1926.201 Signaling	X						
1926.300 General requirements (hand and power tools)					X		
1926.303 Abrasive wheels and tools					X		
1926.353 Ventilation and protection in welding, cutting, and heating	X				X		
1926.354 Welding, cutting and heating in way of preservation coatings						X	
1926.400 General requirements (electrical)		X				X	
1926. 403 Battery rooms and battery charging	X	X			X		
1926.551 Helicopters	X			X	X		
1926.605 Marine operations and equipment	X						
1926.650 General protection requirements (excavations, trenching, and shoring)	X	X	X	X	X	X	
1926.800 Tunnels and shafts		X			X	X	
1926.950 General requirements (power transmission and distribution)		X					
1926.951 Tools and protective equipment		X		X			

Table 2-1. (continued)

Section and Title	Personal Protective Equipment (PPE) Referenced in Regulation						
	Protective Clothing	Gloves or Handwear	Footwear	Headwear	Eye or Facewear	Respirators	Hearing Protectors
Maritime Standards							
1915.2 Precautions before entering						X	
1915.32 Toxic cleaning solvents	X	X			X	X	
1915.33 Chemical paint and preservative removers	X	X			X	X	
1915.34 Mechanical paint removers	X	X			X	X	
1915.35 Painting	X	X	X		X	X	
1915.51 Ventilation and protection in welding, cutting, and heating	X			X	X	X	
1915.53 Welding, cutting, and heating in way of preservative coatings						X	
1915.73 Guarding of deck openings and edges	X						
1915.77 Working surfaces	X						
1915.97 Health and sanitation					X		
1915.134 Abrasive wheels					X		
1915.135 Power actuated fastening tools					X		
1915.154 Lifesaving equipment	X						
1917.22 Hazardous cargo	X						
1917.23 Hazardous atmospheres and substances						X	
1917.25 Fumigants, pesticides, insecticides, and hazardous preservatives						X	

Table 2-1. (continued)

Section and Title	Personal Protective Equipment (PPE) Referenced in Regulation						
	Protective Clothing	Gloves or Handwear	Footwear	Headwear	Eye or Facewear	Respirators	Hearing Protectors
Maritime Standards (continued)							
1917.71 Terminals handling intermodal containers or roll-on roll-off operations	X						
1917.73 Terminals handling menhaden and similar species of fish						X	
1917.152 Welding, cutting, and heating (hot work)	X				X	X	
1918.86 Hazardous cargo			X				
1918.93 Ventilation and atmospheric conditions						X	
1918.95 Longshoring operations in the vicinity of repair and maintenance work					X		
1918.96 First aid and life saving equipment	X						
1918.106 Protection against drowning	X						

- EEC/89/686—Council Directive on the Approximation of the Laws of the Member States Related to Personal Protective Equipment

☐ EEC/89/656 addresses employer responsibilities for selection and use of PPE and is similar to OSHA regulations contained in Title 29 of Code of Federal Regulations (CFR) Subpart I for the United States.

☐ EEC/89/686 applies to the manufacture of PPE, including textile-based apparel. The requirements of this directive include:

- PPE must provide general protective qualities balanced against ease of use and comfort.
- PPE shall be classified into one of three types:
  - *Simple design*—PPE where serious injury is not likely and user can judge the level of protection provided
  - *Intermediate design*—PPE where serious injury is possible.
  - *Complex design*—PPE intended to protect against serious irreversible injury or mortal danger
- PPE shall be marked with the CE mark and manufacturer identity.

☐ This type classification of PPE has significance in obtaining CE marking (known as EC Type Examination):

- PPE classified as being simple design are examined for conformance.
- PPE classified as being of intermediate design must be tested and the manufacturer is required to provide a technical file and copies of customer information to ensure compliance. PPE classified as intermediate design are only tested once.
- PPE classified as being of complex design must meet the requirements of intermediate design PPE and manufacturers must demonstrate an ongoing quality system (often through outside monitoring).

☐ European type classification is performed by testing and certification organizations, called "Notified Bodies," which are approved by respective EC countries.

2.2 General Requirements for PPE Use under OSHA 29 CFR 1910.132

☐ PPE includes protective devices for the eyes, face, head, and extremities, protective clothing, respiratory devices, and protective shields and barriers.

- ☐ PPE shall be provided, used and maintained in a sanitary and reliable condition, when necessary for:
  - Process or environment hazards
  - Chemical hazards, radiological hazards, or mechanical irritants capable of causing injury or impairment to any part of the body through absorption, inhalation, or physical contact
- ☐ PPE shall be of safe design and construction for work to be performed.
- ☐ Defective and damaged PPE shall not be used.
- ☐ The employer must ensure the adequacy, proper maintenance, and sanitation of PPE even when owned and provided by employees.

2.3 General Employer Selection Responsibilities under OSHA 29 CFR 1910.132

- ☐ The employer is responsible for conducting a hazard assessment of the work place to determine if hazards requiring PPE are present or are likely to be present.
- ☐ If hazards are present, the employer must:
  - Select and have each affected employee use the types of PPE that will protect the affected employee from the hazards identified in the hazard assessment.
  - Communicate selection decisions to each affected employee.
  - Select PPE that properly fits each affected employee.
- ☐ The employer must verify that the required workplace hazard assessment has been performed through a written certification that:
  - Identifies the workplace evaluated
  - Identifies the person certifying that the evaluation has been performed
  - Identifies the date(s) of the hazard assessment
  - Is itself clearly identified as the documentation of certification of the hazard assessment
- ☐ For selection of PPE for protection against respiratory and electrical hazards, the employer should refer to Section 1910.134 for respiratory protection and Section 1910.137 for electrical protection.

2.4 General Employer Training Responsibilities under OSHA 29 CFR 1910.132

☐ The employer must provide training to each employee who is required to use PPE with each employee instructed in the following:

- When PPE is necessary
- What PPE is necessary
- How to properly don, doff, adjust, and wear PPE
- The limitation of PPE
- The proper care, maintenance, useful life, and disposal of PPE

☐ The employer must have each affected employee demonstrate an understanding of the required training and the ability to use PPE properly before being allowed to perform work requiring the use of PPE.

☐ If the employer has reason to believe that any affected employee who has already been trained does not have the required understanding, then the employer must retrain each such employee. The employer must conduct retraining under circumstances that include, but are not limited to, situations in which:

- Changes in the workplace render previous training obsolete
- Changes in the types of PPE to be used render previous training obsolete
- Inadequacies in an affected employee's knowledge or use of assigned PPE indicate that the employee has not retained the requisite understanding or skill

☐ The employer must verify that each affected employee has received and understood the required training through a written certification that:

- Lists the name of each employee trained
- Indicates the date(s) of training
- Identifies the subject of the certification

☐ For selection of PPE for protection against respiratory and electrical hazards, the employer should refer to Section 1910.134 for respiratory protection and Section 1910.137 for electrical protection.

2.5 Overview of Requirements in OSHA 29 CFR 1910.133 through 1910.140

☐ Section 1910.133 on eye and face protection:

- Requires use of eye and face protection for specific hazards, including:
  - -- Flying particles
  - -- Molten metal
  - -- Liquid chemicals
  - -- Acids or caustic liquids
  - -- Chemical gases or vapors
  - -- Potentially injurious light radiation
- Requires side protection for flying object hazards, provision for prescription lenses, PPE marking, and use of filter lenses for protection against injurious light radiation
- Requires compliance of protective eye and face devices with ANSI Z87.1-1989, *American National Standard Practice for Occupational and Educational Eye and Face Protection (Devices purchased before July 4, 1994 must meet ANSI Z87.1-1968)*

☐ Section 1910.134 on respiratory protection:

NOTE: Section 1910.134 was updated on January 8, 1998.

- Requires use of engineering controls where feasible and use of respirators when necessary to protect employee health against occupational diseases caused by contaminated air
- Requires establishment of a respiratory protective program
- Requires selection of respirators based on an evaluation of respiratory hazards and relevant workplace and user factors which affect respirator performance and reliability
- Requires medical exams for employees who must wear respirators
- Requires fit testing of employees who must wear respirators
- Requires that employers implement procedures for use of respirators that:
  - -- Provide proper facepiece seal protection
  - -- Continuing respirator effectiveness
  - -- Protection in Immediately Dangerous to Life and Health (IDLH) atmospheres

-- Protection during interior structural fire fighting

- Requires specific maintenance and care of respirators, including
  - -- Inspection for defects
  - -- Cleaning and disinfection
  - -- Repair
  - -- Storage
- Requires meeting minimum air quality standards
- Requires identification of respirator filters, cartridges, and canisters
- Requires employers to provide effective training to employees who must use respirators
- Requires evaluation of the workplace to ensure that provisions of the respirator program are being carried out
- Requires recordkeeping of medical examinations and fit testing

☐ Section 1910.135 on head protection:

- Requires use of protective helmets in areas where potential exists for head injury from falling objects
- Requires use of protective helmets designed to reduce electric shock when employees are near exposed electrical conductors that could contact the head
- Requires that protective helmets comply with ANSI Z89.1-1986, *American National Standard for Personnel Protection-Protective Headwear for Industrial Workers-Requirements (Protective helmets purchased before July 5, 1994 must meet ANSI Z89.1-1969: A new edition of this standard exists as ANSI Z89.1-1997)*

☐ Section 1910.136 on foot protection:

- Requires use of protective footwear in areas where potential for foot injury from:
  - -- Falling or rolling objects
  - -- Objects piercing the sole
  - -- Exposure to electrical hazards
- Requires that protective footwear comply with ANSI Z41-1991, *American National Standard for Personal Protection-Protective Footwear (Protective footwear purchased before July 5, 1994 must meet ANSI Z41.1-1967)*

- ☐ Section 1910.137 on electrical protective equipment:
  - Addresses insulating electrical protective equipment made from rubber that includes:
    - -- Blankets
    - -- Mattings
    - -- Covers
    - -- Line hose
    - -- Gloves
    - -- Sleeves
  - Sets specific design requirements for electrical protective equipment that include:
    - -- Manufacture and marking
    - -- Electrical requirements
    - -- Workmanship and finish
  - Sets requirements for in-service care and use for electrical protective equipment
- ☐ Section 1910.138 on hand protection:
  - Requires employers to select and require "appropriate" hand protection when employees' hands are exposed to hazards from:
    - -- Skin absorption of harmful substances
    - -- Severe cuts or lacerations
    - -- Severe abrasions
    - -- Punctures
    - -- Chemical burns
    - -- Thermal burns
    - -- Harmful temperature extremes
  - Requires employers to base selection of hand protection on:
    - -- Performance of hand protection relative to task(s) to be performed
    - -- Conditions present
    - -- Duration of use
    - -- Hazards and potential hazards identified

- ☐ Section 1910.139 provides a list of standards on which the requirements in sections 1910.133 through 1910.138 are based.
- ☐ Section 1910.140 lists the American National Standards Institute (ANSI) as the recognized for obtaining referenced standards.

NOTE: *OSHA 29 CFR 1910.133 through 1910.138 cover PPE in varying levels of detail but do not specifically address all hazards or types of PPE. For example, no specific requirements are provided for overall skin or body protection. This book is intended to provide a more comprehensive list of hazards and types of PPE.*

2.6 Limitations and Shortcomings of Regulations

- ☐ Regulations specifying the selection and use of PPE are often:
  - General in scope,
  - Limited to specific applications, or
  - Do not provide specific guidance.
- ☐ Only in a few areas are regulations specific in recommending particular types of PPE in terms of design, performance properties and service life.
- ☐ The majority of OSHA and other governmental regulations simply specify the use of general types of PPE (e.g., "use protective clothing") without indicating a particular configuration and required performance.
  - The principle exceptions to generic requirements exist when national standards exist, such as in the case of:
    - -- Protective footwear (ANSI Z41.1)
    - -- Protective face and eyewear (ANSI Z87.1)
    - -- Protective headwear (ANSI Z89.1)
    - -- Respirators (ANSI Z88.2 and 42 CFR Part 84)
  - There are no national standards for protective garments or gloves except for very specific applications (e.g., emergency response).
- ☐ Employers and end users are faced with a variety of choices for PPE selection for meeting regulatory requirements and must decide on appropriate PPE design, performance and service life through a risk assessment and determination of protection needs.

Section 3

The Risk Assessment-Based Approach to Selecting PPE

The purposes of this section is describe OSHA requirements for selecting PPE as based on a risk assessment and to provide a specific multi-step approach for carrying out risk assessment-based selection of PPE.

3.1 OSHA Requirements for Performing a Risk Assessment

☐ OSHA 29 CFR 1910.132 requires that employers use a hazard assessment to determine the need for and then to select PPE.

☐ Appendix B to Subpart I of OSHA 29 CFR provides non-mandatory guidelines for conducting hazard assessment and for selecting personal protective equipment; the following subjects are addressed:

- Controlling hazards
- Assessment guidelines
- Selection guidelines
- Fitting the device
- Devices with adjustable features
- Reassessment of hazards
- Selection guidelines for eye and face protection
- Selection guidelines for head protection
- Selection guidelines for foot protection
- Selection guidelines for hand protection
- Cleaning and maintenance

(Appendix B to OSHA 29 CFR Subpart I is repeated as Appendix B in this text)

3.2 Recommended Risk Assessment Approach

3.2.1 Overall Approach

☐ The risk assessment-based approach for selecting PPE in this text uses the following steps:

- Conducting a hazard assessment of the workplace
- Determining the risk of exposure and rank protection needs
- Evaluating available PPE designs, design features, performance, and applications against protection needs
- Specifying appropriate PPE

☐ The overall approach for selecting PPE based on a risk assessment is illustrated in Figure 3-1.

3.2.2 General Approach for Conducting the Hazard Assessment

☐ The hazard assessment consists of the following steps:

- Defining the workplace and tasks to be evaluated
- Identifying the hazards in the workplace
- Determining areas of the body or body systems which are affected the hazards
- Estimating the potential for employee contact with hazards in the workplace
- Estimating the consequences of employee contact with hazards

☐ Section 4 provides general guidelines for conducting a thorough hazard assessment. Sections 5 through 13 assist in the identification of specific hazards.

3.2.3 General Approach for Determining Relative Risk and Establishing Protection Needs

☐ The risk of exposure is determined by the hazard assessment. Protection needs are ranked by the relative risk.

☐ Section 14 provides details on how relative protection needs are established.

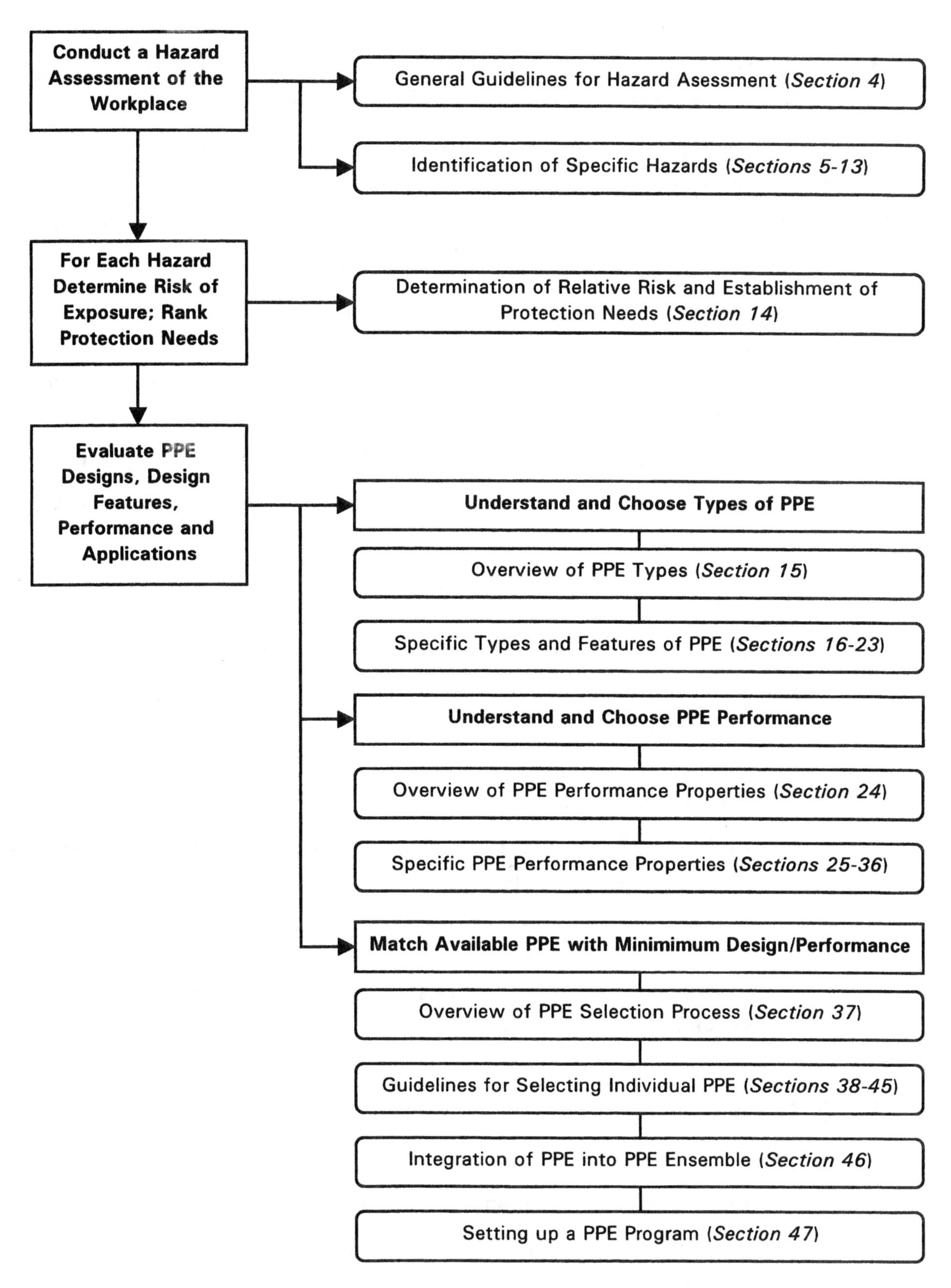

Figure 3-1. Overview of PPE Selection Process

3.2.4 General Approach for Evaluating PPE Designs, Design Features, Performance and Applications

☐ The evaluation of PPE designs, design features, performance, and applications encompasses the following steps:

- Understanding and choosing the types of PPE available for protection:
  - Section 15 provides an overview of PPE types
  - Sections 16 through 23 discuss specific types and features of different PPE
- Understanding and choosing relevant performance properties of PPE to consider during PPE selection:
  - Section 24 provides an overview of PPE performance properties.
  - Sections 25 through 34 provide specific details on different types of PPE performance properties.
  - Section 35 describes PPE service life and estimating life cycle costs.
  - Section 36 discusses PPE labels and user information.
- Selection of PPE by matching available PPE with minimum design and performance needs:
  - Section 37 provides an overview of the selection process on how to match design and performance needs to PPE.
  - Sections 38 to 45 establish specific guidelines for selecting individual types of PPE.
  - Section 46 describes the integration of individual PPE to provide complete protective ensemble.
  - Section 47 provides guidance on setting up a program covering selection, care, and maintenance of PPE.

3.2.5 General Approach for Specifying PPE

☐ Specifying appropriate PPE entails:

- Referencing appropriate standards or
- Developing a comprehensive product design and performance specification,
- Establishing acceptance criteria.

Guidelines for writing PPE specifications are provided generally within Section 37 and for specific PPE items in Sections 38 through 45.

Section 4

Conducting the Risk Assessment

The purpose of this section is describe how to conduct a risk assessment based on the steps of defining the workplace and the tasks to be evaluated, identifying hazards, determining affected body areas, estimating likelihood of exposure, and estimating the consequences of exposure.

4.1 Specific Steps in the Risk Assessment

4.1.1 Define the Specific Workplace and Tasks to Be Evaluated

- ☐ Define each workplace to be evaluated. The workplace should include the area that encompasses the range of hazards that may be encountered.
- ☐ Examples of a workplace can be:
  - The specific work locations for a particular employee
  - A laboratory
  - A part of a production process
- ☐ List the work tasks that are performed at the work place. Work tasks should be defined as those worker activities which:
  - Involve unique hazards
  - Are accomplished by a single individual or group of individuals within a given period of time

4.1.2 Identify Hazards

- ☐ Identify the hazards associated with each work task. Refer to the sections 5 through 13 for assistance in determining which hazards may be present.

☐ Table 4-1 lists general hazard categories that include:

- Environmental hazards
- Physical Hazards
- Environmental Hazards
- Chemical hazards
- Biological hazards
- Thermal hazards
- Electrical hazards
- Radiation hazards
- Person-position hazards
- Person-equipment hazards

4.1.3 Determine Affected Body Area or Body System

☐ For each hazard, determine which portion of the body can be affected by the hazard. Sections 5 through 13 indicate the areas of the body typically affected by specific hazards.

☐ Table 4-1 provided in this section identifies general body areas and body systems that are typically affected by workplace hazards, including:

- Head
- Eyes and face
- Hands
- Arms
- Feet
- Legs
- Trunk or torso
- Entire body
- Respiratory system
- Hearing

Table 4-1. Categories and Ratings for Risk Assessment Model

General Hazard Categories		*Body Areas and Body Systems Potentially Exposed*	
1	Physical hazards	1	Entire body
2	Environmental hazards	2	Trunk or torso
3	Chemical hazards	3	Head
4	Biological hazards	4	Eyes and face (including vision)
5	Thermal hazards	5	Arms
6	Electrical hazards	6	Hands
7	Radiation hazards	7	Legs
8	Person-position hazards	8	Feet
9	Person-equipment hazards	9	Respiratory system
		10	Hearing

Likelihood of Exposure Ratings	
0	Exposure cannot occur
1	Exposure very unlikely
2	Exposure possible, but unlikely
3	Exposure likely
4	Multiple exposures likely
5	Continuous exposure likely

Rating for Consequences of Exposure	
0	No Effect
1	Temporary effect on employee (such as discomfort) with no long term consequences
2	Exposure results in temporary, treatable injury
3	Exposure results in serious injury with loss of work time
4	Exposure results in permanent debilitating injury
5	Exposure results in likely death

4.1.4 Estimate the Likelihood of Employee Exposure to Identified Hazards

☐ For every identified hazard affecting a specific portion of the body (or the whole body), indicate the likelihood of exposure.

☐ Table 4-1 provides a rating scale of 0 to 5 based on both the risk and the frequency of exposure. These ratings include:

- 0 Rating: Exposure cannot occur
- 1 Rating: Exposure very unlikely
- 2 Rating: Exposure possible, but unlikely
- 3 Rating: Exposure likely
- 4 Rating: Multiple exposures likely
- 5 Rating: Continuous exposure likely

4.1.5 Estimate the Possible Consequences of Exposure to Identified Hazards

☐ For every identified hazard affecting a specific portion of the body (or the whole body), indicate the consequences of exposure.

☐ Table 4-1 provided a rating scale of 0 to 5 based on the "worst case" effects on the worker, if exposed. These ratings include:

- 0 Rating: No effect
- 1 Rating: Temporary effect on employee (such as discomfort) with no long-term consequences
- 2 Rating: Exposure results in temporary, treatable injury
- 3 Rating: Exposure results in serious injury with loss of work time
- 4 Rating: Exposure results in permanent debilitating injury
- 5 Rating: Exposure results in likely death

4.2 Sources of Information

☐ The principal source of information for conducting a hazard assessment comes from an inspection of the workplace with actual observation of specific tasks being carried out.

- ☐ Additional information can be obtained by interviewing the affected employee and asking about:
  - The types and frequency of hazards encountered in the task
  - Specific instances in which hazards have been encountered in the past
  - The past effectiveness or ineffectiveness of any PPE used in the task
- ☐ Another source of information is a review of the log and summary of all occupational illnesses and injuries at the workplace. This may be accomplished by examining OSHA No. 200 or equivalent forms.
  - The specific hazard and nature of any accidents or exposures should be evaluated to determine the possible preventative role of using PPE or improving PPE if involved.
- ☐ In some cases, it will be necessary to measure hazard levels using special instrumentation. Examples include:
  - Portable sampling devices to measure airborne concentrations of chemicals
  - Globe, dry and wet bulb thermometers to measure wet bulb globe temperatures
  - Geiger counters to measure radiation levels

4.3 Use of the Hazard Assessment

- ☐ A completed hazard assessment will provide a list of hazards, which parts of the worker body may be affected, how likely exposure will occur, and what the probable consequences of exposure might be.
  - This list can then be used to determine which parts of the body must be protected for which hazards.
  - This list can also be used to set priorities for providing worker protection based on exposure likelihood and possible consequences.
- ☐ Table 4-2 provides a general matrix of hazards versus body areas and body systems that may be affected by these hazards.
- ☐ Section 14 addresses how to determine relative risk and protection needs based on the hazard assessment.

Table 4-2. General Risk Assessment Matrix

Hazard Category	Body Area of Body System Potentially Affected									
	Entire Body	Trunk or Torso	Head	Eyes/ Face?	Arms	Hands	Legs	Feet	Respiratory System	Hearing
Physical										
Environmental										
Chemical										
Biological										
Thermal										
Electrical										
Radiation										
Person-position										
Person-equipment										

Section 5

Identifying Physical Hazards

The purpose of this section is assist in the identification of workplace/task physical hazards. Several categories of physical hazards are discussed, including falling objects, flying debris, projectiles, abrasive or rough surfaces, sharp or jagged edges, pointed objects, slippery surfaces, or excessive vibration.

5.1 Types of Physical Hazards

☐ Physical hazards include those hazards that are mechanical in nature or involve contact with an object that causes harm in some way. General categories of physical hazards include:

- Falling objects
- Flying debris
- Projectiles
- Abrasive or rough surfaces
- Sharp or jagged edges
- Pointed objects
- Slippery surfaces
- Excessive vibration

☐ Depending on the activity and orientation of the worker, some portions of the body may be more likely to be exposed to physical hazards than others.

☐ Table 5-1 summarizes the types of physical hazards, the body areas or body systems affected, and how these hazards are prevented or minimized.

Table 5-1. Overview of Physical Hazards

Physical Hazard	Body Areas Usually Affected	Relevant Occupations	Mitigation Methods
Falling objects	Head Hands Feet	Material handling; Construction; Job involving multiple levels	PPE which resists impact/attenuates transmitted energy
Flying debris	Face and eyes Respiratory system	Abrasive wheel grinding Sandblasting Machine cutting operations	Enclosed chambers or guards on equipment; Puncture/impact-resistant PPE
Projectile/ ballistic	Entire body	Law enforcement Military Hunting Security services	PPE which resists impact, absorbs and dissipates energy
Abrasive/rough surfaces	Hands Arms Feet Legs	Abrasive wheel grinding; Jobs on rough surfaces; Jobs with moving belts or machinery	Equipment guards; PPE reinforcement and abrasion-resistant materials
Sharp edges	Hands Arms Feet Legs	Jobs with sharp tools; Material handling with glass or metal objects; Forestry; Construction Law enforcement Emergency response	Automated equipment; Equipment guards; Cut-resistant PPE
Pointed objects	Hands Feet	Jobs with pointed tools, syringes, or involving drills or punches; Law enforcement Emergency response	Automated equipment; Equipment guards; Puncture-resistant PPE
Slippery surfaces	Hands Feet	Wet operations; Work on smooth surfaces; Emergency response	Non-skid work surfaces; Slip-resistant footwear; Good "grip" gloves
Excessive vibration	Entire body; Hands	Driving vehicles Operation of large machines Use of power towers	Modification of vibration frequency; Reduction of exposure time; PPE which attenuates vibration energy

5.2 Falling Object Exposures

- ☐ Falling object hazards typically involve heavy objects falling from above the worker onto his or her head, hands, or feet, or heavy objects falling from a work surface onto the worker's feet.
- ☐ Falling object hazards are most common in the following activities:
  - Material handling and lifting
  - Construction work
  - Tasks involving different work levels
- ☐ In the absence of engineering or administrative controls, protection from falling objects is best provided by providing PPE that resists impact and attenuates the imparted energy.

5.3 Flying Debris Exposures

- ☐ Flying debris hazards typically involve very small objects that are subject to forced air circulation through machinery or other unnatural means.
- ☐ While any part of the body may be struck by flying debris, the face and eyes are the most susceptible to exposure. Flying debris may also present a significant hazard to the respiratory system in conjunction with other hazards.
- ☐ Flying debris hazards are most common in the following activities:
  - Abrasive wheel grinding
  - Sandblasting
  - Machine cutting operations
- ☐ Reducing exposure to flying debris can be done by:
  - Using enclosed chambers or guards (shields) for equipment or processes that produce flying debris
  - Changing to processes that create less debris
  - Using protective clothing that protects against flying debris

5.4 Projectile/Ballistic Exposures

- ☐ Hazards from projectiles primarily include exposure to bullets or other high-speed objects such as fragments from explosions.
  - The sudden pressure forces from an explosion also constitute a hazard.

- All portions of the body are at risk, but the head and torso are considered most vulnerable.
- The worker's hearing can also be affected from the overpressure of the explosion.

☐ The relative hazard is based on the size and shape of the projectile, its velocity, and the angle of its impact.

- Most bullet projectiles weigh from a few grains to several ounces.
- Velocities of projectiles can reach several thousand feet per second.
- An item of PPE may be able to stop projectile at one velocity but the same projectile can penetrate at a higher velocity.

☐ Applications where projectile hazards exist include:

- Law enforcement
- Military
- Hunting
- Security services

☐ The principal types of protection used are vests or similar garments that are constructed of materials capable of absorbing and dissipating the energy of the impacting projectile.

☐ Anti-fragmentation clothing is used in some special applications.

5.5 Abrasive or Rough Surface Exposures

☐ Abrasive or rough surface hazards typically involve potential for worker contact with a surface that will abrade or lacerate the skin either by worker movement or by movement of the rough surface.

☐ Hands, feet, arms, and legs are most likely to be affected.

☐ Abrasive or rough surface hazards are most common in the following activities:

- Abrasive wheel grinding
- Tasks performed on asphalt or other rough surfaces
- Tasks involving moving belts or machinery

☐ Minimization of abrasive or rough surface hazards can be best achieved by:

- Installing automated equipment

- Reducing operations involving abrasives or rough surfaces
- Providing guards on equipment
- Using reinforcements on areas of protective clothing subject to repeated abrasive or rough surface contact

5.6 Sharp Edge Exposures

- ☐ Sharp edge hazards typically involve sharp tools/machinery or materials that have a cut or sharp edge that moves across the body, usually in a slicing action. The working environment may also contain sharp edges in the form of rough edges or unprotected machinery.
- ☐ Most sharp edge exposures involve the worker's use of his or her hands; arms, feet, or legs may also be affected by this hazard.
- ☐ Sharp edge hazards are most common in the following activities:
  - Any kind of cutting operation (e.g. metal working, forestry)
  - Material handling involving metal, glass, or other objects with sharp edges
  - Construction work
  - Law enforcement
  - Emergency response
- ☐ Minimization of sharp edge hazards can be best achieved by:
  - Installing automated equipment
  - Reducing operating involving sharp edges
  - Providing guards on equipment
  - Using cut-resistant protective clothing

5.7 Pointed Object Exposures

- ☐ Pointed object hazards involve a sharp point, which is capable of puncturing PPE or the worker.
- ☐ Pointed object exposures differ from sharp edge exposures by the way that the force is applied behind the object (puncturing versus slicing action).
- ☐ Hands and feet are the most likely portions of the body to be affected by pointed object hazards.

☐ Pointed object hazards are most common in the following activities:

- Tasks involving the use sharp, pointed tools
- Tasks involving handling of syringes
- Tasks involving use of machinery such as punches or drills
- Law enforcement
- Emergency response

☐ Minimization of pointed object hazards can be best achieved by:

- Installing automated equipment
- Reducing operations involving pointed objects
- Providing guards on equipment
- Using puncture-resistant protective clothing

5.8 Slippery Surface Exposures

☐ Slippery surface hazards are typically presented by handling of smooth objects or walking surfaces that may also be wet or covered with some other liquid.

☐ Hand grip and foot traction are generally affected by slippery surface hazards.

☐ Hands and feet are the most likely portions of the body to be affected by slippery surface hazards.

☐ Slippery surface hazards are most common in the following activities:

- Wet operations
- Handling wet or smooth objects
- Walking on smooth surfaces which may become wet
- Emergency response

☐ Protection from slippery surfaces can be addressed:

- Providing non-skid work surfaces
- Using protective gloves with improved grip
- Using slip-resistant footwear which has a high level of traction

5.9 Excessive Vibration Exposures

- ☐ Excessive vibration hazards exist from exposure to vibrating machinery in the absence of any cushioning or energy absorbing materials.

- ☐ Worker tolerance to vibration is defined as the level of vibration at which effects on performance of a motor or visual task are detectable. Worker tolerance to vibration is affected by the vibration's:

  - Frequency
  - Acceleration
  - Duration

- ☐ Worker tolerance to vibration in terms of duration generally decreases with increasing acceleration.

  - Whole body tolerance to vibration is lowest for vibrations of 4 to 8 Hertz.
  - For hands, vibrations at frequencies from 8 to 500 Hertz are of concern.

- ☐ Hands and the whole body are most often affected by excessive vibration.

- ☐ Repeated or prolonged exposure to excessive vibration can cause Cumulative Trauma Disorders (CTDs).

  - Vibration is also known to cause other whole body physiological effects such as:

    -- Abdominal pain

    -- Loss of equilibrium

    -- Nausea

    -- Muscle contractions

    -- Chest pain

    -- Shortness of breath

    -- Excessive vibration for hands causes:

    -- Stiffness

    -- Numbness

    -- Pain

    -- Loss of strength

- ☐ The most common sources of excessive vibration include:

- Whole body vibration associated with prolonged driving of trucks and other vehicles, especially those with poor suspension or when traveling on rough surfaces
- Whole body vibration associated with the operation of large production machinery, such as presses
- Segmental vibration associated with the operation of power tools, such as chipping hammers, chain saws, or grinders

☐ The American Conference of Governmental Industrial Hygienists (ACGIH) establishes Threshold Limit Values (TLVs) for whole body and hand-arm (segmental) vibration in its annual publication, *Threshold Limit Values for Chemical Substances and Physical Agents and Biological Exposure Indices*.

- TLVs are expressed in frequency, weight, component accelerations, or gravitational units based on the daily duration exposure

☐ Protection against excessive vibration is met by:

- Substituting automated processes that remove the workers from areas of excessive vibration
- Modifying the frequency of vibration source
- Decreasing the duration of exposure
- Using protective clothing with materials which attenuate the energy of the vibration source

Section 6

Identifying Environmental Hazards

The purpose of this section is assist in the identification of workplace/task environmental hazards. Several categories of environmental hazards are discussed, including high heat and humidity, cold temperatures, wetness, insufficient or extreme light, and excessive noise.

6.1 Types of Environmental Hazards

- ☐ Environmental hazards are encountered in the work environment, typically from existence of extreme conditions.
- ☐ Environmental hazards include:
  - Exposure to high heat and humidity conditions (not extreme heat or thermal hazards)
  - Exposure to cold temperatures
  - Exposure to wetness
  - Exposure to bright light
  - Exposure to excessive noise
- ☐ Table 6-1 summarizes the types of environmental hazards, the body areas or body systems affected, and how these hazards are prevented or minimized.

6.2 High Heat and Humidity Exposures

- ☐ Exposure to high temperature and humidity encompasses worker exposure to temperatures and relative humidities that may be either natural or unnatural (such as those encountered in some industrial processes).

NOTE: *Excessively hot temperatures associated with high heat and flame exposures are covered under Section 9, Thermal Hazards.*

Table 6-1. Overview of Environmental Hazards

Environmental Hazard	Body Areas Usually Affected	Relevant Occupations	Mitigation Methods
High heat/humidity	Entire body	Outdoor work; Industrial operations producing heat (smelting, drying, steam cleaning, ironing, molding, chemical processes); Emergency response	Provide shelter or air conditioning; Allow worker acclimization to environment; Control work/rest cycles; Provide appropriate worker rehabilitation to heat exposure; Use cooling devices in conjunction with PPE; Use 'breathable' PPE
Ambient cold	Entire body Face Hands	Outdoor work; Refrigeration operations; Construction in cold climates; Emergency response	Providing heating; Limit cold exposure times; Use insulated PPE
Wetness	Entire body Hands	Outdoor work; Cafeteria work; Plumbing; Cleaning; Emergency response; Underwater construction	Provide shelter; Use water-resistant PPE
High wind	Entire body	Outdoor work; Emergency response	Provide shelter; Use insulated PPE
Insufficient or bright light	Eyes (vision)	Outdoor work; Emergency response; Jobs with excessive or insufficient illumination	Provide adequate lighting; Use anti-glare surfaces; Provide eyewear with shaded lenses
Excessive noise	Hearing	Construction; Work around heavy machinery; Use of power tools	Limit workplace noise through engineering controls; Use ear protectors

☐ Common significant heat and humidity exposures include:

- Work in buildings which are not air conditioned or work outside during summer months in warm climates, especially if heavy physical effort is required
- Work in the following industrial operations that produce heat:
  - Smelting
  - Drying
  - Boiler cleaning and maintenance
  - Steam cleaning
  - Ironing
  - Extruding and molding plastics
  - Chemical manufacturing involving heat-producing reactions

6.2.1 Body Heat Balance

☐ Exposure to temperature extremes affects the worker's body heat balance.

☐ Body temperature must be maintained within a narrow range for worker comfort and safety.

☐ The body heat balance is affected by:

- Metabolic heat gain from the physical work load on the worker
- Convective heat gain or loss represented by the air velocity and the difference in temperature between the air and the person's skin
- Radiative heat gain or loss resulting from the difference between the person's skin temperature and the temperature of surfaces in the environment
- Evaporative heat loss as determined by the difference between the water vapor pressure of a person's skin and the relative humidity of the environment as well as air velocity
- Conductive heat gain or loss based on the difference of the person's skin temperature and the temperature of surfaces contacted

6.2.2 Types of Heat Stress

- ☐ When rises in body temperature cannot be offset by loss of body heat through other means, heat stress may occur. This includes the following heat-related disorders:
  - heat rash
  - heat cramps
  - heat exhaustion
  - heat stroke
- ☐ **Heat rash** may result from continuous exposure to heat or humid air.
- ☐ **Heat cramps** are caused by heavy sweating with inadequate electrolyte replacement. Signs and symptoms include:
  - Muscle spasms
  - Pain in the hands, feet, and abdomen
- ☐ **Heat exhaustion** occurs from increased stress on various body organs including inadequate blood circulation due to cardiovascular insufficiency or dehydration. Signs and symptoms include:
  - Pale, cool, moist skin
  - Heavy sweating
  - Dizziness
  - Nausea
  - Fainting
- ☐ **Heat stroke** is the most serious form of heat stress. Temperature regulation fails and the body temperature rises to critical levels. Immediate action must be taken to cool the body before serious injury or death occurs. Competent medical help must be obtained. Signs and symptoms are:
  - Red, hot, usually dry skin
  - Lack of or reduced perspiration
  - Nausea
  - Dizziness and confusion
  - Strong, rapid pulse
  - Coma

6.2.3 Factors Affecting Heat Stress

☐ The risk for heat stress increases with:

- Higher temperature and relative humidity; several indices have been established regarding safe working limits in different environmental conditions.
- Increased level of work activity; higher work activities produce earlier onset of heat-stress-related symptoms for given environmental conditions and work/rest cycles.
- Increasing amounts of clothing, particularly heavier clothing or clothing which prevents moisture vapor transfer

☐ Individuals vary in their susceptibility to heat stress. Factors that affect individual susceptibility include:

- Physical condition (more physically fit persons general have better tolerance)
- Level of acclimization to work environment
- Age
- Gender
- Weight

☐ The American Conference of Governmental Industrial Hygienists (ACGIH) establishes Threshold Limit Values (TLVs) for heat exposure in its annual publication, *Threshold Limit Values for Chemical Substances and Physical Agents and Biological Exposure Indices*.

- TLVs are expressed as the maximum Wet Bulb Globe Temperature (WBGT) of the work environment for a given work load and work-rest regimen.
- TLVs must be corrected for different types of clothing.
- Other specific guidelines are offered for minimizing exposure to hot conditions.

☐ The potential for heat stress is best managed by:

- Modifying the work environment (providing shade, adding air ventilation or air conditioning, moving work to a controlled indoor environment)

- Modifying the task (reducing the level of effort required by employing automation or special tools or by introducing appropriate work/rest cycles)
- Employing protective clothing that provides the maximum achievable breathability
- Using cooling devices, such as static ice vests, supplied air ventilation, and circulating water cooling systems, in conjunction with other protective clothing
- Improving the physical conditioning of the individual

6.3 Exposure to Cold Temperatures

☐ Exposure to cold temperatures can occur in environments that are either natural or unnatural (such as those encountered in some industrial processes).

☐ Exposure to cold can occur in the following industrial situations:

- Refrigeration or cold storage work
- Construction in poorly heated buildings or outdoors during the winter months in cold climates
- Outdoor maintenance and service work in the winter months in cold climates
- Cleaning with cold water

☐ The major concern about whole body exposure to the cold is the development of serious hypothermia and subsequent death from exposure.

- The body defends core temperature by intense shivering to increase metabolic heat.
- Exhaustion of this resource for generating heat is implied when body temperature falls below 95°F (35°C).
- Frostbite of the face or extremities may result from exposure to extreme cold, often in combination with high air velocities, or from prolonged exposure to less severe cold but with high humidity.
- Exposures may result in cold injury to exposed flesh at equivalent temperatures of -32°F (-35 °C).
- For individuals working outdoors in the cold, body heat losses associated with high winds can be very significant.

- ☐ Local cold discomfort, most often in the hands, feet, and face, is usually the major cause of complaints in the cold discomfort zone.
  - The hands begin to exhibit some loss of flexibility and manipulation skills at ambient dry bulb temperatures of 60°F (16°C) over a few hours of exposure.
  - A 20% decrement in performance is not unusual in manual tasks at ambient temperatures of 7°F (-14°C), dry bulb.
  - Extended exposure to cold conditions leads to frostbite that, in turn, can lead to permanent injuries if left untreated.
- ☐ Susceptibility to cold is primarily a function of:
  - Environmental conditions
    - Temperature
    - Humidity
    - Wind velocity
  - Length of exposure
  - Level of work activity
  - Amount of clothing worn
- ☐ The American Conference of Governmental Industrial Hygienists (ACGIH) establishes Threshold Limit Values (TLVs) for cold exposure in its annual publication, *Threshold Limit Values for Chemical Substances and Physical Agents and Biological Exposure Indices*.
  - TLVs are expressed as maximum work period and the number of necessary breaks based on air temperature and the wind velocity.
  - Other specific guidelines are offered for minimizing exposure to cold temperatures.
- ☐ In the absence of engineering or administrative controls, protection from cold is best achieved through the addition of clothing with several layers of insulation.

6.4 Wetness Exposures

- ☐ Some tasks include exposure to water temperatures that range from comfortable to very uncomfortable.

- ☐ Water may be contacted in the process of doing a production, maintenance, or clean-up task, such as the following:
  - Cafeteria work, including operation of dishwashers
  - Chemical-making operations
  - Building cleaning
  - Plumbing, heating, and refrigeration work
  - Hosing-down tasks or cleaning of facilities and equipment
  - Outside work on rainy or snowy days
  - Underwater construction work
- ☐ In most industries, water contact is confined to the extremities, except for work in inclement weather.
- ☐ The effect of working in wet clothing on tolerance times for high and low temperatures is strongly influenced by:
  - Duration of exposure
  - Area of contact of the skin
  - Level of work activity
  - Air velocity
- ☐ Wetness may reduce the insulative value of clothing, so that in cold temperatures, skin temperature is reduced and heat loss from the body is increased.
- ☐ When there is airflow, water contacting the skin and clothing in hot environments helps increase evaporative heat loss without sweating. Thus worker tolerance is higher in the heat when the clothing is wet than when it is dry.
- ☐ Loss of finger flexibility and tissue injury may result when the hands are safely immersed in hot or cold water. Temperatures of liquids in which a worker may immerse his or her hands are seldom below 34°F (1°C) or above 158°F (70°C).
- ☐ Protection from wetness requires protective clothing that prevents the penetration of water under the conditions of use and for the expected exposure.

6.5 High Wind

- ☐ High winds pose hazards by disrupting tasks or making it difficult for works to function; high winds may also be associated with wetness (*see 6.4*) or flying debris (*see 5.3*).
- ☐ High winds are most likely to affect the whole body or face and eyes when exposed.
- ☐ High winds may be encountered in outdoor operations or emergency response during poor weather (thunderstorms, tornados, and hurricanes).
- ☐ Protection against high winds can be achieved:
  - Limiting work activity in adverse conditions
  - Using rugged shelter
  - Wearing barrier protective clothing (such as wind breakers or rain suits)
  - Wearing eye and face protection to prevent debris from blowing into eyes.

6.6 Insufficient or Extreme Light Exposures

- ☐ Sufficient light is needed to carry out most tasks. When there is insufficient illumination for specific work functions, task efficiency may be affected and workers may be subjected to eye strain.
- ☐ Certain kinds of light may be hazardous to the worker's eyes (*see 11.3 on Non-ionizing radiation*)
- ☐ The direction of provided light might also result in glare that can also affect task productivity and cause worker eyestrain.
- ☐ Strategies for eliminating insufficient or extreme light include:
  - Providing proper lighting of the work place
  - Using anti-glare surfaces
  - Providing employees with shaded glasses or safety glasses
- ☐ PPE may interfere with worker vision or reduce field of vision *(see Section 13)*.

6.7 Excessive Noise Exposures

- ☐ Excessive noise on workers may:
  - Contribute to hearing loss

- Interfere with communication
- Annoy or distract individuals
- Alter the performance of some tasks.

☐ Hearing loss represents irreversible damage to the inner ear. The degree to which hearing is affected depends on:

- Noise intensity
- Noise frequency spectrum
- Noise duration
- Individual susceptibility

☐ Excessive noise is likely to be encountered:

- In work around heavy machinery
- When using power tools
- During emergency response (sirens and alarms)

☐ Although noise levels below 85 (decibels) dBA probably do not contribute to hearing loss problems, they may contribute to performance decrements due to distraction or annoyance

- Noisy equipment can reduce the effectiveness of communications and make it difficult for the worker to concentrate on some types of tasks.
- Speech interference by noise is fairly common around noisy machinery or vehicles. The criticalness of communications should determine the steps to be taken to improve the noise levels in the environment.

☐ Noise may affect performance in a number of ways. Noise features that are most likely to degrade performance include:

- Variability in level or content
- Intermittency
- High-level repeated noises
- Frequencies above 2000 Hz
- Any combination of these features

☐ OSHA 29 CFR 1910.95 sets specific requirements for protecting workers against the effects of noise exposure. These include:

- Defining the need for noise protection by the noise exposure:
  - -- Octave band sound pressure levels (in decibels)

  - -- Band center frequency in cycles per second
- Setting permissible noise exposure levels in terms of duration and sound level in decibels
- Establishing a noise conservation program
- Monitoring of noise exposure
- Notifying employees when noise exposures exceed 85 dBa
- Allowing employees to observe noise exposure monitoring
- Conducting audiometric testing of employees
- Providing hearing protectors
- Maintaining records

☐ Ear protectors must be used for noise in excess of OSHA levels.

☐ The American Conference of Governmental Industrial Hygienists (ACGIH) establishes Threshold Limit Values (TLVs) for both continuous/intermittent and impulsive/impact noise exposures in its annual publication, *Threshold Limit Values for Chemical Substances and Physical Agents and Biological Exposure Indices*.

- Continuous or intermittent noise TLVs are expressed in permissible maximum sound level in decibels based on a specific daily duration.
- Impulsive or impact noise TLVs are expressed in the permitted number of noise impulses or impacts per day for a given sound level.

☐ Ear protectors may diminish the worker's ability to communicate or hear work-related signals.

Section 7

Identifying Chemical Hazards

The purpose of this section is assist in the identification of workplace/task chemical hazards. Several categories of physical hazards are discussed, including inhalation, skin absorption or contact, injection, ingestion, flammability, and reactivity.

7.1 Types of Chemical Hazards

☐ Each chemical has different levels of hazards associated with exposure.

☐ Different chemicals present a variety of general hazards, which include:

- Toxicity (poisonous)
- Corrosiveness
- Irritation
- Flammability
- Reactivity (explosiveness or oxidation)

☐ Table 7-1 summarizes the types of chemical hazards, the body areas or body systems affected, and how these hazards are prevented or minimized.

7.2 Nature of Chemical Hazards

☐ Most chemicals present multiple hazards and have to potential to affect the entire worker's body or specific areas of the worker's body.

☐ Exposures in turn are affected by the chemical's properties. Key properties include:

- Density
- Vapor density (determines whether chemicals will be lighter or heavier than air, thus affecting rate of dispersal)

Table 7-1. Overview of Chemical Hazards

Chemical Hazard	Body Areas Usually Affected	Relevant Occupations	Mitigation Methods
Inhalation	Respiratory system	Chemical industries; Emergency response; Any job with possible exposure to chemicals	Limit worker exposure by engineering/ administrative controls; Use respirators
Skin absorption or contact (vapor contact)	Entire body; Hands	Chemical industries; Emergency response; Any job with possible exposure to chemicals as vapors or gases	Limit worker exposure by engineering/ administrative controls; Use vapor-tight protective clothing
Skin absorption or contact (liquid contact)	Entire body; Hands	Chemical industries; Emergency response; Any job with possible exposure to chemicals as liquids	Limit worker exposure by engineering/ administrative controls; Use liquid-tight protective clothing
Skin absorption or contact (particulate contact)	Entire body	Chemical industries; Emergency response; Any job with possible exposure to chemicals as liquids	Limit worker exposure by engineering/ administrative controls; Use particle-tight protective clothing
Ingestion	Digestive system	Chemical industries; Any job with possible exposure to chemicals	Separate work areas from eating/rest areas; Practice good hygiene; Cover mouth with PPE
Injection	Entire body	Chemical industries; Any job with possible exposure to chemicals	Limit chemical work around sharp or pointed objects; Use cut- and puncture-resistant protective clothing
Chemical flashover	Entire body	Chemical industries, particularly petrochemical industry; Emergency response; Any job with possible exposure to chemicals	Keep employees from entering area with vapors in flammable range; Wear flame-resistant, heat insulative protective clothing
Chemical explosions	Entire body	Chemical industries; Emergency response; Any job with possible exposure to chemicals	

- Specific gravity (determines whether chemical will be lighter or heavier than water, thus affecting how it will behave when in water)
- Viscosity (a liquid's resistance to flowing)
- Odor (affecting ease of detection)
- Flash point (method for rating chemical flammability)
- Flammable limits (indicates chemical concentration in air needed for ignition)
- Melting point (temperature at which solid changes to a liquid)
- Boiling point (temperature at which liquid changes to a gas)
- Solubility (ability of chemical to dissolve or completely mix with a solvent)

☐ One of the best sources of information on a chemical's hazards and properties is its Material Safety Data Sheet (MSDS). OSHA 29 CFR 1910.1200 requires that employers obtain and maintain MSDS's for each chemical at the work place.

☐ The potential for exposure to chemical hazards exists in any workplace where chemicals are manufactured, used, or stored.

☐ The toxic, corrosive, and irritant effects of chemicals can collectively be considered health effects.

☐ Chemical health hazards are rated by the National Fire Protection Association ratings:

- 0: Materials that on exposure under ordinary conditions would offer no perceivable health hazard
- 1: Materials that an exposure would cause irritation but only minor residual injury even if no treatment was given
- 2: Materials that an intense or continued exposure could cause temporary incapacitation or possible residual injury unless prompt medical treatment was given
- 3: Materials that on short exposure could cause serious temporary or residual injury even though prompt medical treatment was given
- 4: Materials that on very short exposure could cause death or major residual injury even though prompt medical treatment was given

☐ Chemicals may enter the body and cause adverse health effects the body through:

- Inhalation

- Skin absorption
- Injection
- Ingestion

7.3 Inhalation Exposures

- ☐ Any of the normal constituents of atmospheric air in greater than normal concentrations or any other substance present in the atmospheric air may be regarded as an air contaminant.
- ☐ Displacement of normal air below its normal concentration of 20.9% constitutes an oxygen deficiency.
  - Oxygen contents less than 19.5% are considered extremely hazardous and require respiratory protection.
  - Oxygen levels can be measured and monitored with oxygen indicators or meters calibrated at the pressure and temperature of the use area.
- ☐ Air contaminants vary in form (gas, vapor, liquid, and solid) and composition (atomic particles, ions, and molecules). The two primary classes of air contaminants are aerosols and gaseous air contaminants.

7.3.1 Aerosols

- ☐ Solid or liquid particles suspended in air are known as aerosols.
- ☐ Aerosols may be classified by the process by which they are formed, their physical type, their particle size, and their physiological effects on the body.
- ☐ Aerosols are created and dispersed by:
  - Mechanical means (such as grinding, crushing, drilling, blasting, or spraying)
  - Condensation (produced by physicochemical reactions such as combustion, vaporization, condensation, sublimation, calcination, and distillation)
- ☐ Physical types of aerosols include:
  - Dusts (solids dispersed in the air mechanically)
  - Sprays (liquids dispersed in the air mechanically)
  - Fumes (solids dispersed in the air by condensation)

- Mists (liquids dispersed in the air by condensation)
- Fogs (a mist which obscures vision)
- Smokes (a combination of liquids and solids caused by the incomplete combustion of organic substances)

☐ Aerosol particle sizes range from fractions of a micrometer to several micrometers.

- The range of interest for respiratory hazards is usually from 0.1 to 10.0 µm in diameter
    - -- Particles smaller than 10 µm generally remain suspended in air and are exhaled
    - -- Particles larger than 0.1 µm are generally removed from the inspired air in the nasal cavity and are carried in the mucous to the pharynx and then to the mouth by ciliary action where they are either expectorated or swallowed.
- Some smaller or larger particles are still hazardous for skin and eye contact depending on other physical or chemical hazards.
- Since the shape of particles varies from round to non-uniform shapes, particles are often characterized by their mass median aerodynamic diameter (MMAD).

☐ Aerosols may be classified by their physiological effect on the body:

- Nuisance and/or relative inert (may cause discomfort or irritation without injury; examples include marble and gypsum dusts)
- Pulmonary fibrosis producing (produce nodules and fibrosis in lungs; examples include silica dust and asbestos fibers)
- Carcinogens (cause cancer in some person after long latent period; examples are chromate dusts and asbestos fibers)
- Chemical irritants (produce irritation, inflammation, and ulceration of respiratory tract; examples include acidic or alkaline particles)
- Systemic poisons (produce toxic pathological reactions in various parts of the body; examples are lead and cadmium dusts)
- Allergy producing (cause allergic, hypersensitive type reactions in the body; examples include tobacco)
- Febrile reaction producing (cause chills followed by fever; examples include zinc and copper fumes)
- Pneumoconiosis (deposition of any dust in lungs)

☐ Concentrations of particles are represented in two different ways:

- Weight of particles per unit volume of air (e.g., mg/m^3)
- Number of particles per unit volume of air

7.3.2 Gaseous Contaminants

☐ Gaseous air contaminants are gases or vapors mixed in air. Gaseous air contaminants may be classified by their chemical composition or their physiological effects on the body:

☐ Chemical classifications include:

- Inert (do not reaction with other substances or harm body, but can displace air; examples are helium, neon, and argon)
- Acidic (produce low pH substances or acids in water; examples include hydrogen chloride, hydrogen fluoride, sulfur dioxide, hydrogen sulfide and hydrogen cyanide)
- Alkaline (produce high pH substances or bases in water; examples include ammonia, amines, phosphine, and arsine)
- Organic (contain carbon atoms; examples are several types of structures such as hydrocarbons, alcohols, ketones, aldehydes, ethers, esters, halides, amides, nitriles, isocyanates, amines, epoxies, and aromatics)
- Organometallic (contain metals that are chemically bonded to organic groups; examples include tetraethyl lead and organophosphates)
- Hydrides (contain hydrogen bonded to metals; examples include diborane and pentaborane)

☐ Physiological effects include:

- Asphyxiation (contaminant interferes with supply or utilization of oxygen in the body; simple asphyxiants include nitrogen, methane, and helium; chemical asphyxiants include carbon monoxide, hydrogen cyanide, and nitriles)
- Irritation (contaminant causes irritation or inflammation of respiratory tract; examples are ammonia, hydrogen chloride, formaldehyde, chlorine, and phosgene)
- Anesthesia (contaminant causes loss of feeling or sensation; examples are nitrous oxide, hydrocarbons, and ethers)

- Systemic poisoning (contaminant produces injury to specific organs or specific parts of the body; examples include mercury vapor, carbon tetrachloride, and hydrogen sulfide)
- Carcinogeneity (contaminant causes cancer; examples include vinyl chloride, benzene, and hydrazine)

☐ The concentration of gaseous contaminants in air is represented in two ways:

- Volume of contaminant per unit volume of air (e.g., parts per million)
- Weight of contaminant per unit volume of air (e.g., mg/m^3)

☐ OSHA 29 CFR 1910.1000 provides the permissible exposure limits (PELs) of several air contaminants for an 8-hour work shift:

- The 8-hour time-weight average concentration is provided as the PEL for most chemicals. The time weighted average equals cumulative exposure divided by 8 hours.
- Some substances are have PELs preceded by a "C' which denotes a ceiling value. Employee exposure to these substances can at no time exceed the ceiling value.
- For a small number of substances, an acceptable maximum peak above the acceptable ceiling concentration for an 8-hour shift is provided in terms of concentration and maximum duration.

☐ The American Conference of Governmental Industrial Hygienists (ACGIH) annually publishes *Threshold Limit Values for Chemical Substances and Physical Agents and Biological Exposure Indices*. This publication also establishes specific limits for several substances, including:

- Threshold Limit Value-Time Weighted Average (TLV-TWA); the time-weighted average concentration for a normal 8-hour workday and a 40-hour to which nearly all workers may be repeatably exposed, day after day, without adverse effect.
- Threshold Limit Value-Short Term Exposure Limit (TLV-STEL); the concentration to which workers can be exposed continuously for a short period of time without suffering from irritation, chronic or irreversible lung damage, or narcosis of sufficient degree as to increase the likelihood of accidental injury, impair self-rescue, or materially reduce work efficiency. The TLV-STEL is based on a 15-minute exposure.
- Threshold Limit Value-Ceiling (TLV-C); the concentration that should not be exceeded during any part of the working exposure.

  - Indications whether the substance is a confirmed human carcinogen, suspected human carcinogen, or animal carcinogen.

- ☐ Levels of air contaminants may be measured and monitored using either fixed or portable sensing instruments. Common detectors include:
  - Photoionization detectors
  - Flame ionization detectors
  - Electrochemical detectors
  - Colorimetric indicator tubes

- ☐ The primary strategy for protecting against inhalation hazards is the use of respirators that remove contaminants from ambient air or provide an alternative air source.

7.4 Skin Absorption Hazards

- ☐ The skin acts as a two-way barrier, preventing entry of environmental toxins while preventing the loss of water, electrolytes, and other substances necessary to maintain homeostasis.

- ☐ Four possible actions can occur when a chemical contacts the skin:
  - The skin can act as a barrier preventing the chemical from causing injury or penetrating into the body.
  - The chemical can act directly on the skin surface to cause injury.
  - The chemical can penetrate the surface of skin and injure the skin tissues beneath the surface.
  - The chemical can penetrate the skin, enter the blood stream, and be dispersed throughout the body where it can produce injury to various parts of the body.

- ☐ Chemical exposure to the skin can cause several skin disorders, including:
  - Irritant dermatitis, a localized inflammatory reaction following a single or repeated exposure to a chemical, which may be of three types:
    - -- The first type causes skin redness and swelling of the exposure area to a single chemical exposure.
    - -- The second type shows no visible reaction to the first exposure, but upon repeated exposure, produces redness, chapping, fissures, and cracks in the skin.

  - -- The third type occurs on repeated exposure but results in a chronic dermatitis, which is not easily reversible.
  - -- Common irritants include detergents, various organic solvents, acids, and bases.
- Allergic contact dermatitis, similar to irritant dermatitis, but occurring through skin sensitization of a long period.
  - -- Occupational skin sensitizers include chromium, cobalt, and nickel.
- Phototoxicity, analogous to irritant dermatitis, but requiring UV or visible light for activation.
  - -- Examples of phototoxic chemicals include polycyclic aromatic hydrocarbons and coat-tar derivatives.
- Follicular diseases, disorders of the hair follicles, such as chloracne.
  - -- Follicular diseases can be cause by dioxin, petroleum derivatives, and halogenated hydrocarbons.
- Skin cancers or tumors
- Pigmentation responses, changes in skin coloration produced by chemical exposure
  - -- Hydrogen fluoride and alkyl phenols are known to cause pigmentation responses.
- Systemic toxicity; effects on various organs of the body from the permeation of hazardous chemicals into the body.

☐ Many chemicals can permeate through the skin without any apparent skin damage.

☐ The skin is susceptible to permeation by chemicals to varying degrees depending on the properties of the chemical:

- Simple, polar non-electrolytes penetrate the skin at rates similar to water.
- Electrolytes do not penetrate the skin readily.
- Dilute solutions of anionic and cationic surfactants penetrate the skin readily.
- Many non-aqueous solutions do not alter skin integrity, but do permeate the skin.
- Some low-weight volatile organic solvents alter skin properties allowing permeation.

- ☐ Both OSHA 29 CFR 1910.1000 and the ACGIH *Threshold Limit Values for Chemical Substances and Physical Agents and Biological Exposure Indices* provide a skin designation or notation for those chemicals where the overall exposure to the individual by the cutaneous route (through skin absorption) is significant.
- ☐ There are currently no Permissible Exposure Limits (PELs) or Threshold Limit Values (TLVs) in use for determining acceptable skin exposure to chemicals.
- ☐ Protective clothing that provides a barrier to the specific chemical(s) is used to prevent skin absorption.

7.5 Ingestion Exposures

- ☐ While the ingestion of hazardous chemicals is not a likely occupational exposure, ingestion of hazardous chemicals can occur from:
  - Accidental swallowing
  - Contamination of food, tobacco, cosmetics, or other personal items which may contact the mouth
  - Inhalation of particles which are soluble in the mucous membranes of the respiratory tract and are carried to the mouth cavity and digestive system by the motion of cilia lining the respiratory tract
- ☐ Chemicals that make it into the digestive tract can:
  - Fail to absorb into the blood and be eliminated from the body
  - Be diluted by food and liquid, decreasing the substance's toxicity
  - React with food and liquid to become toxic
  - Pass through the walls of the digestive tract and be absorbed by the blood to be carried to other parts of the body.
- ☐ Strategies for preventing ingestion chemical exposures involve:
  - Clear separation of work areas from eating or rest areas
  - Good personal hygiene (hand washing, proper decontamination, storage and disposal of PPE)
  - Wearing of PPE that covers the mouth

7.6 Injection Exposures

- ☐ Injection can occur when a syringe or other sharp object-containing chemical penetrates the skin and releases chemical directly into the blood stream.
- ☐ Because injection bypasses the body's natural defenses to chemical, exposure to chemicals by injection can result in the more rapid onset of symptoms produced by chemicals compared to skin absorption.
- ☐ Injection of chemicals can occur in any process involving potential for chemical exposure.
- ☐ Strategies for preventing injection chemical exposures include:
  - Limiting contact with sharp or pointed objects in chemical-containing environments
  - Use of cut or puncture-resistance chemical protective clothing

7.7 Chemical Flammability Exposures

- ☐ Many chemicals may be flammable as gases, liquids, or solids.
- ☐ Principal properties used to determine the potential for flame hazards from chemicals include:
  - Flammability limits: the range of chemical concentration of air that will support combustion; usually represented in terms of:
    - -- Lower flammability or explosive limit
    - -- Upper flammability or explosive limit
  - Flash point: the minimum temperature at which a substance produces sufficient vapors to form an ignitable mixture
  - Autoignition temperature: the temperature at which a material will self-ignite (spontaneous combustion)
- ☐ The U.S. Department of Transportation provides several classifications and categories of flammable substances:
  - Gases are classified as flammable when their Lower Explosive Limit is less than 13% and the gas has a flammable range of more than 12% (examples include hydrogen, methane, and acetylene).
  - Liquids are classified as flammable when they have a flash point below 100°F [38°C] (examples include methanol, acetone, and gasoline).

- Liquids are classified as pyrophoric when they will spontaneously ignite in air at or below 140°F [60°C] (examples are pentaborane and aluminum alkyls).
- Liquids are classified as combustible when they have a flash point between 100°F and 200°F [38–94°C] (examples include kerosene and diesel fuel).
- Solids which cause fires through friction, retained heat during manufacturing or processing, or which can be readily ignited are classified as flammable (examples include magnesium and titanium).
- Spontaneously combustible solids either ignite within 5 minutes in coming in contact with air or are liable to self-heat when in contact with air but without an energy supply.

☐ Chemical flammability hazards are rated by the National Fire Protection Association ratings:

- 0: Materials that will not burn
- 1: Materials that must be preheated before ignition can occur
- 2: Materials that must be moderately heated or exposed to relatively high ambient temperature before ignition can occur
- 3: Liquids and solids that can be ignited under almost all ambient temperature conditions
- 4: Materials that:
  - -- Rapidly and completely vaporize at atmospheric pressure and normal ambient temperatures and burn readily, or
  - -- Are readily dispersed in air and burn readily

☐ Flammable chemicals create specific hazards resulting in:

- Ignition and continued combustion of chemical and surrounding materials
- Chemical flash fires

☐ Continued combustion of chemicals produces intense heat with fuel flame front at temperatures up to 1800°F (980°C), and primarily presents thermal hazards *(see Section 9, Thermal Hazards)*.

☐ The ignition of a flammable atmosphere may cause a chemical flash fire. Chemical flash fires are characterized by:

- Rates of heat release of 100-350 kW/m^2 (2.4–8.4 cal/cm^2s)
- Short exposure times ranging from 0.1 to several seconds

- Ignition of combustibles in the area of the chemical flash fire

☐ Flammable hazards associated with chemicals should be avoided. When the potential exists for flame exposure:

- Provisions should be made to use reflective, flame-resistant protective clothing *(see Section 9, Thermal Hazards)* over chemical protective clothing.
- Flame resistant underclothing should be used.

7.8 Chemical Reactivity Exposures

☐ Chemicals may pose reactivity hazards:

- As explosives
- By interacting with other substances to undergo violent reactions (e.g., oxidizers and water-reactive substances)
- By interacting with itself under specific conditions to undergo a violent reaction (e.g., polymerization)

☐ Explosives are materials that function by a rapid release of energy that is highly destructive. These materials are generally designed to explode and have the following characteristics:

- Heat and shock sensitive
- Sensitive to contamination
- Thermal/mechanical effects

☐ The U.S Department of Transportation provides several categories of explosives:

- Class 1.1 materials present the maximum hazard and function by detonation with shock waves that travel faster than the speed of sound (examples include dynamite and trinitrotoluene).
- Class 1.2 materials function by rapid burning, known as deflagration, with shock waves slower than the speed of sound (an example is rocket propellent)
- Classes 1.3 and 1.4 materials contain small amounts of Class 1.1 or 1.2 explosives (examples are fire works or ammunition).
- Class 1.5 materials are harder to detonate than Class 1.1 or 1.2 explosives but produce the same level of destruction (examples are ammonium nitrate and fuel/oil mixtures).

- ☐ Oxidizers are chemicals that yield oxygen readily to stimulate combustion of organic materials. Oxidizers have the following characteristics:
  - Heat and shock sensitive
  - Unpredictable behavior
  - Problems with contamination
  - Reactivity
- ☐ The U.S. Department of Transportation provides two categories of oxidizers:
  - General oxidizers react chemically or by heating to evolve oxygen and heat (examples include sodium chlorate and perchloric acid).
  - Organic peroxides contain a bivalent oxygen to oxygen bond which readily release oxygen when in contact with combustible materials and under high temperature conditions (an example is benzoyl peroxide).
- ☐ Water-reactive chemicals violently react when contacted by water (examples include sodium metal, potassium superoxide, and some organic metallic compounds).
- ☐ Some chemicals may undergo rapid polymerization with detonation or the release of heat under high temperature or pressure conditions unless inhibited (examples include acrylonitrile and methyl methacrylate).
- ☐ Many chemicals are incompatible with other chemicals or substances. For example, acrylonitrile will react violently with bases, amines, and copper-based metals.
- ☐ Chemical reactivity hazards are rated by the National Fire Protection Association ratings:
  - 0: Materials that are intrinsically stable, even when exposed to fire, and that do not react with water
  - 1: Materials that are intrinsically stable, but which can:
    - -- Become unstable at elevated temperatures, or
    - -- React with water with some release of energy, but not violently
  - 2 Materials that are intrinsically unstable and may:
    - -- Readily undergo violent chemical change, but do not detonate, or
    - -- React violently with water, or
    - -- Form explosive mixtures with water

- 3: Materials that are intrinsically capable of detonation or explosive reaction, but:
    - -- Require a strong initiating source, or
    - -- Must be heated under confinement before initiation, or
    - -- Must be combined with water
- 4: Materials that and pressures are intrinsically capable of detonation or of explosive decomposition or reaction at normal temperatures

☐ Proper handling and prevention of chemical contact with incompatible substances is necessary to avoid reactivity hazards of chemicals.

- Some forms of protective clothing and equipment may offer limited protection against reactive substances.
- For some substances, there is no PPE available to offer acceptable protection to violent reactions or explosions.

Section 8

Identifying Biological Hazards

The purpose of this section is assist in the identification of workplace/task biological hazards. Several categories of biological hazards are discussed, including bloodborne pathogens, tuberculosis, biogenic toxins, and biogenic allergens.

8.1 Types of Biological Hazards

☐ Principal biological exposures include:

- Bloodborne pathogens
- Tuberculosis and other airborne pathogens
- Biogenic toxins
- Biogenic allergens

☐ Table 8-1 summarizes the types of biological hazards, the body areas or body systems affected, and how these hazards are prevented or minimized.

8.2 Nature of Biological Hazards

☐ Biological hazards involve exposure to biological or etiological agents. Etiological agents are agents capable of causing disease.

8.2.1 Types of Biological Agents

☐ Biological agents include microorganisms or other living matter, which are pathogenic to humans, including:

- Viruses, a submicroscopic pathogen consisting of a single nucleic acid enclosed by a protein coat, able to replicate only within a living cell

Table 8-1. Overview of Biological Hazards

Biological Hazard	Body Areas Usually Affected	Relevant Occupations	Mitigation Methods
Bloodborne pathogens	Entire body, primarily exposed mucous membranes	Health care industry Medical waste disposal Medical equipment repair Correctional facilities Law enforcement Emergency response	Universal precautions; Proper personal hygiene (handwashing); Hepatitis B vaccinations; Use of PPE to prevent blood/body fluid contact with mucous membranes or skin; Puncture-resistant PPE to prevent accidental inoculation
Tuberculosis or airborne pathogens	Respiratory system	Health care industry Correctional facilities Law enforcement Emergency response	Isolation of affected persons; Room filtering systems; Proper personal hygiene; Use of particulate filtering respirators
Biological toxins	Entire body Respiratory system	Agriculture production and processing; Meat processing; Natural fiber processing; Biotechnology and fermentation; Office work with indoor air contaminated by fungi or bacteria; Road maintenance work	Isolation from biological hazards by use of engineering controls; Proper personal hygiene; Use of PPE (respirators, gloves, and clothing)
Biogenic allergens	Entire body Respiratory system	Agricultural workplaces; Natural fiber processing; Sewage treatment facilities; Humidified office buildings	Isolation from biological hazards by use of engineering controls; Proper personal hygiene; Use of PPE

- Bacteria, a unicellular microorganism, existing as a free-living organism or as a parasite
- Chlamydiae, an intracellular parasite that multiplies by means of a unique development cycle
- Rickettsiae, small microorganism that can only multiply within a host, usually carried by ticks and fleas
- Mycoplasmas, the smallest cells capable of independent existence differing from bacteria by not having a cell wall
- Fungi, either as unicellular yeasts or branching filaments of cells in molds

☐ Each type of microorganism requires a set of parameters that affects its growth, metabolism, development, and reproduction. The microorganism must have an environment that provides favorable conditions for its survival and growth.

☐ Environmental factors affecting survival of microbes include:

- Moisture content
- Temperature
- pH balance
- Osmotic pressure
- Oxygen tension
- Nutrients
- Lighting

8.2.2 Routes of Exposure

☐ The primary routes of entry for infectious biological agents include:

- Ingestion
- Inhalation
- Inoculation
- Skin and mucous membrane penetration
- Animal and insect bites

☐ Ingestion occurs frequently as the result of poor personal hygiene and poor laboratory practice, such as through:

- Oral pipeting
- Handling infectious materials without gloves and/or failing to wash contaminated hands before eating
- Eating, drinking, smoking, and applying cosmetics or contact lenses in laboratories

☐ Inhalation exposure occurs when aerosol-generating procedures are conducted in an open area without containment. Possible inhalation hazards exist during the following operations:

- Centrifugation
- Sonification
- Homogenization
- Mixing

☐ Inoculation most frequently occurs as the result of accidental injections with contaminated needles or cuts with contaminated sharp instruments.

☐ Penetration of skin or contact of mucous membranes with microorganisms is usually the result of poor hygiene or failure to use protective devices. Subcutaneous exposures may occur through existing abrasions, cuts, or other areas of non-intact skin.

☐ Failure to properly control laboratory animals can result in bites. Wild animals and insects can also transfer infectious agents to workers through bites.

8.2.3 General Principles of Biosafety

☐ Biosafety in the laboratory is affected by use of appropriate:

- Laboratory practice and technique
- Safety equipment (primary barriers)
- Facility design (secondary barriers)

☐ Four biosafety levels are established which combine laboratory practices and techniques, safety equipment, and laboratory facilities specifically appropriate for different types of biological hazards:

- Biosafety Level 1
    - -- Applies to work performed with defined and characterized strains of viable microorganisms not known to cause disease in healthy

adult humans (e.g., *bacillus subtilis, Naegleria gruberi*, and infectious canine hepatitis virus

-- Represents a basic level of containment that relies on standard microbiological practices with no special primary or secondary barriers recommended other than a sink for handwashing.

- Biosafety Level 2

-- Applies to work done with the broad spectrum of indigenous moderate-risk agents present in the community and associated with human disease with varying severity (e.g., Hepatitis B virus, salmonellae, and *Toxoplasma*

-- Includes primary hazards of accidental percutaneous or mucous membrane exposure, or ingestion of infectious materials

-- Requires use of splash shields, face protective devices, gowns, and gloves as primary barriers with hand washing and decontamination facilities as secondary barriers

- Biosafety Level 3

-- Applies to work with indigenous or exotic agents with a potential for respiratory transmission, and which may cause serious and potentially lethal infection (e.g., *Mycobacterium tuberculosis*, St. Louis encephalitis virus, and *Coxiella burnetii*)

-- Includes primary hazards of autoinoculation, ingestion, and exposure to infectious aerosols

-- Requires use of respirators, full body protective clothing, Class I or II biological safety cabinets, controlled laboratory access, specialized ventilation systems

- Biosafety Level 4

-- Applies to work with dangerous and exotic agents which pose a high individual risk of life-threatening disease, which may be transmitted via the aerosol route, and for which there is no available vaccine or therapy (e.g., Congo-Crimerian hemorrhagic fever)

-- Includes primary hazards of respiratory exposure to infectious aerosols, mucous membrane exposure to infectious droplets, and autoinoculation

-- Requires use of complete isolation of infectious agents by working in Class III Biological Safety Cabinets or a full, body, air-supplied positive pressure personnel suit

8.3 Bloodborne Pathogen Exposures

☐ Common bloodborne pathogens include:

- Hepatitis A
- Hepatitis B
- Hepatitis, non-A, non-B
- Human Immunodeficiency Virus

☐ While health care workers are at the greatest risk, the following workers also represent high-risk occupations:

- Morticians
- Firefighters and emergency response personnel
- Law enforcement personnel
- Correctional facility personnel
- Personnel involved in infectious waste disposal
- Personnel involved in the repair or cleaning of medical equipment

☐ The primary modes of transmission of Hepatitis B and HIV in occupational settings are:

- Direct inoculation through the skin
- Contact with an open wound
- Contact with non-intact skin (chapped, abraded, weeping skin)
- Needle sticks and cuts with sharp instruments
- Mucous membrane exposure (eyes and mouth) with blood or body fluids containing blood

☐ OSHA 29 CFR 1910.1030 regulates occupational exposure to bloodborne pathogens. These regulations require:

- Developing an exposure control plan
- Using methods for eliminating or minimizing employee contact with bloodborne pathogens, including

    -- Use of universal precautions
    -- Use of PPE which keeps bloodborne pathogens off the worker's skin or underclothing

- Keeping the work site clean and sanitary with proper cleaning and decontamination
- Providing Hepatitis B vaccinations for affected employees
- Communicating hazards to employees
- Conducting training
- Keeping appropriate records

8.4 Tuberculosis and Airborne Pathogen Exposures

☐ Tuberculosis (TB) is caused by the tubercle bacillus (i.e., *Mycobacterium tuberculosis*) and is spread by airborne particles expelled when persons with pulmonary TB sneeze, cough, speak, or sing.

☐ TB is not equally prevalent throughout the United States.

☐ Susceptibility to infection depends on the health of the individual's immunity system.

☐ Protection from transmission of TB can be accomplished by:

- Administrative controls for isolating those patients with TB infection or active TB
- Engineering controls in the form of high-efficiency particulate air (HEPA) filters or germicidal ultraviolet (UV) lamps in combination with control of ventilation for recirculated air
- Use of personal protective equipment including surgical masks and respirators

    -- Surgical masks may limit exposure but are not an effective PPE item against airborne pathogens because these masks do not provide an air-tight seal to the face of the wearer.
    -- The National Institute for Occupational Safety and Health (NIOSH) recommends NIOSH-certified, powered, half mask respirators equipped with HEPA filters.

☐ Other airborne pathogens, such as influenza, are similarly transmitted. Microorganisms must remain viable for transmission to occur. Strategies for preventing exposure involve isolation and respiratory protective equipment and sometimes other forms of protective equipment.

8.5 Biological Toxin Exposures

☐ Biological toxins include:

- Biogenic Toxins
- Endotoxins

8.5.1 Biogenic Toxins

☐ Biogenic toxins are naturally occurring substances that can cause acute toxic disease in addition to long-term reproductive and carcinogenic effects.

☐ Biogenic toxins include those of bacterial, algal, fungal, plant, or animal origin. Examples of biogenic toxic effects are:

- Botulism
- Tetanus
- Diphtheria
- Dysentery
- Poison ivy
- Animal bites or stings

☐ The primary route of exposure is by inhalation, but some toxins are spread by dermal contact (e.g., poison ivy) or injection (animal bites or stings).

☐ Workplace exposure to biotoxins can occur:

- In any indoor environment where there is extensive growth of microorganisms
- Outdoors from direct contact with plants, animals, or their products

☐ Specific occupations at risk include:

- Agriculture production and processing
- Poultry processing
- Natural fiber technology
- Biotechnology and fermentation

- Office work with indoor air contaminated by fungi or bacteria
- Road maintenance workers

☐ Protection from biogenic toxins can be provided by:

- Isolating biological hazards
- Using engineering controls to limit exposure
- Use of respirators equipped with appropriate filtering material
- Use of clothing to cover skin to protect from animal bite or insect stings

8.5.2 Endotoxins

☐ Endotoxin makes up part of the cell envelope of Gram-negative bacteria.

Endotoxin is produce in a variety of work environments, including:

- Agricultural workplaces
- Processing of vegetable fiber
- Swine confinement buildings
- Grain storage facilities
- Poultry houses
- Cotton mills
- Flax mills
- Sewage treatment facilities
- Humidified office buildings

☐ Endotoxins primary affect the lungs causing difficulty in breathing, coughing, and fever.

☐ Repeated exposures cause hyperreactive airways, which may increase susceptibility of workers to other toxic exposures.

☐ Protection from endotoxins can be provided by:

- Isolating biological hazards
- Using engineering controls to limit exposure
- Use of respirators equipped with appropriate filtering material
- Use of protective clothing to protect exposed skin

8.6 Biogenic Allergen Exposures

☐ Biogenic allergens are substances produced or derived from plants, animals, or microorganisms, which can elicit an immune response from certain individuals.

☐ Normally the immune system has a protective role, but repeated exposure to antigens (produced by the body in response to exposure by biogenic allergens) can cause excessive immune reactions (hypersensitivity), such as:

- Allergic asthma
- Rhinitis (inflammation of the nasal mucous membranes)
- Pneumonitis (inflammation of the lungs)
- Contact dermatitis

☐ Hypersensitivity reactions are of four types:

- Type I (immediate hypersensitivity) occur immediately after a second contact with the original sensitizing antigen.
- Type II (cytotoxicity reactions) causes destruction of affected cells.
- Type III (immune complex reactions) occurs when a soluble antigen binds with antibodies and activates a number of local immune complexes or circulating immune complexes.
- Type IV (cellular immunity) begins with an initial nonspecific inflammation followed by an immune reaction.

☐ In occupational settings, biogenic allergens arise from:

- Animal products and insects
- Plants
- Wood dust
- Microorganisms and products

☐ Workplace engineering and administrative controls are typically used to reduce exposure. However, some activities may require personal protective equipment.

Section 9

Identifying Thermal Hazards

The purpose of this section is assist in the identification of workplace/task thermal hazards. Several categories of thermal hazards are discussed, including high heat, flame, hot liquid or gas, and molten substances.

9.1 Types of Thermal Hazards

☐ Thermal hazards include exposure to thermal energy from:

- High heat sources
- Flame
- Hot liquids and gases
- Molten substances

NOTE: *Thermal exposure may also be present from chemical flash fires (see 6.5) and electrical arc exposures (see 10.3).*

☐ Table 9-1 summarizes the types of physical hazards, the body areas or body systems affected, and how these hazards are prevented or minimized.

9.2 Nature of Thermal Hazards

☐ Thermal hazards will affect any exposed area of the human body including the respiratory system when hot gases are breathed in.

☐ The principal injury resulting from exposure to high amounts of thermal energy is burn injury.

- Burn injury occurs when the skin absorbs a sufficient amount of heat energy to damage the skin or underlying tissue.
- First degree burn injury occurs when the skin temperature reaches 111°F (44°C); the skin appears reddened and is tender to the touch.

Table 9-1. Overview of Thermal Hazards

Thermal Hazard	Body Areas Usually Affected	Relevant Occupations	Mitigation Methods
High heat	Entire body, especially exposed skin	Fire fighting; Work around ovens or kilns; Foundries and smelting; High temperature manufacturing processes	Provide insulative shielding to individual; Limit time of exposure; Use flame-resistant protection clothing which insulates from convective, radiative, and conductive heat
Flame	Entire body, especially exposed skin	Fire fighting; Work around ovens or kilns with open flames; Operations with open flames	Isolate worker from flame contact; Use flame-resistant, non-melting, heat insulative protective clothing
Hot gas or liquid	Entire body, especially exposed skin; Respiratory system	Fire fighting; Steam plants and boilers; Industrial processes involving hot liquids or gases	Isolate worker from hot gas or liquid contact; Use flame-resistant, non-melting, heat insulative protective clothing which provide barrier to hot liquids and gases Use respirators to protection against hot gases
Molten substances	Entire body, especially exposed skin	Fire fighting Foundries and smelting Welding	Isolate worker from flame contact; Use flame-resistant, non-melting, heat insulative protective clothing which sheds molten metals

- Second degree or partial thickness burn injury occurs when the skin temperature reaches 119°F; 2nd degree burns produce blistering.
- Third degree or full thickness burn injury occurs when the skin temperature reaches 131°F; 3rd degree burns involve destruction of the skin.
- Burn injury may be also referred to in terms of its depth.

☐ Severe burn injury (2nd or 3rd degree burns) may be preceded by the sensation of pain.

☐ Burn injury is a function of incident thermal energy and time or heat dose.

- Higher incident thermal energies cause burns earlier
- Given a sufficient time, exposure to elevated thermal energy will cause burn injury

☐ Burn injuries may be sustained by:

- Direct contact with a heat source (solid, liquid, or vapor)
- Heat transfer through the clothing
- Ignition and combustion of the clothing and equipment which is worn by the individual

9.3 High Heat Exposures

☐ Heat transfer may occur by:

- Conduction
- Convection
- Thermal radiation

9.3.1 Conduction

☐ Conduction involves heat transfer as the result of direct contact of the individual or his or her clothing with a hot surface.

☐ Heat flows through the resulting continuity of surfaces.

☐ Heat conduction is affected when the protective clothing is wet or compressed.

- Water can provide a bridge between surfaces that might not otl touch, increasing the chances of heat conduction by dis insulating air between and within the layers of clothing.
- Water can also act as an insulator since water has a relative heat capacity increasing the overall mass of the material syst extending the time it can absorb heat.
- Compression may bring surfaces closer together, permitting transfer of heat between clothing layers.

☐ Contact with surfaces that are above 140°F (60°C) will cause pa tissue damage.

9.3.2 Convection

☐ Convection entails heat transmission through the movement an density of surrounding gases or liquids (normally air or water).

☐ Convection also affects the transfer of heat within layers of clothi between these layers and the body.

☐ Some convective heat loss occurs by evaporation of wearer's swea

☐ Spaces within the clothing or between clothing layers, if filled w provide convective air currents affecting heat transfer through pro clothing.

☐ Tolerance times for specific heat exposures depend on the exposu energy and duration of exposure.

9.3.3 Thermal Radiation

☐ Thermal radiation depends upon the temperature difference betwe surfaces, the distance between two surfaces, and the reflectivity surface.

☐ Heat exchange by radiation does not depend on the temperature o between each surface.

☐ Thermal radiation does not always depend on the clothing color darker color in the visible spectrum of light may actually be more re in the near infrared region than a white color, depending on the p properties of the textile and dyes).

9.3.4 Effects of High Heat Exposures

☐ High heat exposures are most common in the following industries:

- Fire fighting
- Work around ovens or kilns
- Foundries and smelting
- High temperature manufacturing processes

☐ Clothing material response to heat typically proceeds through the following stages:

- Temperature rise and subsequent heat transfer through the material
- Decomposition and change in physical form
- Ignition
- Combustion

☐ In the absence of engineering or administrative controls, protection of the individual from thermal injuries is provided by PPE that offers:

- Thermal resistance
- Thermal capacity
- Reflectivity

9.4 Flame Exposures

☐ A flame exposure is a specialized form of heat exposure, which generally involves the combination of convection and thermal radiation; a typical flame exposure involves:

- Incident thermal energies of 15 to 110 Btu/ft^2 (4 to 30 cal/cm^2)
- 30-50% radiant energy and 50-70% convective energy

☐ Depending upon the distance from the fire, thermal loads vary from that of the radiant heat energy of a JP-4 fuel flame front (982°C or 1800°F) to the mixture of radiant heat and convective heat typical in a smoky structural fire (93 to 315°C or 200 to 600°F).

☐ Other than burn injury, the principal hazard from exposure to flame is the ignition of clothing.

☐ Flame exposures are most common in the following applications:

- Fire fighting
- Work around ovens or kilns with open flames
- Operations with open flames

☐ Protection strategies for minimizing hazards from flame exposure is based on the use of PPE which is flame resistant; flame resistance may be imparted by:

- Treating PPE materials with flame retardants
- Using materials which are intrinsically flame resistant

9.5 Hot Liquid or Gas Exposures

☐ Contact with hot liquids or gases provide primarily convective thermal exposures.

☐ Hot liquids or gases may penetrate clothing causing increased heat transfer to the individual.

- Steam is an example of a hot vapor, which transfers heat by penetration and direct contact with the skin.

☐ Hot liquid or gas exposures are most likely in the following environments:

- Fire fighting
- Steam plants and boilers
- Industrial processes involving hot liquids or gases.

☐ Hot gases may affect all areas of the body while hot liquids are most likely to affect hands and feet. Hot gases also pose a respiratory system hazard.

☐ Protection from hot liquid or gas exposure involves principles similar to those used for protection from high heat but also requires the provision of a barrier material, which prevents penetration of the liquid or gas.

9.6 Molten Substance Exposures

☐ Common molten substances that pose thermal hazards include:

- Molten metals
- Molten bath electrolyte ("crysotile" used in smelting aluminum)
- Molten cast iron
- Welding splatter

- ☐ The principle occupations involving molten substance exposures include:
  - Fire fighting
  - Foundries and smelting
  - Welding
- ☐ Molten substances pose hazards from heat conduction and ignition of clothing fabrics.
- ☐ When needed for protection, the selected PPE must be flame resistant, limit heat transfer of molten substances, and allow molten substances to readily run off.

Section 10

Identifying Electrical Hazards

The purpose of this section is assist in the identification of workplace/task electrical hazards. Several categories of electrical hazards are discussed, including electric shock objects, electric arc, and static electricity.

10.1 Types of Electrical Hazards

☐ Electricity may creates three possible form of hazards:

- Exposure to electrical shock
- Exposure to electrical arcs
- Exposure to static electricity

☐ Table 10-1 summarizes the types of electrical hazards, the body areas or body systems affected, and how these hazards are prevented or minimized.

☐ OSHA 29 CFR 1910.137 addresses specific concerns for electrical hazards as related to PPE.

10.2 Electrical Shock Exposures

☐ Line-to-ground electrical hazards are the most common type of hazard.

☐ Electrical shock hazards are difficult to detect because the electric device may continue to operate normally.

☐ Examples of electrical shock hazards include:

- Ground wire of a power cord broken or not connected
- Use of a two-prong adapter plug with equipment having a three-wire power cord
- Ungrounded wall receptacles

- Polarity reversed on an equipment power cord or a wall receptacle
- Electrical insulation failure in a heating element
- Power lines short circuit to the equipment case

Table 10-1. Overview of Electrical Hazards

Electrical Hazard	Body Areas Usually Affected	Relevant Occupations	Mitigation Methods
Electric shock	Entire body Head Hands Feet	Electrical utilities and repair; Any exposure to poor or improper wiring; Emergency response	Insulate and maintain wiring and electrical equipment in accordance with local codes; Insulate and guard high voltage terminals; Use grounded equipment; Avoid use of metallic jewelry or conductive items; Use PPE with electrical insulation qualities
Electric arc	Entire body Head Face and eyes Arms Hands	Public utilities and power generation; Facility plant generation	Isolate workers from potential electrical arcs; Use flame resistant PPE which provide thermal and electrical insulation from electrical arcs
Static charge generation	Entire body	Cleanroom operations; Work in potentially flammable or explosive atmospheres; Semiconductor electronics manufacturing and repair; Emergency response	Adjust environmental conditions to reduce potential of static charge build-up; Use static removing treatments on clothing; Use conductive or static charge generation resistant protective clothing

- ☐ Electrical shock hazards also exist for workers who repair electrical equipment or lines, or those in the utility industries.
- ☐ Individuals have different body resistance to electrical shock, which are affected by:
  - Body size
  - Gender
  - Age
- ☐ The severity of the electric shock can be related to the amount of voltage or current involved in the exposure.
- ☐ The susceptibility of an individual to and seriousness of an electrical shock is also influenced by the presence of moisture on the individual's skin.
- ☐ Electrical shock can affect processes in the body that are controlled by electrical activity such as:
  - Muscle contraction
  - Sensory processes
  - Heart action
- ☐ Applied external voltage can create the following effects on the body:
  - Increased threshold perception (tingling or warm sensation)
  - Pain
  - Sustained muscle contraction
  - Ventricular fibrillation
  - Cardiac arrest
  - Convulsions
  - Burns
- ☐ Strategies for reducing the possibility of electrical shock include:
  - Installing and maintaining electrical circuits, equipment, and devices to electrical codes
  - Insulating and guarding high voltage terminals and components that are subject to accidental contact
  - Locating test points away from exposed high-voltage circuits
  - Prohibiting use of metallic jewelry and rings on hands and wrists while working

- Providing PPE for the wearer's head, hands, and feet that insulate from electrical shock

10.3 Electrical Arc Exposures

☐ An electric arc is produced by the passage of an electrical current between two electrodes in ionized gases and vapors.

☐ In industrial power systems, electrical arcs:

- Are normally short in duration, 1 second or less
- Produce very high energy levels, primarily radiant energy (2 to >100 cal/cm^2)
- Have temperatures as high as 20,000°F

☐ Electrical energy from the arc can be converted to other forms of energy and create other hazards:

- Intense thermal radiation
- Damaging noise levels
- Explosive expansion of the air surrounding the arc due to rapid heating
- Melting/vaporization of arc electrodes (producing molten metal hazards)
- Significant damage of electrical equipment
- Ignition or melting of normal wearing apparel in the vicinity of the arc

☐ Injuries are typically second or third degree burns; ignited fabrics can cause burn injuries over a high percentage of the body.

☐ Electrical hazards are most likely encountered in the power generation facilities of utilities and major plants.

☐ In the absence of engineering or administrative controls, approaches to protecting personnel from electrical arcs involve the use of PPE including flame resistant clothing, hooded visors, gloves, and footwear.

10.4 Static Charge Exposures

☐ Protection from static charge and discharge is preferable and may be vital in many environments.

☐ Consequences from static discharge range from minor physical discomfort to the destruction of sensitive electronic devices or the entire facility.

- ☐ Static charge is characterized by its lack of warning; it has no odor, color, or sound and is usually only detected by clothing clinging to skin. It is typically not detected until it discharges.
- ☐ Commercial applications where static discharges are a concern include:
  - Cleanrooms: garments must not allow particles to cling to the exterior of the fabric making the garment collect contamination.
  - Flammable/explosive atmospheres: the discharge of a spark will ignite a flammable environment; charges over 12.5 kV/in will typically ignite a gas requiring 0.2 milliJoules for ignition.
  - Electronic assembly/repair: sensitive electronic devices can be damaged by contact with garments and other materials sustaining relative low voltages (400 volts).
- ☐ Methods for controlling static in clothing include:
  - Controlling environmental conditions to reduce likelihood of static charge generation
  - Using topical antistatic treatments for increasing the surface conductivity of fabrics; these treatments help by spreading the charge generated or induced by the fabric
  - Using conductive fibers for transferring the charge
  - Using fibers which promote air ionization

Section 11

Identifying Radiation Hazards

The purpose of this section is assist in the identification of workplace/task radiation hazards. Several categories of radiation hazards are discussed, including ionizing and non-ionizing radiation.

11.1 Types of Radiation Hazards

- ☐ Radiation hazards are classified as either:
  - Ionizing radiation, or
  - Non-ionizing radiation
- ☐ Table 11-1 summarizes the types of radiation hazards, the body areas or body systems affected, and how these hazards are prevented or minimized.

11.2 Ionizing Radiation

- ☐ Ionizing radiation include X-rays and radiation emitted from radioactive materials.
  - Radiation associated with radioactive materials is produced by the spontaneous transformation of the element to other elements or isotopes produced either naturally or unnaturally from man-made process (usually by neutron bombardment).
  - Radioactivity is not affected by the physical state or chemical combination of the element.
- ☐ The time associated with the radioactive disintegration process is characteristic of the specific element. The time period during which an amount of a radioactive isotope decays to half the value is called the half-life.
- ☐ The energy of this process is emitted in the form of:

- Alpha particles
- Beta particles
- Gamma rays

Table 11-1. Overview of Radiation Hazards

Radiation Hazard	Body Areas Usually Affected	Relevant Occupations	Mitigation Methods
Ionizing radiation (alpha/beta particles)	Entire body Respiratory system	Nuclear power plants	Shielding and distance from radioactive source; PPE which provide protection from particulates
Ionizing radiation (gamma rays)	Entire body	Nuclear power plants; Use of radioisotopes; Equipment with radioisotope sources; Special sterilization or polymerization processes	Shielding and distance from radioactive source; Specialized PPE with heavy shielding
Ionizing radiation (X-rays)	Entire body	Nuclear power plants Medical treatments Use of radioisotopes; Equipment with radioisotope sources; Special sterilization or polymerization processes	Shielding and distance from radioactive source; Specialized PPE with heavy shielding
Non-ionizing radiation	Entire body; Eyes	Arc welding; Hot metal operations; Microwave ovens and transmission towers; Radar; Electric power transmission lines	Shield or isolation from some non-ionizing radiation sources; Use of PPE to attenuate non-ionizing radiation energy; Use of eyewear which attenuates non-ionizing radiation energy

☐ Ionizing radiation is likely to be encountered in:

- Nuclear power generating plants
- Military facilities
- Laboratories which use radioactive substances or conducting experiments involving ionizing radiation
- Health care and other specialized equipment which uses ionizing radiation

11.2.1 Alpha Particles

☐ Alpha particles are doubly ionized nuclei of helium having the following characteristics and hazards:

- Relatively little penetrating power
- Considered an internal hazard that can be absorbed into the body through the respiratory tract, ingestion, open wounds, or body orifices

11.2.2 Beta Particles

☐ Beta particles are negatively charged particles identical to electrons having the following characteristics and hazards:

- Wide range of energies
- Considered to be intermediate between alpha and gamma radiation in their penetration
- Capable of penetrating the outer layers of the skin, but are easily shielded against
- Primarily (although not always) an internal hazard

11.2.3 Other Ionizing Radiation

☐ Gamma rays are electromagnetic radiation of extremely short wavelengths and intensely high energy. Gamma rays:

- Can penetrate many objects, including the human body.
- Are best absorbed by dense materials like lead and depleted uranium

☐ Another form of ionizing radiation is X-rays. X-rays are electromagnetic radiation of extremely short wavelengths emitted as the result of electron transitions in the inner orbits of heavy atoms bombarded by cathode rays in a vacuum tube. X-rays have the following characteristics and hazards:

- Penetrate solids of moderate density (such as human tissue, but not bone)
- Affect photographic plates and fluorescent screens
- Ionize gases through which they pass
- Able to destroy or damage diseased tissue
- Overexposure permanently destroys cells and tissue structures; damage is cumulative

☐ Ionizing radiation may also include:

- Neutrons
- High speed electrons
- High speed protons
- Other atomic particles

11.2.4 Effects of Radiation

☐ Ionizing radiation hazards are likely to be encountered in the following applications:

- Nuclear power generation on research facilities
- Use of radioactive isotopes as tracers in biochemical, metallurgical, medical, geochemical, and archeological research
- Irradiation processes for sterilization and polymerization
- Medical diagnostic or therapeutic treatments

☐ Radiation can damage body tissue in various ways. Radiation primarily affects the cells of living tissues. The energy emitted from the radioactive source is deposited in tissues and disrupts cells either directly or indirectly:

- Large doses of radiation directly destroy cell components and cause death of the affected cells or prevent reproduction.
- The genetic material in cells is particularly sensitive to radiation effects which disrupting reproduction.
- Radiation can interact with cell chemical components and can lead to the development of cancer.

- ☐ Specific whole body biological effects of short-term radiation exposure include:
  - Doses of 100 to 200 rem cause nausea and vomiting.
  - Doses of 200 to 600 rem affects the levels of circulating blood cells.
  - Doses of 300 rem or more cause hair loss.
  - Doses between 600 and 1000 rem cause infection and hemorrhage, and other symptoms of decreased bone marrow functioning, which may take months to occur.
  - Doses over 1000 rem result in irreversible damage of the small intestine lining causing death with weeks.
  - Doses over 5000 rem have severe effects on gastrointestinal, cardiovascular, and central nervous systems with death occurring in hours.
- ☐ Specific acute local biological effects of short-term radiation exposure include:
  - Large doses may also cause skin burns that are similar to thermal burns but take longer to develop.
  - Doses of 10 to 15 rem cause temporary sterility in men; Permanent sterility occurs in men and women at doses over 200 rem.
  - Depending on the stage of development, fetuses have different sensitivities to radiation.
- ☐ Long term exposures to radiation can cause:
  - Cancer of the skin, bone marrow, lungs, breast, stomach, and thyroid
  - Genetic effects or mutations of the genes in reproductive cells, producing birth defects in offspring or succeeding generations

11.2.5 Radiation Exposure Levels

- ☐ Radioactivity is measured in several different ways:
  - A *roentgen* (R) is the amount of radiation that produces sufficient ion pairs in a cubic centimeter of air to carry one electrostatic unit of electrical charge (used for measuring gamma rays and X-rays).
  - A *radiation absorbed dose* or *rad* is the amount of radiation that results in the absorption of 100 ergs of energy by 1 gram of a material.

- A *curie* is the amount of radioactivity that decays at the same rate as 1 gram of Radium 226, equivalent to 3.7 x 10^{10} disintegrations per second.
- Biological effects of radiation depend on the type of radiation in addition to the dose:
  - The *relative biological effectiveness* (RBE) is the ratio of the absorbed dose of gamma radiation to the absorbed dose of the given radiation which gives the same biological effect (RBE are tabulated for various types of radiation).
  - A *roentgen equivalent man* or *rem* is RBE multiplied by the dose in rems and indicates the amount of biological damage from radiation.

☐ The U.S. Department of Transportation classified radioactive materials into three categories:

- *Radioactive I*: Materials that register 0.5 millirems per hour or less on the external surface of the container
- *Radioactive II*: Materials that register less than 1 millirem per hour at 3 feet from external points of the container
- *Radioactive III*: Materials which register more than 50 millirems per hour at the external surface or more than 1 millirem 3 feet away from the package surface

☐ Normal (natural) background exposure rates for gamma radiation are 0.01 to 0.02 milliroentgen per hour (mR/hr) and may vary from region to region.

☐ Permissible exposure levels to ionizing radiation are set by two organizations:

- OSHA 29 CFR 1910.96 prohibits exposure of individuals to ionizing radiation at the following levels:
  - Whole body—head and trunk, active blood forming organs, lens of eyes, or gonads at 1¼ rems per calendar quarter
  - Hands and forearms—feet and ankle at 18¾ Rems per calendar quarter
  - Skin of whole body at 7½ rems per calendar quarter
- The International Commission on Radiation Protection (ICRP) and the National Council on Radiation Protection and Measurement (NCRP) recommend that radiation doses be kept as low as reasonably achievable, but below the federal OSHA limits.

- The American Conference of Governmental Industrial Hygienists (ACGIH) establishes a Threshold Limit Value (TLVs) for particulate or electromagnetic ionizing radiation having an energy of 12.4 electron volts in its annual publication, *Threshold Limit Values for Chemical Substances and Physical Agents and Biological Exposure Indices*.

☐ The principal method for protecting individuals from ionizing radiation involves applying the principles of time, distance, and shielding:

- Reduced exposure reduces dose
- Increased distance from the radiation source reduces dose
- Shielding reduces dose
- Protective clothing effective against against particulates will also be effective against alpha and beta particles

11.3 Non-Ionizing Radiation

☐ Non-ionizing radiation is electromagnetic energy that does not change the structure of atoms of elements that it contacts.

☐ Categories of non-ionizing radiation include:

- Ultraviolet and visible light
- Infrared light
- Microwaves
- Radio frequencies
- Extremely low frequency radiation

☐ Sources of non-ionizing radiation may be natural, such as the sun, or may be from industrial products or processes, including:

- Arc welding
- Hot metal operations
- Microwave ovens and transmission towers
- Radar
- Electric power transmission lines

☐ Health effects of non-ionizing radiation include:

- Alteration of biochemical structures in the skin and eyes, leading to inflammation, from ultraviolet radiation (an example is sunburn)

- Dissipation of microwave and infrared energy in the form of heat

☐ The effects of radio frequencies are still being studied, but new regulations are forthcoming which limit the amount of exposure, particularly from electric power transmission lines.

☐ The American Conference of Governmental Industrial Hygienists (ACGIH) in its annual publication, *Threshold Limit Values for Chemical Substances and Physical Agents and Biological Exposure Indices*, establishes Threshold Limit Values (TLVs) for exposure to:

- Lasers
- Light and near-infrared radiation
- Radio frequency/microwave radiation
- Electromagnetic pulses and radio frequency radiation
- Static magnetic fields
- Sub-radio frequency magnetic fields
- Sub-radio frequency and static electric fields
- Ultraviolet radiation

☐ OSHA 29 CFR 1910.133 sets requirements for shielding the eyes from the light associated with non-ionizing radiation caused by:

- Shielded metal arc welding
- Gas metal arc welding and flux cored arc welding
- Gas tungsten arc welding
- Air carbon arc cutting
- Plasma arc welding and cutting
- Torch brazing, soldering
- Carbon arc welding
- Gas welding
- Oxygen cutting

Section 12

Identifying Person-Position Hazards

The purpose of this section is assist in the identification of workplace/task person-position hazards. Several categories of person-position hazards are discussed, including hazards from the worker not being visible, hazards from drowning, and hazards from falling off of elevated platforms.

12.1 Types of Person-Position Hazards

- ☐ The position of the person in performing work can also create hazards, such as:
  - Hazards from the worker not being visible
  - Hazards from drowning
  - Hazards from falling off of elevated surfaces
- ☐ Table 12-1 summarizes the types of person-position hazards, the body areas or body systems affected, and how these hazards are prevented or minimized.

12.2 Worker Visibility

- ☐ Workers who perform tasks outdoors or in areas near vehicular traffic face the potential for being struck and must remain highly visible under a number of circumstances both during the day and at night.
  - Daytime visibility is different from nighttime visibility.
  - High visibility means conspicuity of the individual against their background.
- ☐ Tasks involving low visibility risks include activities in:
  - The transportation industry
  - Emergency response

Table 12-1. Overview of Person-Position Hazards

Person-Position Hazard	Body Areas Usually Affected	Relevant Occupations	Mitigation Methods
Low visibility	Entire body	Law enforcement Emergency response Road maintenance Rail operations	Use signs, flashing lights or flares to alert traffic to personnel near roadway or railroad tracks; Use fluorescent material on PPE to enhance daytime visibility; Use retroreflective material on PPE to enhance night time visibility
Drowning	Entire body	Any work on or next to water or waterways	Provide guards to prevent workers from falling in water; Require workers to wear U.S. Coast Guard approved Personal Flotation Devices (PFDs)
Falling from elevated platforms	Entire body	Construction Material handling Outdoor cleaning Any work on elevated platforms	Provide guards on elevated platforms, scaffolding, stairs, or manholes; Limit worker traffic on elevated platforms; Use personal fall arrest equipment

☐ Work visibility is best enhanced by the use of high visibility materials:

- Fluorescent or phosphorescent materials should be used to provide enhanced daytime visibility:
  -- Phosphorescent materials absorb radiant energy and reradiate light after the energy source has been removed

-- Fluorescent materials are unnatural in color and provide high contrast during the day

- Retroreflective materials should be used to provided enhanced nighttime visibility (retroreflective materials have built in optical systems that redirect incoming light back to the source).

12.3 Drowning

☐ Workers operating in marine areas may be laden with equipment and may be subject to drowning.

☐ Increased risk for drowning exists when water temperatures are near freezing; hypothermia can quickly set in and limit the individual's attempt to self-rescue.

☐ Prevention of drowning is best accomplished by:

- Moving operations away from water, if possible
- Providing railings which prevent persons from falling in the water
- Using personal flotation devices

☐ OSHA 29 CFR 1926.106 requires that employees working over or near water be:

- Be provided with U.S. Coast Guard-approved personal flotation devices or buoyant work vests
- Have ring buoys with at least 90 feet of line available
- Have at least one lifesaving skiff available

12.4 Falling from Elevated Surfaces

☐ OSHA 29 CFR Subpart M on Fall Protection requires employers to provide fall protection when the worker is 6 feet or more above a lower level, without a guardrail system or safety net.

☐ The potential for fall hazards exists:

- On surfaces with unprotected sides or edges
- In hoist areas
- Around uncovered holes
- Near excavations
- During elevated construction operations

- On top of roofs

☐ Fall protection can be provided by employing:

- Guardrail systems
- Safety net systems
- Personal fall arrest systems
- Positioning device systems
- Warning line systems
- Controlled access zones
- Safety monitoring systems

☐ Personal fall arrest systems include lifelines, ropes, straps, body belts, harness, and related hardware components.

NOTE: *This reference does not provide detailed guidance for selection of personal flotation devices or personal fall arrest systems.*

Section 13

Identifying Person-Equipment Hazards

The purpose of this section is assist in the identification of workplace/task person-equipment. The specific categories of person-equipment hazards presented in this section include hazards from the PPE that the worker is wearing.

13.1 Types of Person-Equipment Hazards

☐ Person-equipment hazards are hazards that are created by the person or the equipment they are wearing.

☐ Examples of person-equipment hazards include:

- Creation of dangerous or destructive environments
- Decrease in worker function
- Increase in worker's potential for heat stress
- Reduction of PPE performance through wear and use (e.g., durability)

☐ Person-equipment hazards must be evaluated after the PPE has been chosen since the hazard created by the clothing or equipment will depend on the specific item chosen.

☐ Table 13-1 summarizes the types of person-equipment hazards, the body areas or body systems affected, and how these hazards are prevented or minimized.

Table 13-1. Overview of Person-Equipment Hazards

Person-Equipment Hazard	Body Areas Usually Affected	Relevant Occupations	Mitigation Methods
PPE creation of particles	Effect on environment	Cleanroom operations Semiconductor manufacturer	Use of controlled environments coupled with PPE which prevents particles coming from workers
PPE creation of static electricity	Effect on environment	Cleanroom operations Semiconductor manufacturer	Use of controlled environments coupled with use of static treatments or conductive material in PPE
Sensitization or allergic reaction from PPE contact	Entire body Hands	Any operation involving incompatible materials in PPE	Limit need for PPE; Use alternative materials
PPE retention of contamination	Entire body Respiratory system	Any operation where contamination occurs	Use contamination resistant PPE; Use disposable PPE; Apply proper PPE decontamination
PPE with loose straps or material	Entire body Arms Hands	Operations where loose clothing or straps are used	Use low profile PPE; Secure loose material or straps
PPE with poor interfaces	Neck Head/Face Wrists Ankles	Any operation where exposed or poor protected PPE interface areas can occur	Use PPE designed to work together; Use auxiliary PPE to cover interfaces
PPE reduction in mobility	Entire body Head Arms Legs	Any use of full or partial body PPE	Select PPE which offers least restriction of movement without loss of needed protection
PPE reduction in hand function	Hands	Any use of gloves or other handwear	Select gloves or other handwear which still permits acceptable hand function (dexterity, tactility, grip) without loss of needed protection

Person-Equipment Hazard	Body Areas Usually Affected	Relevant Occupations	Mitigation Methods
PPE impairment of vision	Eyes	Any use of face or eyewear; or other PPE items shielding eyes	Select eyewear, or other PPE which shield eyes to provide acceptable vision clarity and field of vision without loss of needed protection
PPE impairment of communications	Hearing/Speaking	Any use of PPE which affects wearer hearing	Select PPE which permit intelligible communications; Use alternative devices (e.g., radios) to assist communications
Lack of footwear ankle support	Feet	Any use of footwear in operation with uneven ground	Select footwear which provides good support in ankle region without loss of needed protection
Lack of PPE back support	Torso	Any use of PPE where significant lifting is required	Train workers in proper lifting techniques; Ergonomically design work tasks; Use PPE with auxiliary back support or separate back support belts
Difficulty in PPE use	All areas	Any use of PPW where difficulty is encountered in donning, adjusting, using, or doffing	Select PPE which is easy to use; Train workers in proper use
Increase in potential for heat stress	Entire body	Any use of full or partial body PPE under high heat or humid conditions	See Table 6 for Environmental Hazards (High heat/humidity)
Decrease in PPE performance by wear and use	Entire body	Any use of PPE	Select PPE with sufficient durability to provide needed protection during intended service life
Difficult serviceability	Entire body	Any use of PPE	Select PPE which is easy to inspect/service

13.2 Creation of Hazardous Conditions or Environments

☐ In some cases, the PPE itself can create a hazardous environment.

☐ Examples of hazards created by PPE include:

- Generation of particles in a particle-sensitive environment
- Generation of static charge in a static charge-sensitive environment
- Sensitization and allergic reactions from contact with PPE materials
- Absorption of hazardous materials which then contact the skin (ease of contamination or failure of decontamination to remove hazardous substances)
- Excess clothing or straps that can get caught in machinery or other obstructions

13.2.1 PPE Creation of Particles

☐ Some work environments such as cleanrooms for electronic and pharmaceutical manufacture require air free of particulate contamination to protect the purity of the product.

☐ In these situations, PPE must possess certain surface characteristics in which external surface are particle-free and non-linting.

☐ Processes for maintaining and cleaning PPE for use in cleanroom environments must also leave PPE particle free.

☐ In some applications, PPE contact with workplace solvents must not allow leaching (extraction) of chemicals from the PPE materials into the environment causing contamination of the manufacturing process.

13.2.2 PPE Creation of Static Electricity

☐ Creation of static electricity from PPE can destroy sensitive electronic equipment (for manufacturer of semiconductors) or ignite flammable or explosive atmospheres (hazardous environments).

☐ Static electricity can be controlled by providing surface treatments (anti-static) of PPE or by using conductive fiber material based PPE.

13.2.3 Sensitization or Allergic Reaction from PPE Contact

- ☐ Materials may contain substances that irritate, cause allergic reactions of the skin, or create toxic reactions.
- ☐ Common examples of PPE skin irritation problems include:
  - Latex rubber protein sensitization
  - Chromium content in leather
  - Organic solvents in adhesives used for construction of PPE items
- ☐ Materials should be tested for biocompatibility before use in PPE (*see 32.2*)
- ☐ Skin irritation or allergic reactions can be prevented by substitution of materials or use of PPE materials that have demonstrated biocompatibility.

13.2.4 PPE Retention of Contamination

- ☐ Some reusable PPE can retain chemicals, biological materials, and other hazardous substances and must be decontaminated or sterilized before reuse.
- ☐ Decontamination or sterilization processes may not always be effective in removing all contaminants.
- ☐ Remaining contamination in PPE can continue to contact skin or become aerosolized or made airborne to create hazards to wearer of PPE upon subsequent wearings.
- ☐ Residual contamination in PPE ('matrix' contamination) can be avoided by using:
  - Contamination resistant PPE
  - Disposable or single use PPE
  - Proper cleaning, decontamination, and sterilization procedures demonstrated to be effective for specific contaminants

13.2.5 PPE with Loose Straps or Material

- ☐ Extra material or straps associated with PPE can create hazards by getting caught on rough surfaces or in machinery.
- ☐ Hazards from loose materials or straps can be minimized by:

- Choosing form-fitting PPE which has a low profile
- Properly wearing PPE by securing loose material or straps (some applications may permit duct tape for this purpose, however, duct tape must be suitable for the full range of expected hazards.)

13.2.6 Lack of Appropriate PPE Interfaces

☐ The use of multiple items of PPE for protection against specific hazards may leave "gaps" in the wearer's protection if the interfaces between PPE items are poor.

☐ PPE interfaces include:

- Upper and lower torso garment overlap (mid-torso)
- Upper torso garment (collar) to hood interface (neck)
- Hood to respirator or face/eyewear interface (face)
- Garment sleeve end to handwear interface (wrist)
- Garment trouser cuff to footwear interface (ankle)

☐ Potential PPE interface problems include:

- Leakage of hazardous substances through interface areas
- Lower insulation or performance of interface areas
- Accidental separation of PPE item interfaces during use created exposed areas on the wearer's body

☐ PPE interface performance is particularly critical in environments where hazards affect all parts of the body or parts of the body that are covered by multiple PPE items.

☐ PPE interface problems can be avoided by:

- Design of PPE items such that different items fit together or can be properly integrated in an ensemble (*see Section 46*)
- Use of anciliary items on PPE to assist in proper interface area protection (such as tape, when permitted, or accessory garments)

13.3 Decrease in Worker Function

- ☐ By virtue of its protective qualities, usually for isolating the worker from the hazard or hazards, PPE creates burdens on the wearer which decrease work function.
- ☐ Examples of PPE effects on worker function include:
  - Reduction of worker's mobility or range of motion
  - Decline in worker's hand function (e.g., dexterity, tactility, and grip)
  - Impairment of worker's vision
  - Impairment of worker's hearing or ability to communicate
  - Lack of adequate ankle support for work on uneven surfaces
  - Lack of adequate back support for lifting operations
  - Difficulty in donning (putting on) or doffing (taking off) PPE
- ☐ These person-equipment hazards must be assessed by observing workers using current or intended PPE in actual or simulated tasks.
- ☐ The sizing and fit of PPE is critical to its optimum performance and to limit effects on worker function.
- ☐ Often optimization of worker function represents a tradeoff with PPE protective performance.

13.3.1 PPE Reduction in Mobility

- ☐ PPE clothing and equipment designs may restrict movement by design or use of thick, bulky, or heavy materials that encumber worker movement or range of motion.
- ☐ Special applications such as confined space entry may require low profile PPE which enables workers to easily enter and exit confined spaces through restricted doors, holes, or hatchways.
- ☐ Restricted movement may create falling or tripping hazards or lessen worker ability to escape a hazardous situation.
- ☐ Reductions of mobility usually come with the provision of specialized protection and therefore represent tradeoffs between worker protection and ergonomics.

☐ Minimization of PPE impact on mobility should be accomplished by selecting PPE that provides the best range of movement without sacrificing needed protection.

13.3.2 PPE Reduction in Hand Function

☐ Gloves and other handwear often diminish hand function in terms of:

- Dexterity (ability to manipulate objects)
- Tactility (ability to sense objects by touch)
- Grip (ability to grasp and hold onto objects)

☐ Reductions of hand function results from providing specialized protection and therefore represent tradeoffs between worker protection and ergonomics.

☐ Minimization of glove/handwear impact on hand function should be accomplished by selecting PPE which provides the best dexterity, tactility, and grip required for worker function without sacrificing needed protection.

13.3.3 PPE Impairment of Vision

☐ Facewear, eyewear, and other items of PPE that cover the eyes or provide a shield of hazards often restrict worker vision in terms of:

- Clarity
- Field of vision

☐ Impairment of vision results from providing specialized protection and therefore represents a tradeoff between worker protection and ergonomics.

☐ Minimization of PPE impact on worker vision should be accomplished by selecting PPE that provides acceptable clarity and field of vision without sacrificing needed protection.

13.3.4 PPE Impairment of Communications

☐ PPE that cover the ears and mouth may often restrict worker speaking and hearing.

- ☐ Impairment of worker communications results from providing specialized protection and therefore represents a tradeoff between worker protection and ergonomics.
- ☐ Minimization of PPE impact on worker communications should be accomplished by:
  - Selecting PPE which provides acceptable understanding communications from speaking and hearing
  - Using alternative methods for enhancing communication:
    - -- Radios with special microphones
    - -- "Speaking" diaphragms in full facepiece respirators

13.3.5 Lack of Footwear Ankle Support

- ☐ Looseness or poor fit of footwear can create hazards by not properly supporting the ankle leading to possible strains or sprains when walking on uneven or rough surfaces.
- ☐ Proper ankle support is provided by choosing footwear with good fit, especially in the ankle area without sacrificing need protection.

13.3.6 Lack of PPE Back Support

- ☐ Operations involving repeated heavy lifting or other strenuous activity in addition to the wearing of heavy or bulky PPE may create back strain.
- ☐ Back strain may be avoided by:
  - Training workers in proper lifting techniques
  - Ergonomically designing work tasks
  - Use of PPE which provides back support
  - Use of auxiliary back support belts
- ☐ Controversy over the effectiveness of back supports continue with one group advocating their use and an opposing group indicating that they do little to prevent back strain.

13.3.7 Difficulty in Donning or Doffing PPE

- ☐ Difficulty in putting on (donning), adjusting, using, or taking off (doffing) PPE can discourage workers from properly wearing PPE.
- ☐ PPE should be selected so that workers can easy don, adjust, use, and doff PPE.
- ☐ Employers must ensure that workers are properly trained in using PPE including its donning, adjustment, wearing, and doffing.

13.4 Increase in Worker Potential for Heat Stress

- ☐ The wearing of some forms of PPE causes physiological stress. This physiological stress is increased in hot and humid working environments.
- ☐ Heat stress hazards are described in detail in *6.2, High Heat and Humidity Exposures*.

13.5 Reduction of PPE Performance through Wear and Use

- ☐ Even if appropriate PPE is selected, the ability for correctly used PPE to provide adequate protection is dependent on:
  - Continued protective performance of PPE over its intended service life
  - The ability to inspect and service PPE.

13.5.1 Poor Durability

- ☐ Durability-related person-equipment hazards are avoided by carefully selecting PPE that maintains its performance properties throughout the period of intended use and from the proper care and maintenance of PPE.
- ☐ PPE should maintain sufficient performance properties during use period and for any successive period. PPE may be intended for:
  - Multiple reuses
  - Limited reuse
  - Disposable after a single use
- ☐ The life cycle of PPE is dependent on several factors including durability. Durability is demonstrated when PPE performance properties do not significantly decline following:

- Use or wearing
- Cleaning, decontamination, or sterilization
- Maintenance
- Storage

☐ Additional information on PPE durability is provided in Section 34.

13.5.2 Limited Serviceability

☐ Since most PPE is recognized to have finite service life, equipment service life can be extended by proper care and maintenance, or serviceability.

☐ The inability to properly service PPE can create a hazard; these hazards arise for reusable PPE when:

- PPE cannot be adequately inspected
- PPE cannot be maintained or repaired

☐ Inability to inspect PPE can occur because:

- In some cases, decreases in PPE performance cannot be readily discerned by the wearer or end user organization.
- PPE is designed so that inspections cannot be made (e.g., a liner material with the barrier material facing inside the garment that can only be examined if the inspector disassembles the PPE).

☐ PPE may be difficult to maintain or repair because:

- Proper instructions (readily understood and detailed) are not provided by the manufacturer (*see Section 36 on Labeling*)
- Maintenance materials or repair supplies are not available for servicing PPE

☐ PPE reservicing hazards are overcome by selecting PPE that is easily inspected, maintained, and repaired (or serviced).

Section 14

Determining Relative Risk and Establishing Protection Needs

The purposes of this section are to assist in the determining relative risk and to establish protection needs to be met by PPE. This section provides a form and instructions for completing a risk assessment for a particular workplace setting.

14.1 Determination of Relative Risk

☐ Establish a Risk Assessment Form for listing:

- Hazards identified for the specific workplace/task
- Body areas or body systems affected by the respective hazard
- The rating associated with the likelihood or frequency of the respective hazard
- The rating associated with the severity or consequences of the respective hazard

☐ Table 14-1 provides a generalized, sample Risk Assessment Form which allows marking the respective body areas or body systems affected by each hazard, but does not permit individual rankings for each hazard by body area. *A modification of this form, appearing in Appendix C, can be tailored to specific work applications or occupations.*

☐ Fill in the Risk Assessment Form using each row to indicate a separate hazard with the associated information and ratings.

☐ For each identified hazard, multiply the rating for the likelihood (or frequency) of the hazard by the rating for the severity (or consequences) of the hazard to determine the risk rating. Risk ratings of 0, 1, 2, 3, 4, 5, 6, 8, 9, 10, 12, 15, 16, 20, and 25 are possible.

Table 14-1. Sample Risk Assessment Form

Hazard Category/ Specific Hazards	Primary Body Area(s) or Body System(s) Affected										Probability Rating(P)	Severity Rating (S)	Risk (P x S)
	Whole Body	Torso	Head	Face or Eyes?	Arms	Hands	Legs	Feet	Resp. System	Hearing			
Physical Hazards													
Falling objects													
Flying debris													
Projectile/ballistic													
Abrasive/rough surfaces													
Sharp edges													
Pointed objects													
Slippery surfaces													
Excessive vibration													
Environmental Hazards													
High heat/humidity													
Ambient cold													
Wetness													
High wind													
Insufficient/bright light													
Excessive noise													
Chemical Hazards													
Inhalation													
Skin contact (vapor)													
Skin contact (liquid)													
Skin contact (particles)													
Ingestion													
Injection													
Chemical flashover													
Chemical explosions													

Table 14-1. (continued)

Hazard Category/ Specific Hazards	Primary Body Area(s) or Body System(s) Affected										Probability Rating(P)	Severity Rating (S)	Risk (P x S)
	Whole Body	Torso	Head	Face or Eyes?	Arms	Hands	Legs	Feet	Resp. System	Hearing			
Biological Hazards													
Bloodborne pathogens													
Airborne pathogens													
Biological toxins													
Biological allergens													
Thermal Hazards													
High heat (contact)													
High heat (convective)													
High heat (radiant)													
Flame impingement													
Hot liquids and gases													
Molten substances													
Electrical Hazards													
Electric shock													
Electrical arc flashover													
Static charge buildup													
Radiation Hazards													
Ionizing radiation													
Non-ionizing radiation													
Person-Position Hazards													
Lack of visibility													
Drowning													
Falling from elevation													
Creation of particles	*												
Creation of static electricity	*												

Table 14-1. (continued)

Hazard Category/ Specific Hazards	Primary Body Area(s) or Body System(s) Affected										Probability Rating(P)	Severity Rating (S)	Risk (P x S)
	Whole Body	Torso	Head	Face or Eyes?	Arms	Hands	Legs	Feet	Resp. System	Hearing			
Person-Equipment Hazards													
Material incompatibility													
Retained contamination													
Loose straps/material													
Unprotected interfaces													
Reduced mobility													
Reduced hand function													
Impaired vision													
Poor communications													
Lack of ankle support													
Lack of back support													
Difficult use													
Heat stress													
Wear through use													
Difficult serviceabilty													

? Includes vision; * Impact on environment; Shaded squares represent unlikely affected body areas or body systems for specific hazards

14.2 Establishment of Protection Needs

- ☐ Using risk determinations from the Risk Assessment Form, rank all hazards associated with the workplace/task:
  - Those hazards with the highest amount of risk should be assigned higher priority for prevention or minimization.
  - Those hazards with "0" risk or low risk should be assigned lower priority for prevention or minimization.
- ☐ Examine possible engineering or administrative controls for those hazards with the highest risk:
  - Engineering controls can encompass changes in the task or process or use of protective shields or other designed measures that eliminate or reduce possible exposure to hazards.
  - Administrative controls can include changes in tasks or work practices to eliminate or limit employee exposure time to a hazard.
- ☐ If engineering or administrative controls are not possible, examine different types of PPE for elimination or reduction of exposure to hazards
- ☐ From information about affected body area or body systems, decide which type of PPE can be used to eliminate or minimize the hazard.
  - In many cases, the type of PPE to be used will be obvious and limited to a single general type (for example, inhalation hazards can be protected against from a respirator).
  - In other cases, there may be several types of PPE that can be used to provide the needed protection.
  - Table 14-2 provides some assistance for selecting general types of PPE for protecting specific body areas and body systems by general hazards (*a more specific form can be created using the blank form provided in Appendix C*).
- ☐ List the engineering controls, administrative controls, or general type(s) of PPE needed on the Protection Strategy Form shown in Table 14-3 (*a blank sample form is provided in Appendix C*).
- ☐ Proceed with an evaluation of PPE design features, performance properties, and service life as described in the relevant following sections.

Table 14-2. Recommended PPE by Type of Hazard? SEE WARNING

Hazard Category/ Specific Hazards	Recommended Personal Protective Equipment (PPE)							
	Full Body Garments	Partial Garments	Gloves or Handwear	Footwear	Headwear	Eye or Facewear	Respirators	Hearing Protectors
Physical Hazards								
Falling objects				X	X			
Flying debris						X	X	
Projectile/ballistic		X						X
Abrasive/rough surfaces		X	X	X				
Sharp edges		X	X	X				
Pointed objects		X	X	X				
Slippery surfaces			X	X				
Excessive vibration			X					
Environmental Hazards								
High heat/humidity	(1)							
Ambient cold	X	X	X	X	X	X		
Wetness	X	X	X	X	X			
High wind	X	X				X		
Insufficient/bright light					X			
Excessive noise								X
Chemical Hazards								
Inhalation							X	
Skin contact (vapor)	X	X	X	X	X	X		
Skin contact (liquid)	X	X	X	X	X	X		
Skin contact (particles)	X	X	X	X	X	X		
Ingestion						X		
Injection	X	X	X	X	X	X		

Table 14-2. (continued)

Hazard Category/ Specific Hazards	Recommended Personal Protective Equipment (PPE)							
	Full Body Garments	Partial Garments	Gloves or Handwear	Footwear	Headwear	Eye or Facewear	Respirators	Hearing Protectors
Chemical Hazards (continued)								
Chemical flashover	X	X	X	X	X	X	X	
Chemical explosions	X	X	X	X	X	X	X	X
Biological Hazards								
Bloodborne pathogens	X	X	X	X		X		
Airborne pathogens							X	
Biological toxins	X	X	X	E		X	X	
Biological allergens		X	X	X		X	X	
Thermal Hazards								
High heat (contact)	X	X	X	X	X	X		
High heat (convective)	X	X	X	X	X	X	X	
High heat (radiant)	X	X	X	X	X	X	X	
Flame impingement	X	X	X	X	X	X		
Hot liquids and gases	X	X	X	X	X	X	X	
Molten substances	X	X	X	X	X	X		
Electrical Hazards								
Electric shock		X	X	X				
Electrical arc flashover	X	X	X	X	X	X		
Static charge buildup	X	X		X				
Radiation Hazards								
Ionizing radiation	X	X	X	X	X	X	X	
Non-ionizing radiation	X	X	X	X	X	X		

Table 14-2. (continued)

Hazard Category/ Specific Hazards	Recommended Personal Protective Equipment (PPE)							
	Full Body Garments	Partial Garments	Gloves or Handwear	Footwear	Headwear	Eye or Facewear	Respirators	Hearing Protectors
Person-Position Hazards								
Lack of visibility		X						
Drowning		X(2)						
Falling from elevation		X(2)						
Person-Equipment Hazards								
Creation of particles	X	X	X	X				
Creation of static elect.	X	X	X	X				
Material incompatibility	X	X	X	X	X	X	X	X
Retained contamination	X	X	X	X	X	X	X	X
Loose straps/material	X	X	X	X	X	X	X	X
Unprotected interfaces	X	X	X	X	X	X	X	
Reduced mobility	X	X			X	X	X	
Reduced hand function			X					
Impaired vision	X	X			X	X	X	
Poor communications	X	X			X	X	X	X
Lack of ankle support				X				
Lack of back support	X	X						
Difficult use	X	X	X	X	X	X	X	X
Heat stress	X	X	X	X	X	X	X	
Wear/serviceability	X	X	X	X	X	X	X	X

(1) Wearing of PPE does not protect against high heat and humidity. (2) Specialized equipment items not generally considered partial body garments

WARNING?This table provides generalized PPE recommendations for starting purposes only. PPE must be selected based on a detailed risk assessment and analysis of PPE design, performance, and service life.

Table 14-3. Protection Strategy Form

Hazard Category/ Specific Hazards	Principal Body Areas or Body Systems Affected	Risk Rating	Priority Area?	Engineering or Administrative Control?	Recommended Protection Strategy (Type of PPE, Design, and Performance)
Physical Hazards					
Falling objects					
Flying debris					
Projectile/ballistic					
Abrasive/rough surfaces					
Sharp edges					
Pointed objects					
Slippery surfaces					
Excessive vibration					
Environmental Hazards					
High heat/humidity					
Ambient cold					
Wetness					
High wind					
Insufficient/bright light					
Excessive noise					
Chemical Hazards					
Inhalation					
Skin contact (vapor)					
Skin contact (liquid)					
Skin contact (particles)					
Ingestion					
Injection					
Chemical flashover					

Table 14-3. (continued)

Hazard Category/ Specific Hazards	Principal Body Areas or Body Systems Affected	Risk Rating	Priority Area?	Engineering or Administrative Control?	Recommended Protection Strategy (Type of PPE, Design, and Performance)
Chemical Hazards (continued)					
Chemical explosions					
Biological Hazards					
Bloodborne pathogens					
Airborne pathogens					
Biological toxins					
Biological allergens					
Thermal Hazards					
High heat (contact)					
High heat (convective)					
High heat (radiant)					
Flame impingement					
Hot liquids and gases					
Molten substances					
Electrical Hazards					
Electric shock					
Electrical arc flashover					
Static charge buildup					
Radiation Hazards					
Ionizing radiation					
Non-ionizing radiation					
Person-Position Hazards					
Lack of visibility					
Drowning					
Falling from elevation					

Table 14-3. (continued)

Hazard Category/ Specific Hazards	Principal Body Areas or Body Systems Affected	Risk Rating	Priority Area?	Engineering or Administrative Control?	Recommended Protection Strategy (Type of PPE, Design, and Performance)
Person-Equipment Hazards					
Creation of particles					
Creation of static elect.					
Material incompatibility					
Retained contamination					
Loose straps/material					
Poor interfaces					
Reduced mobility					
Reduced hand function					
Impaired vision					
Poor communications					
Lack of ankle support					
Lack of back support					
Difficult use					
Heat stress					
Wear through use					
Difficulty serviceability					

Section 15

General Types and Classification of PPE

The purposes of this section are to describe the general roles of PPE available in different applications and to provide a system of classifying PPE by design, performance, and service.

15.1 The PPE Marketplace

- ☐ The types of PPE available in the marketplace and thus the choices available to the end user are rapidly increasing.
- ☐ PPE exists in a variety of designs, materials, and methods of construction, each having advantages and disadvantages for specific protection applications.
- ☐ End users should have an understanding of the different types of PPE and their features in order to make appropriate selections.

WARNING: ***PPE that is similarly designed may offer different levels of performance. Examine PPE performance in addition to design and features.***

- ☐ PPE must be properly sized to provide adequate protection.

WARNING: ***Improperly sized or ill fitting PPE may reduce or eliminate protective qualities of PPE.***

15.2 General Classification of PPE

- ☐ Classification of PPE is useful for understanding its intended use.
- ☐ PPE may be classified by:
  - Design
  - Performance
  - Intended service life

☐ The system used for PPE classification in this book is illustrated in Figure 15-1.

15.2.1 Classification by Design

☐ Classification of PPE by its design usually reflects how the item is configured or the part of the body area or body systems that it protects. For example, footwear by design provides protection to the wearer's feet.

☐ The types of PPE can be generally categorized as:

- Full body garments
- Partial body garments
- Gloves and handwear
- Footwear
- Headwear
- Face and eye wear
- Hearing protectors
- Respirators

☐ Classification of PPE by design may also provide an indication of specific design features that differentiate PPE items of the same type.

- For example, closed-circuit self-contained breathing apparatus are configured with significant design differences when compared to open-circuit self-contained breathing apparatus.

☐ Some designs of PPE may offer varying protection against hazards in different parts of the PPE item.

- For example, the palm material in a glove palm may provide a better grip surface than the glove's back material.

WARNING: ***PPE coverage of a specific body area, in and of itself, does not guarantee protection of that body area.***

☐ An important design feature of all PPE is sizing as it affects fit of the PPE on the individual wearer; Poor fitting PPE can significantly diminish wearer function and create additional hazards.

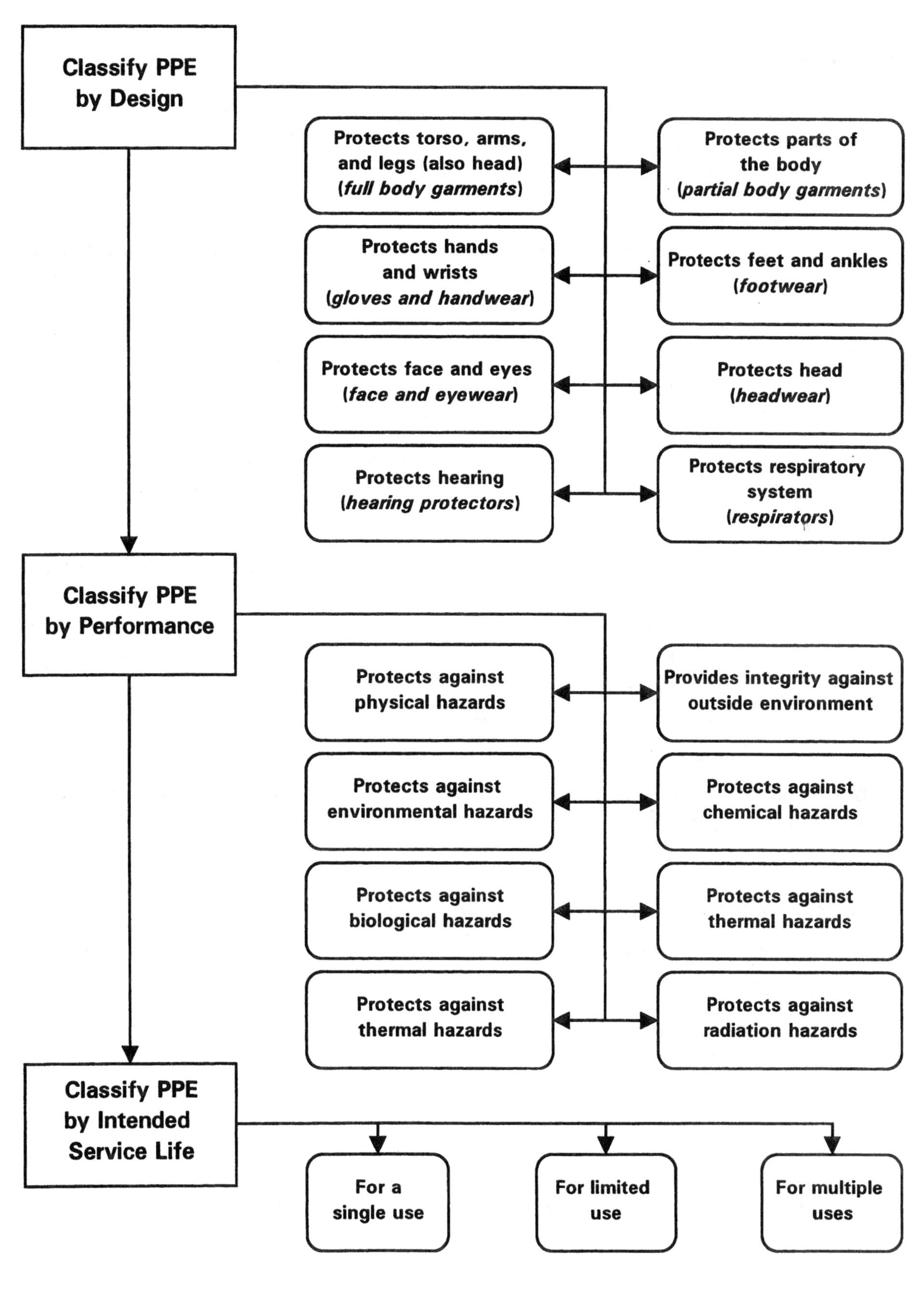

Figure 15-1. System for Classifying PPE

15.2.2 Classification by Performance

- ☐ Classification of PPE by performance indicates the actual level of performance to be provided by the item of PPE. This may include a general area of performance or a more specific area of performance.
  - For example, while two items of PPE might be considered to be chemical protective clothing, one item may provide an effective barrier to liquids but not to vapors while the other item provides an effective barrier to both liquids and vapors.
- ☐ Classification of PPE by performance is best demonstrated by actual testing or evaluations of PPE with a standard test. *(See Sections 24 through 32)*
- ☐ Comparative evaluations of PPE performance against specific hazards will often establish a hierarchy of performance for ranking PPE in that particular application area.
  - For example, in the area of biological resistance, some protective garment materials may demonstrate fluid resistance by prevent most liquid penetration under ordinary conditions, other materials may demonstrate fluid penetration resistance by preventing liquid protection under most conditions, and still other materials may demonstrate viral penetration resistance in which individual organisms are prevented from penetrating garment fabrics.

WARNING: ***Intended or manufacturer-claimed performance does not always match actual performance.***

15.2.3 Classification by Expected Service Life

- ☐ Classification of PPE by expected service life is based on the useful life of the PPE item.
- ☐ PPE may be designed to be:
  - Reusable
  - Used a limited number of times (Limited Use)
  - Disposable after a single use
- ☐ Classification of PPE by expected service life is based on:
  - Durability
  - Life cycle cost

- Ease of reservicing

☐ Durability is determined by evaluating how the item of PPE maintains its original performance properties following the number of expected uses. Assessments of PPE durability should account for:

- Normal "wear and tear" on the PPE item
- Wear created by servicing, cleaning, or maintaining PPE

☐ The life cycle cost of PPE is the total cost for using an item of PPE and is usually represented as the cost per use for a PPE item.

☐ The following costs should be considered in determining the life cycle cost:

- Purchase cost
- Labor cost for selection/procurement of PPE
- Labor cost for inspecting PPE
- Labor and facility cost for storing PPE
- Labor and materials cost for cleaning, decontaminating, maintaining, and repairing PPE
- Labor and fees for retirement and disposal of used PPE

☐ The total life cycle cost is determined by adding the separate costs involved in the PPE life cycle and dividing by the number of PPE items and number of uses per item. *(See Section 35, PPE Service Life and Life Cycle Cost).*

☐ If an item of PPE cannot be reserviced to bring it to an acceptable level of performance, then it cannot be reused

WARNING: ***Expected service life does not always equal actual use life.***

Section 16

Full-Body Protective Garment Designs, Features, and Sizing

The purpose of this section is to describe the different types of full-body protective garments in terms of their general design, materials of construction, design features, and sizing.

16.1 Application

☐ Full-body garments are designed to provide protection to the wearer's upper and lower torso, arms, and legs

☐ Full-body garments may also provide protection to the wearer's hands, feet, and head when auxiliary PPE is integrated with the garment to form a suit

☐ The level of protection varies with garment design; many garment designs do not provide uniform protection for all areas of the body covered by the garment.

☐ Full-body garments may offer protection against any one or combination of the following hazards:

- Physical
- Environmental
- Chemical
- Biological
- Thermal
- Electrical
- Radiation
- Person-Position

- Person-Equipment

16.2 General Design

☐ Full-body garments may be single or multi-piece clothing items:

- Full body suits
- Jacket and trouser combinations
- Jacket and overall combinations
- Coveralls

☐ Examples of full body garments are illustrated in Figure 16-1.

16.2.1 Full Body Suits

☐ Full-body suits are one-piece garments that may offer a variety of entry options depending on the type and placement of the closure.

☐ Full-body suits, sometimes known as totally encapsulating suits, may encapsulate the wearer and other protective equipment such as the respirator depending on the design and intended application.

☐ Full-body suits may include other items of PPE integrated with or attached to the suit such as gloves and footwear.

16.2.2 Jacket and Trouser Combinations

☐ Jacket and trouser combinations mimic normal wearing apparel:

- Jackets or coats generally provide some overlap of the waist portion of trousers with or without a collar, hood, or wrist protection.
- Trousers usually rely on a fly, drawstring, or snaps and may have openings with closures at the foot end to allow entry while wearing footwear.

16.2.3 Jacket and Overall Combinations

☐ Jacket and overall combinations have a higher "bib" style pants and permit a shorter jacket.

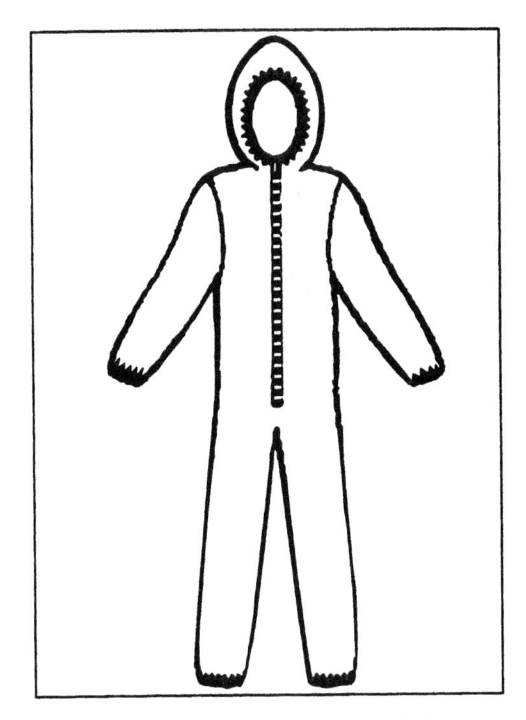

(a) Hooded coverall with front closure and elasticized openings

(b) Bib overall (without jacket)

(c) Firefighter multilayer coat and pants

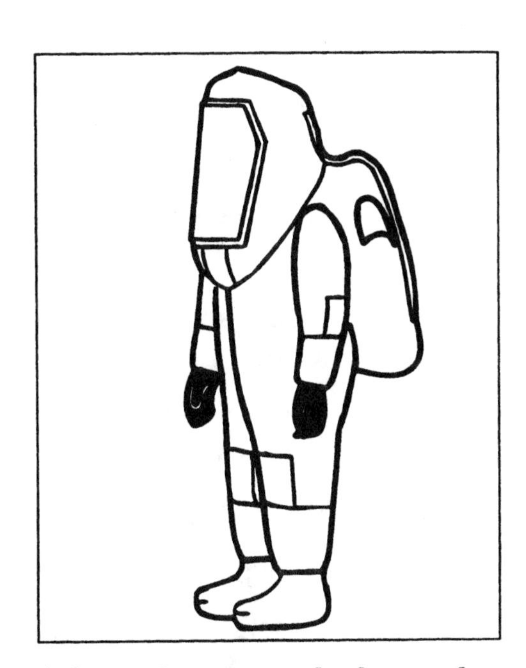

(d) Total encapsulating suit with attached gloves and booties

Figure 16-1. Examples of Full Body Protective Garments

- ☐ Overalls often use straps for adjust garment on lower torso.
- ☐ Area of body covered by overall will vary with manufacturer design.

16.2.4 Coveralls

- ☐ Coveralls are one-piece garments, usually with a front closure, and have options for:
  - Attachment of hoods
  - Type of sleeve end
  - Type of trouser cuff end

16.2.5 Interface Areas of Full Body Garments

- ☐ Hoods on garments may be permanently attached or removable. Removable hoods may be attached by:
  - Buttons
  - Zippers
  - Hook and loop closure tape
- ☐ Sleeve ends may be of multiple designs:
  - Open sleeves allow for no adjustment.
  - Sleeves may be adjusted by buttons, snaps, or hook and loop closure tape.
  - Sleeves may have knit wristlets; wristlets may have thumbholes or bar strips to secure sleeve to hand.
  - Sleeves may have devices for attaching gloves or have flaps or other means for providing protection to the sleeve end to glove interface.
- ☐ Trouser cuff ends may be of multiple designs:
  - Straight cuffs allow for no adjustment.
  - Tapered cuffs are designed to fit securely around the ankles.
  - Cuffs may be adjustable by using buttons, snaps, or hook and loop closure tape.
  - Cuffs with zippers allow expansion of trouser end for donning while wearing footwear and may be covered with protective flaps.

- Trouser ends may be designed for attaching footwear or have flaps or other means for protecting the trouser cuff to footwear interface.

16.3 Materials of Construction

☐ Full-body garments are constructed of materials appropriate for the specific application; typical materials include:

- Woven, non-woven, or knit textiles
- Leather
- Unsupported rubber materials
- Unsupported plastic materials
- Rubber coated fabrics
- Plastic coated fabrics
- Aluminized fabrics

☐ Full-body garments may be constructed of several layers to provide insulation or other properties related to comfort.

- Linings may enhance comfort by wicking away body perspiration.
- Barrier materials may prevent the penetration of liquids.
- Batting materials may provide insulation from severe heat or cold.

16.4 Design Features

☐ Features affecting the design of full-body garments include:

- Type and location of seams; depending on the material involved, seams may be:
  - -- Sewn
  - -- Sewn and bound with a separate material piece
  - -- Heat sealed
  - -- Glued and taped
  - -- Sewn and glued or heat sealed
- Type, length, and location of the closure system(s); common closure systems include:
  - -- Snaps

    - Hooks and dees
    - Zippers
    - Two-track closure systems
    - Hook and loop tape
    - Adhesive strips or tapes
- Amount of overlap for multi-piece garments
- Design of interface areas with other PPE
    - Sleeve to glove
    - Trouser cuff to footwear
    - Collar to hood
- Type, size, and location of pockets
- Types, function, and location of hardware

16.5 Sizing

☐ There are few uniform sizing practices for the design of full-body garments.

- ANSI/ISEA 101-1996 provides a system of sizing for disposable or limited use coveralls.
- ASTM F 1731 provides a practice for determining body measurements and sizing for fire and rescue uniforms.

☐ The availability of sizing often depends on the design of the garment and the relative volume of garments sold by the manufacturer.

☐ Sizing may be based on:

- Individual measurements for custom sizing
- Numerical sizing for wearing apparel
- Alphabetic sizing (e.g., small, medium, and large)
- Two or more wearer dimensions, such as
    - Height and weight
    - Inseam and waist size
    - Collar size and sleeve length
    - Height and chest circumference

- ☐ Sizing between manufacturers is often inconsistent.
- ☐ Sizing of protective garments often does not address the needs of special worker populations.

Section 17

Partial Body Protective Garment Designs, Features, and Sizing

The purpose of this section is to describe the different types of partial body protective garments in terms of their general design, materials of construction, design features, and sizing.

17.1 Application

☐ Partial body garments provide protection to only a limited area of the wearer's body including the upper torso, lower torso, arms, legs, neck, or head.

☐ Partial body garments may offer protection against any one or combination of the following hazards:

- Physical
- Environmental
- Chemical
- Biological
- Thermal
- Electrical
- Radiation
- Person-Position
- Person-Equipment

17.2 General Designs and Materials of Construction

☐ Partial body protective garments vary significantly with the application. Examples of partial body protective garments are:

- Hoods, head covers, and bouffants for head/or face protection
- Aprons, gowns, smocks, lab/shop coat, and vests for front or upper torso protection
- Sleeve protectors for arm protection
- Chaps, legging, or spats for leg protection

☐ Different examples of partial body protective garments are illustrated in Figure 17-1.

☐ Specific design features and sizing vary with the type of partial body garment.

☐ Partial body protective garments use the same materials and construction features as used for full body garments.

17.3 Hoods, Head Covers, Bouffants

17.3.1 Hoods

☐ Hoods cover the wearers head and include either a face/eye opening or may be provided with an integrated faceshield or visor.

☐ Hoods materials may be made of:

- Knit materials which conform to the wearer's head shape
- Woven or non-woven textiles, coated fabrics, rubber or plastic materials that are relatively loose fitting

☐ Hood features and options include:

- Pullover or front closure styles
- Type, size of visor and method of integration if present
- Use of different face opening designs:
    - Elastic
    - Pliable, stretchable material
    - Drawstrings
- Ties, snaps, or hook and loop closures (for closure styles)
- Size of face opening
    - Full face
    - Eye slit

(a) Full length apron

(b) Lab coat

(c) Long style hood

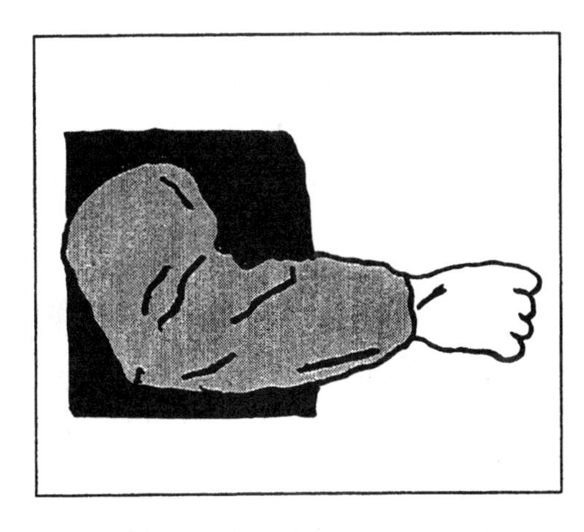

(d) Sleeve protector with elasticated ends

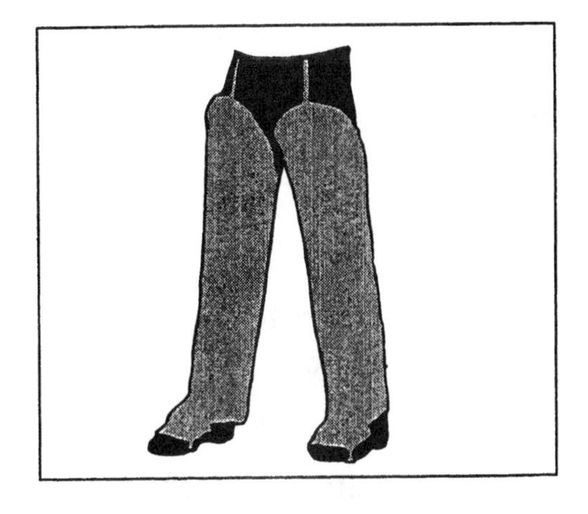

(e) Full leg chaps

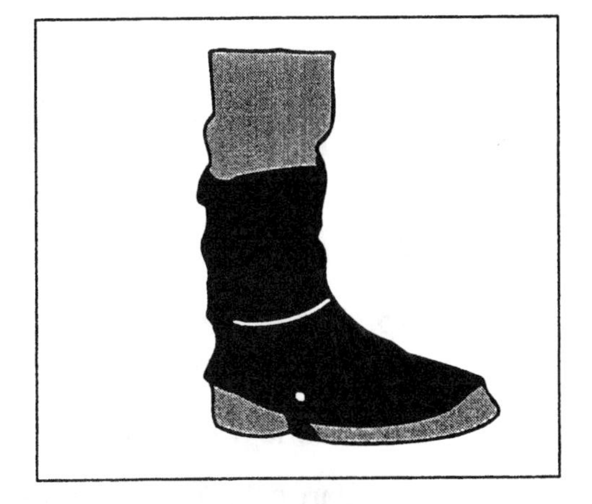

(f) Leather spats

Figure 17-1. Examples of Partial Body Protective Garments

  - -- Eye holes
  - Length below neck:
    - -- Short (to base of shoulders only)
    - -- Long (extends down onto shoulders)
    - -- Extra long (extends over shoulders)
    - -- 'Dickey" style: extends longer in front and back, short on sides

- ☐ Hoods are usually offered in limited number of sizes
- ☐ The hood length and size of hood affects how it will fit with other PPE
  - Length affects integration with upper torso garments
  - Face opening size affects integration with eye/face protection and respirators
  - The bulk of the hood on the wearer's crown may affect the fit of helmets and respirator straps

17.3.2 Head Covers and Bouffants

- ☐ Head covers or bouffants provide protection to the wearer's upper head, but are most often used to contain or cover the wearer's hair.
  - Bouffant style head covers are secured to the wear's head by elastic around the periphery.
  - Head covers may also be used as helmet covers as an aid to preventing contamination.

17.4 Aprons, Gowns, Smocks, Lab/Shop Coats and Vests

17.4.1 Aprons

- ☐ Aprons consist simply of a flat piece of material contoured to the front of the body for providing lower or lower/upper torso protection
  - The lower part extends around the wearer's sides and ties in the back.
  - Larger aprons covering the upper torso are designated 'bib' aprons; a strap around the wearer's neck holds up the 'bib'.
- ☐ Apron sizing depends on areas of protection:

- Aprons are generally available in more than one size, designated by unisex rectangular dimensions, and are adjusted by the tie straps at the top or sides
- Smaller aprons (24" x 24" or less) are classified by some manufacturers as 'waist' aprons and have belt-like tie straps.

☐ "Split-leg" aprons have the bottom portion divided, and separate ties secure each "leg" portion to the wearer.

☐ Aprons may include pockets for carrying tools or other items.

17.4.2 Gowns

☐ Gowns are usually full-length garments that provide protection to the wearer's front torso, upper legs, and arms.

☐ Gowns are typically designed to minimize seams on the front of the garment.

☐ Gowns typically have a partially open back.

☐ Gowns use ties or snaps that close at the back of the garment.

17.4.3 Smocks and Lab/Shop Coats

☐ Smocks and lab/shop coats are partial body garments that provide front torso, arm, and upper leg protection.

☐ Smocks and lab/shop coats can be of varying lengths and are generally offered in alphabetic sizing.

☐ Smocks and lab/shop coats generally have front closures and open neck areas.

☐ The type and location of pockets is usually a design option.

17.4.4 Vests

☐ Vests are upper torso garments that are usually used for physical protection or mounting high visibility materials:

☐ Vests are usually designed with front and back panels with open sides.

☐ Adjustment of vest sizing is generally by side straps or ties.

17.5 Sleeve Protectors

☐ Sleeve protectors are partial protective garments that provide protection from the wearer's hand to the shoulder area.

☐ Sleeve protectors may be secured to the wearer's arm by several means:

- The elasticity of some materials provides fit around the circumference of the wearer's arm.
- Other sleeve protectors use sewn-in elastic, elastic bands, or hook and loop closure material at the garment ends.
- An adjustable belt or strap may beat the top of the sleeve protector.

☐ Sizing of sleeve protectors is affected by length and method of hand end finish:

- Sleeve lengths begin at 10 inches (25 mm) and may extend to up to 24 inches (600 mm).
- There are three styles of hand end finishes:
  -- Open end
  -- Bar-tacked or thumb hole for thumb
  -- Bar-tacked for all fingers with thumb hole

☐ Sleeve protectors extending the entire length of the arm are called 'cape' sleeves. To secure this garment, a strap across the back connects the shoulder ends of the two sleeves. (This style of sleeve protector is used in leather-based products that do not have the elasticity of knit sleeve protectors for attaining proper fit.)

17.6 Chaps, Leggings, and Spats

17.6.1 Chaps

☐ Chaps are partial pants which are open at the sides/back and are intended to provide wearer front leg protection.

☐ Some chaps are also designed to provide protection to the back of the wearer's legs.

☐ Chaps may be flared at the footwear ends to provide upper foot protection.

- ☐ There are two predominate design styles of chaps that affect sizing and fit:
  - The first style incorporates a belt at the waist that is adjusted like a belt used for regular trousers.
  - The second style has includes loops or straps which fit onto the wearers belt.
- ☐ Most styles of chaps use side ties to secure to the chaps to the wearer's legs so that different leg circumferences can be accommodated.
- ☐ Manufacturers of chaps make these products in different lengths, corresponding to different inseam measurements, usually ranging from 27 to 33 inches (680 to 840 mm).

17.6.2 Leggings

- ☐ Leggings are partial protective garments intended to protect the wearer's lower leg and have the following features:
  - Knee height
  - Elastic top or adjustable tops
  - Snap or tie closures
  - A flare for protecting the top of the shoe (in some designs)
  - Adjustable instep straps to secure legging under footwear (in some designs)

17.6.3 Spats

- ☐ Spats are partial protective garments intended to protect the wearer's ankle, and upper foot, having the same features as leggings (except for height).

Section 18

Protective Gloves and Other Handwear Designs, Features, and Sizing

The purpose of this section is to describe the different types of protective gloves and other handwear in terms of their general design, materials of construction, design features, and sizing.

18.1 Application

☐ Handwear provides protection to the wearer's hands and wrists or portions of the wearer's hands and wrists.

☐ Some handwear may have gauntlets to protect the lower arm.

☐ Handwear may offer protection against any one or combination of the following hazards:

- Physical
- Environmental
- Chemical
- Biological
- Thermal
- Electrical
- Radiation

18.2 General Designs

☐ Hand protection may be provided to the wearer in several forms

- Five-fingered gloves

- Two-fingered gloves
- Mittens
- Partial gloves
- Fingerless gloves
- Finger guards
- Finger cots
- Hand pads

☐ Different types of protective handwear are illustrated in Figure 18-1.

18.2.1 Five-fingered Gloves

☐ Five fingered gloves are traditional style gloves providing full coverage to the wearer's hand.

☐ Of PPE affording complete hand protection, five fingered gloves offer the maximum hand function.

☐ Five-fingered gloves are used when complete hand protection must be provided in combination with full hand functionality.

18.2.2 Two-fingered Gloves

☐ Two fingered gloves group the last three fingers inside a mitt-like covering but provide separate digit covers for the wearer's index finger and thumb.

18.2.3 Mittens

☐ Mittens group the wearer's fingers inside one covering but allow for individual covering of the wearer's thumb.

☐ Mittens are most common in applications requiring thermal insulation from excessive heat or cold.

18.2.4 Partial Gloves

☐ Partial gloves generally provide protection to the wearer's fingers and palm, but have an open back.

☐ Partial gloves are generally used for physical protection applications.

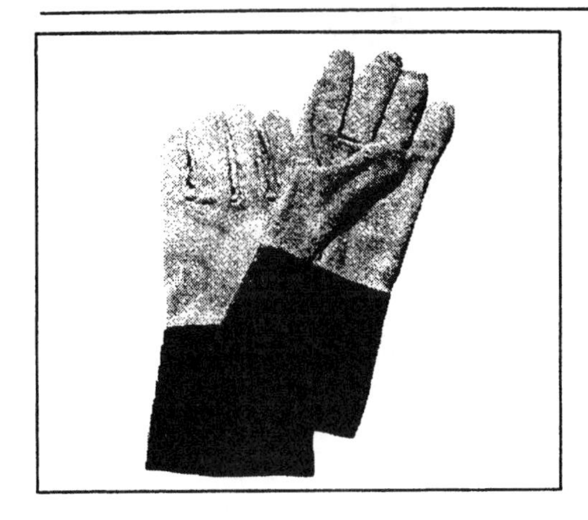

(a) Leather work glove - Gunn cut

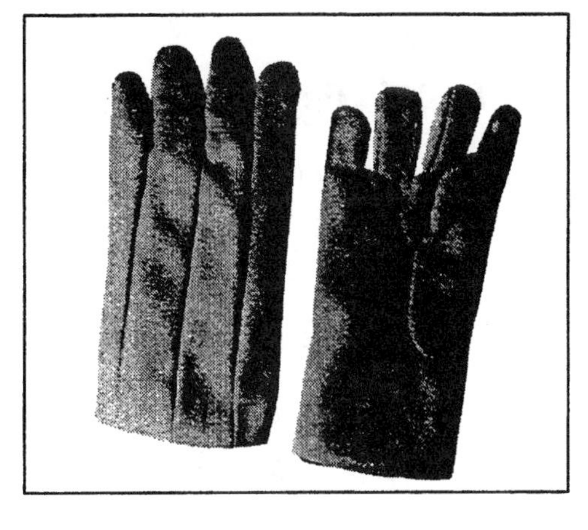

(b) Cotton work glove - Clute cut

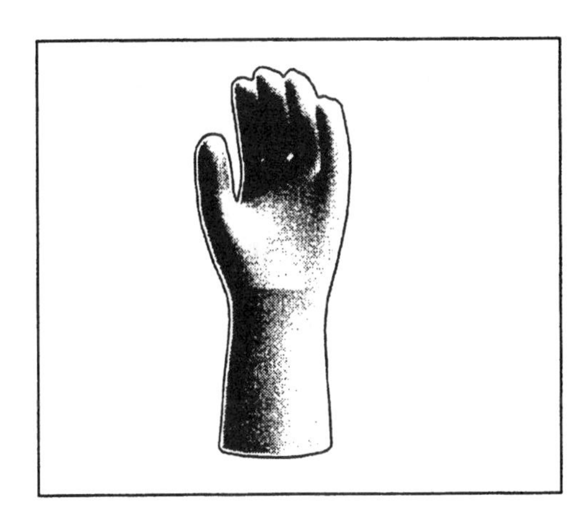

(c) Rubber glove (unsupported)

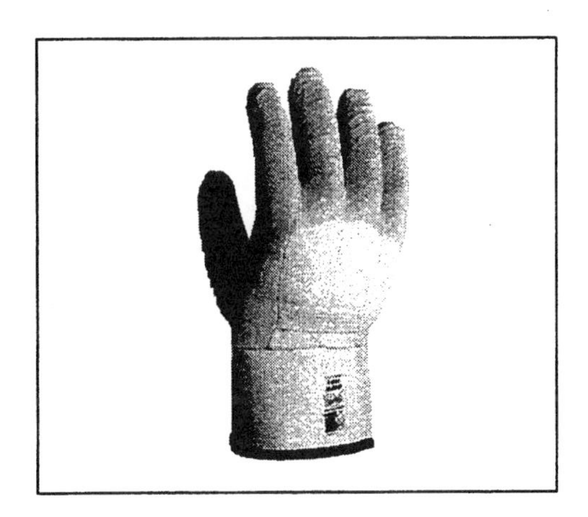

(d) Rubber-coated fabric glove

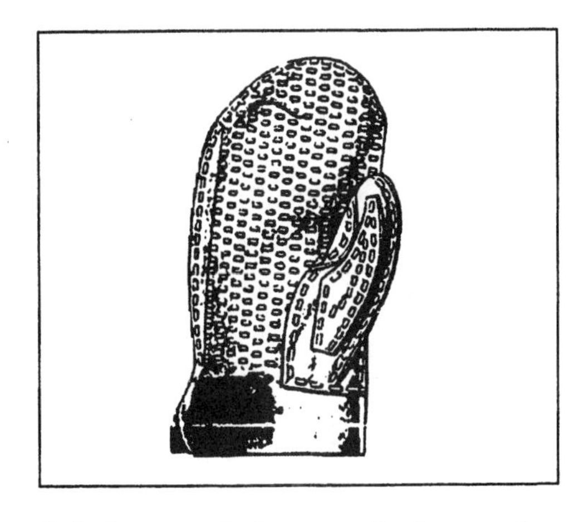

(e) Steel ribbon-reinforced mitten

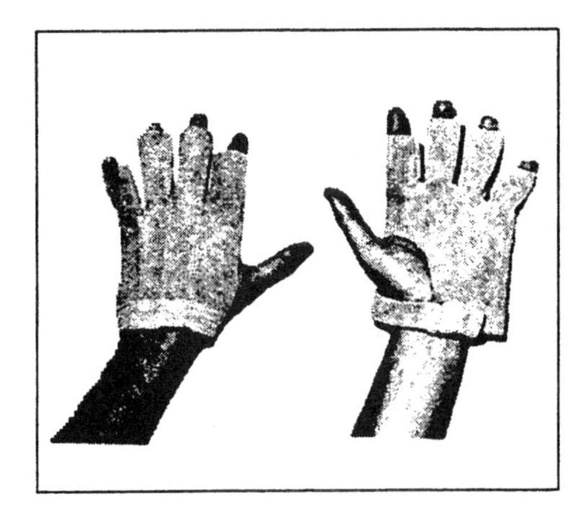

(f) Leather fingerless glove

Figure 18-1. Examples of Protective Gloves and Other Handwear

18.2.5 Fingerless Gloves

- ☐ Fingerless gloves provide protection to the wearer's palm and back of hand, but only partial protection to the lower fingers; leaving the finger tips are left free and unprotected.
- ☐ Fingerless gloves are generally used for physical protection applications.

18.2.6 Finger Guards

- ☐ Finger guards provide protection for the entire finger, and generally are available in three styles:
  - Flat styles that have an elastic back and open end
  - Wrap around styles which have an elastic back, open end, and provide protection around the complete circumference of the wearer's finger
  - Closed end style
- ☐ Finger guards are generally used for physical protection applications.

18.2.7 Finger Cots

- ☐ Finger cots fit around the fingertip and usually extend over the first joint of the wearer's finger.

18.2.8 Hand Pads

- ☐ Hand pads are generally intended to provide protection to the palm side of the hand but some designs may offer protection to the back side of the hand.
- ☐ Designed much like an envelope, many hand pads are rectangular shaped items into which the entire hand fits.
- ☐ Some envelope type hand pads may have a thumb slot.
- ☐ Hand pads offering protection to the palm side of the hand use a wrist strap to help secure the hand pad in place.

18.3 Materials of Construction

- ☐ Gloves and other handwear are constructed of materials appropriate for the specific application; typical materials include:

- Leather
- Cotton and other natural textile knit or woven materials
- Synthetic fiber knit or woven materials
- Aluminized fabrics
- Rubber coated or impregnated fabrics
- Rubber
- Plastic
- Metal mesh

☐ Gloves fabricated from materials using rubber-coated fabrics are known as supported gloves.

☐ Rubber gloves without a supporting fabric are known as unsupported gloves.

☐ Gloves and other handwear may be constructed in several layers to provide insulation or other properties related to comfort.

- Linings may enhance comfort by wicking away hand perspiration.
- Barrier materials may prevent the penetration of liquids.
- Batting materials may provide insulation from severe heat or cold.

18.4 Design Features

☐ Principal glove design features include:

- Type of glove construction
- Cuff designs
- Grip designs

18.4.1 Glove Construction

☐ Glove construction styles include:

- ***Two Piece:*** the glove liner is made from two pieces of material sewn in the general form of a glove.
- ***Clute Cut:*** the front of the glove liner is made from one piece, while the back is sewn from a number of parts. This style provides a formed glove and usually has a knit wrist (see below).
- ***Gunn Cut:*** The back of the glove liner is made from one piece.

-- The second and third fingers are set in with a seam across the bottom of each.

-- The seams in the finger areas extend two-thirds of the way around each finger, to eliminate exposed seams in the common glove wear areas.

- ***Dipped Gloves:*** Gloves are created by dipping a metallic or porcelain glove form into a vat of rubber.

  -- Different processes are used depending on the polymer being used.

  -- Coated gloves are manufactured by dipping hand forms on which woven or knit fabric gloves have first been placed.

- ***Heat-sealed Gloves:*** Gloves are created by sealing the peripheral edges of two-hand silhouette shaped piece of flat material.

18.4.2 Cuff Designs

☐ Textile, leather, or other fabric-based gloves have can have different cuff designs:

- ***Knit wrists cuffs*** are designed to hold the glove in place on the hand and to prevent debris from entering the glove.

  -- As such they provide a relatively tight fit in the form of a wristlet around the arm at or above the wrist.

  -- Because of this construction, knit wrist cuff designs should not be used if rapid glove removal may be necessary.

- ***Safety cuffs*** are an extended piece of material attached to the glove by a seam at the wrist.

  -- Safety cuffs provide additional protection to the wrist area and slide on and off easily.

  -- Safety cuffs are typically made of more rigid material that remains firm even when exposed to perspiration

- ***Slip on (or band top) cuffs*** allow easy donning and doffing and are continuous with the rest of the glove (no seam is used).

- ***Gauntlets*** are similar to safety cuffs but extend farther up the arm to provide protection to the lower forearm.

-- Gauntlets are designed to allow for maximum forearm dexterity at different overall lengths.

-- Glove length is generally measured from the tip of the middle finger to cuff edge of the glove.

- ***Closure ends*** are designs with an open end that is secured around the wrist with a closure, usually a hook and loop closure or an adjustable strap.

- Cuff designs for unsupported gloves include:

  -- Straight edges

  -- Pinked or serrated edges

  -- Rolled cuffs (which usually give greater cuff strength)

18.4.3 Grip Designs

☐ Grip designs enhance wearer gripping power by use of special finishes on the glove palm:

- Finishes offer textures that range from relatively smooth to relatively rough.

- The grip on supported gloves is often affected significantly by the substrate fabric, because the coating conforms to the texture of the liner. Coated or impregnated gloves may include the following possible finishes:

  -- Dipped (unfinished)

  -- Sanded-on

  -- Textured or raised

  -- Wrinkle or crinkled

  -- Rough

  -- Breathable or porous (for impregnated fabric gloves)

- Grip finishes for unsupported gloves generally involve different types of raised patterns (embossing).

18.5 Sizing

☐ Glove sizing is based on either numerical hand sizes or qualitative size ratings, such as small, which may be based on hand size.

- One example of glove sizing system is shown below:

Glove size	XS	S	M	L	XL
Hand size	6-7	7-8	8-9	9-10	10-11

- Hand size is generally based on the circumference, in inches, of the wearer's hand around the palm area.

☐ Glove fit may be based on:

- Hand circumference
- Hand length
- Individual digit circumference
- Individual digit length

☐ Section 6-32 of NFPA 1977, *Standard on Protective Clothing for Wildland Fire Fighting*, provides a method for evaluating glove fit on the hand by comparing predicted fit versus qualitative fit ratings.

☐ Glove length will vary from gloves that end at the wearer wrist crease to those gloves that extend to the upper arm of the wearer.

WARNING: ***Not all gloves offer uniform protection over the entire glove material. For example, some gloves may provide different protection in palm or back areas as compared to glove gauntlets or wristlets.***

☐ The availability for glove sizes will depend on the glove style and relative sales volume:

- Larger volume glove styles may be offered in full and half sizes.
- Gloves sold in small volumes may be offered only in small, medium, and large sizes.

Section 19

Protective Footwear Designs, Features, and Sizing

The purpose of this section is to describe the different types of protective footwear in terms of their general design, materials of construction, design features, and sizing.

19.1 Application

☐ Footwear provides protection to the wearer's feet or portions of the wearer's feet.

☐ Depending on the footwear height, additional protection may be afforded to the wearer's ankles and lower and upper legs.

☐ Footwear may offer protection against any one or combination of the following hazards:

- Physical
- Environmental
- Chemical
- Biological
- Thermal
- Electrical
- Radiation

19.2 General Designs

☐ Foot protection may be provided to the wearer in several forms:

- Shoes
- Boots
- Overshoes or overboots
- Shoe or boot covers
- Metatarsal footwear
- Toe protectors or caps

☐ Different types of protective footwear are illustrated in Figure 19-1.

19.2.1 Shoes

☐ Shoes provide protection to the wearer's feet up to or below the ankles.

19.2.2 Boots

☐ Boots provide protection to the wearer's feet and lower legs.

☐ "Wader" style boots provide protection to the wearer's upper legs as well.

19.2.3 Overshoes and Overboots

☐ Overshoes or overboots are worn over the wearer's street or protective shoes or boots.

☐ Overshoes or overboots are primarily used as a protective cover over regular work shoes or boots and usually do not offer any physical protection to the wearer.

19.2.4 Shoe or Boot Covers

☐ Shoe or boot covers also fit over shoes and boots but are generally made of relative soft textile, rubber, or plastic materials.

19.2.5 Metatarsal Footwear

☐ Metatarsal footwear is footwear specifically designed to provide protection to the wearer's metatarsal area (the region of the foot between the ankle and the toes).

(a) Leather safety shoe with steel toe and laces

(b) Leather work boot with steel toe, laces, and stud posts

(c) Rubber work boot with steel toe and puncture-resistant plate

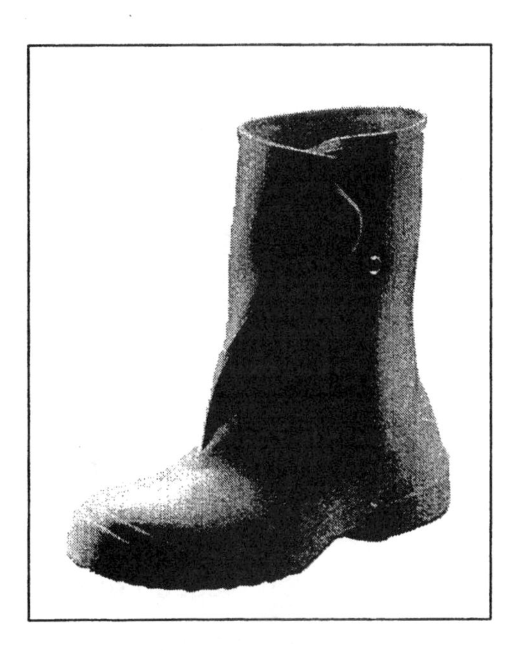

(d) Rubber overboot - no steel toe

Figure 19-1. Examples of Protective Footwear

19.2.6 Toe Caps

- ☐ Toe caps provide protection to the wearer's toes and are generally worn over existing work shoes or boots for physical protection only.

19.3 Materials of Construction

- ☐ The majority of footwear is constructed with leather as the primary material.
- ☐ Other footwear materials include:
  - Rubber or plastic coated fabric
  - Unsupported rubber (for overshoes and overboots)
  - Textiles (for shoe or boot covers)
- ☐ Leather or textile footwear is usually constructed on a last or foot form for creating the desired footwear item size.
- ☐ Rubber or plastic footwear is typically fabricated by injection molding in one or more stages.
- ☐ Boot linings may consist of woven, nonwoven textiles, coated barriers, or foam materials.

19.4 Design Features

- ☐ Footwear consists of the following primary features:
  - Sole (bottom of footwear)
  - Heel (back part of the sole)
  - Ball (front part of the sole)
  - Upper (all portions of the footwear above the sole)
  - Vamp (sides of the footwear)
  - Toe (front section of the footwear)
  - Quarter (back part of footwear that extends up wearer's ankles)
  - Throat (open front area of footwear where the closure system is located)
  - Gusset (tongue or piece of footwear inside of throat)
  - Top line (top of footwear)

19.4.1 Soles

- ☐ Soles provide contact with walking surfaces and are designed to offer different levels of traction (slip resistance) and wear (abrasion) resistance.

- ☐ Soles vary with the material, tread pattern, and depth of tread; Common sole types include:
    - Plain
    - Chevron (for moving debris out while walking)
    - Safety-Loc (for slip resistance on slippery clean surface)
    - Cleated (for extra gripping action on slippery surfaces)

19.4.2 Closures

- ☐ Closure systems for footwear include:
    - None (slip on)
    - Zippers
    - Laces with eyelets
    - Laces with stud and hook posts
    - Straps and buckles
    - Buckles

- ☐ Footwear closures may extend from:
    - The top line of the footwear down to the metatarsal area
    - The top line of the footwear down part of the footwear upper
    - Near the top line (for strap and some bucket or lace closures systems)

19.4.3 Linings

- ☐ Linings may be provided on the interior of footwear to provide:
    - Insulation from extreme heat or cold
    - Barrier protection from penetrating liquids
    - Comfort by absorbing sweat or providing cushioning

19.4.4 Interior Supports

☐ Metatarsal or arch supports provide additional support for the wearer's metatarsal area or arch.

☐ Insoles may be fixed or removable and provide cushioning for the bottom of the wearer's foot.

19.4.5 Protective Hardware

☐ Protective devices are special hardware used in the construction of the footwear to provide specific forms of physical protection to the wearer's foot:

- Interior toe caps provide protection to the top from impact or compression hazards.
- Metatarsal plates provide protection to the metatarsal or top portion of the foot from impact or compression hazards.
- Puncture resistant devices or midsole plates keep sharp objects from cutting or puncturing the bottom of the footwear.
- Ladder shanks keep the footwear sole from bending on narrow steps or ladder rungs.

19.5 Sizing

☐ The majority of protective footwear uses the same footwear sizing system as employed for standard footwear; this system:

- Is based on the Brannock measuring scale, for which foot length and width are the two key dimensions used for choosing footwear
- Requires the individual to measure his or her foot and then to select the corresponding labeled size (e.g., 7D or 9EE)
- Involves designation of sizes by the manufacturer for their products so that appropriate fit with the corresponding dimensions is provided.

☐ However, one manufacturers size may not fit as well as the same size from a different manufacturer because:

- Manufacturers address fit differently and may use different allowances for ease (the difference between PPE measurements and wearer dimensions).
- Individual feet are shaped differently, and a two dimensional system does not fully capture the characteristics of feet to ensure fit.

☐ Availability of sizing is based on the materials of construction, manufacturing methods, design features, and relative production volume:

- Manufacturers of leather-based products generally offer men's and women's safety footwear in sizes based on standard street shoe sizing.
- Rubber boot makers are more constrained by the individual lasts or molds used for making footwear.
- Protective footwear that deviates from common designs is more likely to be available only in full sizes and one width and may not be provided in women's sizes.
- Footwear with laces or other closure devices with snug footwear uppers to the ankle provide better fit and ankle protection.
- Some protective devices affect flexibility and wearer's foot comfort.
- Linings and insoles, particularly removable insoles which can be customized for the individual wearer, provide cushioning in boots and can be used to help adjust fit.

Section 20

Protective Headwear Designs, Features, and Sizing

The purpose of this section is to describe the different types of protective headwear in terms of their general design, materials of construction, design features, and sizing.

20.1 Application

- ☐ Protective headwear provides protection to the wearer's head or portions of the wearer's head.
- ☐ Headwear may or may not provide protection to the wearer's face.

NOTE: *Hoods are considered partial protective garments (see Section 15).*

- ☐ Head protective devices may offer protection against any one or combination of the following hazards:
    - Physical
    - Environmental
    - Chemical
    - Biological
    - Thermal
    - Electrical
    - Radiation
    - Person-Position
- ☐ Industrial headwear is typically designed for physical and electrical protection.

20.2 General Design

- ☐ The basic form of head protection for industry is the helmet.
- ☐ Helmets consist of three primary components:
  - Shell
  - Suspension (absorbs energy within the shell)
  - Harness (secures the helmet to the wearer)
- ☐ Helmets may also have various accessories, including
  - Brackets for lights
  - Faceshield or brackets for other types of face and eye protective wear
  - High visibility materials
- ☐ Headwear designs are illustrated in Figure 20-1.

20.2.1 Helmet Shell

- ☐ The helmet shell is a dome-shaped covering for the head, which may or may not include a brim.

20.2.2 Helmet Suspension

- ☐ The suspension consists of crown straps and protective padding.
- ☐ Crown straps are the part of the suspension that passes over the head.
- ☐ Protective padding is intended to providing cushioning between the shell and the wearer's head.
- ☐ Suspensions are also designed to provide ventilation between shell and the headband.

20.2.3 Helmet Harnesses

- ☐ Harnesses may include:
  - Headbands
  - Sweatbands
  - Chin straps
  - Nape straps

(a) Standard hardhat

(b) Full, wide brim hardhat

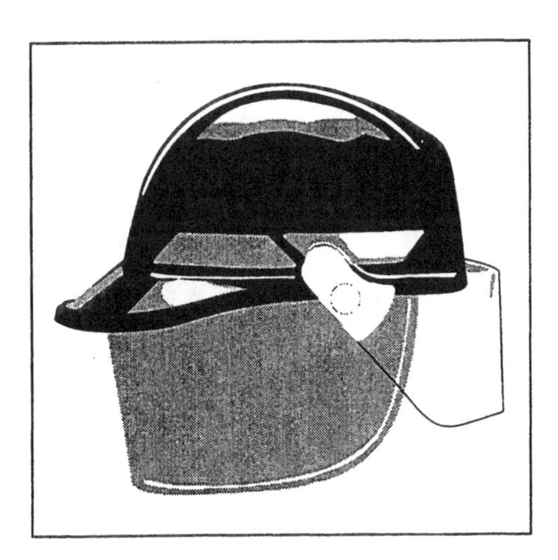

(c) Fire fighter helmet with ear/neck protector

Figure 20-1. Examples of Protective Headwear

- ☐ Headbands are bands in the interior of the shell, which allows for the adjustment of the helmet to fit on the wearer's head.
- ☐ Sweatbands may be integral or removable material attached to the headband for absorbing sweat.
- ☐ Chin straps secure the helmet to the head by means of a strap under the chin.
- ☐ Nape straps go behind the head to secure the helmet.

20.2.4 Accessories

- ☐ Helmets may also include brackets for holding lamps or other accessories such as visors or ear muffs.

20.2.5 Headwear Classification

- ☐ ANSI Z89.1, *Protective Headwear for Industrial Workers—Requirements* defines two impact types and threes electrical classes of protective helmets:
  - Helmet impact types include:
    - -- ***Type 1,*** helmets intended to reduce the force of impact resulting from a blow only to the top of the head
    - -- ***Type 2,*** helmets intended to reduce the force of impact resulting from a blow that may be received off center or to the top of the head
  - Helmet electrical classes include:
    - -- ***Class G (General),*** helmets that are intended to reduce the danger of contact exposure to low voltage conductors
    - -- ***Class E (Electrical),*** helmets that are intended to reduce the danger of exposure to high voltage conductors
    - -- ***Class C (Conductive),*** helmets that are not intended to provide protection against contact with electrical conductors

20.3 Materials of Construction

- ☐ Helmet shells are constructed of hard plastics and composites, typically including:

- Nylon
- High density polyethylene
- Fiberglass
- Vulcanized fiber
- Polycarbonate
- Aluminum

☐ Helmet visors may be:

- Polycarbonate
- Nylon
- Lexan
- Steel mesh
- Materials may include godl-coating for protection against radiant heat.

☐ Most helmet components in the suspension are Nylon, vinyl, or sponge foam for cushioning elements.

☐ Depending on the intended application of these helmets, some materials used in helmets may be heat and flame resistant.

20.4 Sizing

☐ Helmets generally come in a single sized shell; headband adjustments provide individual fit.

- ANSI Z89.1 requires that the headband be adjustable in 1/8-inch (3 mm) hat size increments, with the approximate size range that can be accommodated by the product permanently marked on the helmet.
- The minimum headband size range is 6 to 8 inches (150 to 200 mm).
- Headband sizes are based on circumferential measurements of the head but integral nape straps generally require a further set of measurements for gauging headband sizes.

☐ Manufacturer variations include the style of the headband, its means of adjustment, and the use of accessory straps for securing the helmet. These straps are adjusted with using buckles and sliders.

Section 21

Protective Face/Eyewear Designs, Features, and Sizing

The purpose of this section is to describe the different types of protective face and eyewear in terms of their general design, materials of construction, design features, and sizing.

21.1 Applications and Types

- ☐ Protective face and eyewear provide protection to the wearer's face and eyes or portions of the wearer's face and eyes.
- ☐ Face and eye protection devices may offer protection against any one or combination of the following hazards:
  - Physical
  - Environmental
  - Chemical
  - Biological
  - Thermal
  - Electrical
  - Radiation
- ☐ ANSI Z87.1, *Practice for Occupational and Educational Eye and Face Protection*, defines a number of general types of eye and face protectors. These include:
  - Spectacles (safety glasses)
  - Faceshields
  - Goggles

- Welding helmets
- Hand shields

☐ Protective face and eyewear are illustrated in Figure 21-1.

21.2 Safety Glasses

☐ Safety glasses, also known as spectacles, are protective devices intended to shield the wearer's eyes from certain hazards depending on the spectacle type.

☐ Safety glasses are commonly used to provide protection from impact and optical radiation.

☐ Safety glasses consist of the following components:

- A frame
- A front lens or lenses
- Temples (which secure device on head in front of eyes)
- Bridges (secures device on nose of wearer)
- Sideshields (provide side protection to eyes)

☐ Frames support all parts of the safety glasses and are generally manufactured of:

- Plastic
- Nylon
- Metal
- Titanium

☐ Lens types include:

- Plano, for wearers not requiring vision correction
- Prescription (Rx), for wearers requiring vision correction
- Lenses may be glass or plastic
- Lenses may also have additional treatments:
  - Shaded for absorbing different wavelengths of light (ultraviolet, infrared, laser)
  - surface treatments for resisting fog or glare

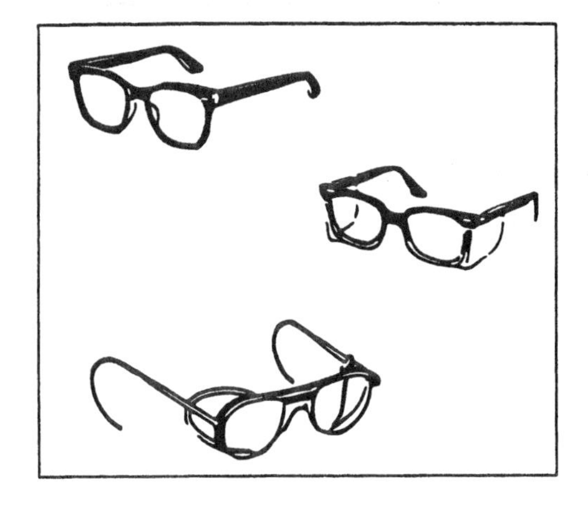

(a) Spectacles

(b) Faceshield

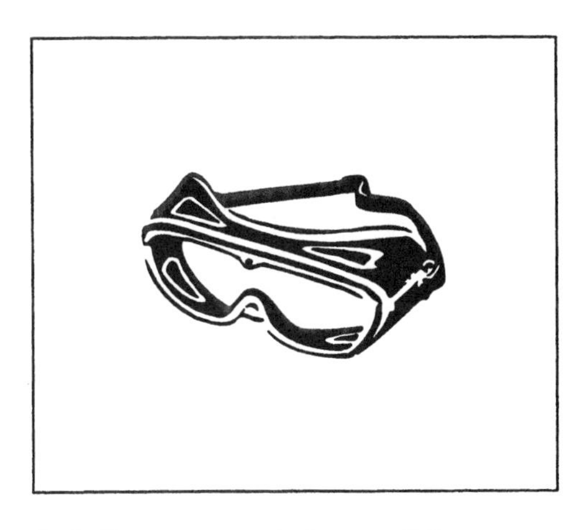

(c) Cover goggles with direct ventilation

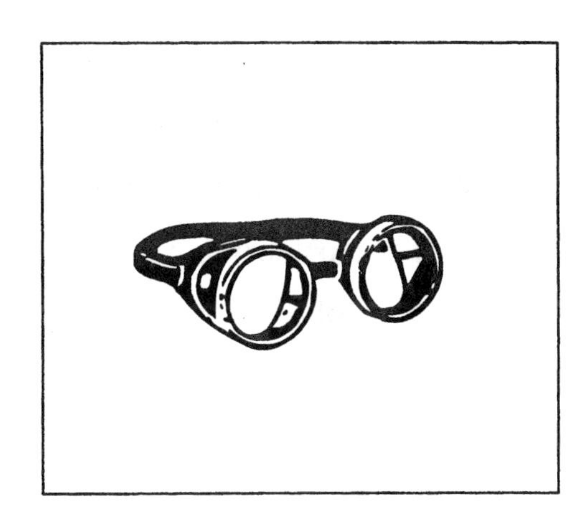

(d) Eyecup goggles

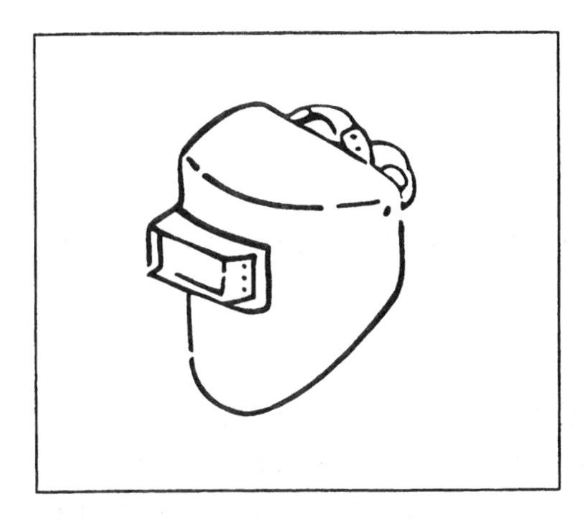

(e) Welding Helmet

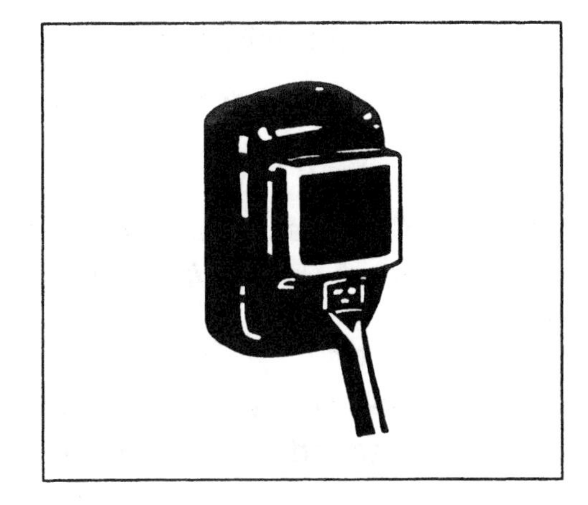

(f) Handshield

Figure 21-1. Examples of Protective Face and Eyewear

☐ Temple types include:

- Spatula (fixed or adjustable): like standard eyeglasses
- Cable (fixed or adjustable): looped temple that fits around ears
- Headband

☐ Front types include:

- Removable lens
- Non-removable lens

☐ Bridge types include:

- Fixed bridge
- Adjustable nose pads
- Adjustable bridge

☐ Sideshields characteristics include:

- Type of surface:
    - Solid
    - Ventilated
- Styles:
    - Flatfold
    - Cup
- Means of attachment:
    - Integrally molded
    - Permanently attached
    - Removable

WARNING: ***Spectacles with removable sideshields are not considered eye protection when the sideshields are detached.***

☐ Special purpose spectacles

- Lift-front spectacles incorporate an additional front, either permanently attached or snap-on, that can be raised or lowered to provide glare and/or optical radiation protection, or to provide for special visual tasks, as needed.

☐ Sizing:

- Most non-prescription safety glasses are available in limited, unisex sizes based on unidentifiable industry "norms."
- Some of the newer styles incorporate a curved design that is contoured to conform to the wearer's face in a one-size-fits-all configuration.
- Manufacturers that offer different sizes generally offer choices of temple size, a typical measurement used in the prescription eyewear industry.
- Safety glasses can be adjusted for individual facial characteristics by adjusting the temple, or in some cases changing the inclination (angle) of the lens (only available on certain types of safety glasses).

☐ Prescription safety glasses must be fitted by a professional.

- The individual is generally sized by a licensed optician in a manner similar to standard eyewear.
- Much of the actual fit process is accomplished by making adjustments at the bridge or by adjusting the temple for fit at the ear by manual bending.

21.3 Faceshields

☐ Faceshields are protective devices intended to shield the wearer's face or portions of the face in addition to the eyes.

☐ ANSI Z87.1 requires that faceshields be used in conjunction with spectacles or goggles. *(Faceshields are not primary protectors).*

☐ Faceshields consist of headgear that supports a window curved to surround and cover the wearer's face.

- Some manufacturers also provide neck and chin protectors.
- The headgear assembly may be provided with or without a crown (top of the head) protector.
- Faceshield windows may also be attached to protective helmets in place of their own headgear.

☐ Sizing:

- Manufacturers offer faceshield windows in a number of sizes, thicknesses, and materials. Therefore, two considerations in sizing this device are:
  - -- The size/shape of the window
  - -- The type of headgear adjustment
- Headgear may include one or two bands for adjustments:

-- One style of band goes around the sides and back of the head.

-- Another style of band goes over the tops and side of the head.

-- Adjustments are typically made by moving the band to the appropriate distance (circumference) which allows the faceshield to stay firmly on the head and in place while being used.

21.4 Goggles

☐ Goggles are face protection devices that are intended to fit the face surrounding the eyes in order to shield the eyes from certain hazards.

☐ Goggles are available in several configurations:

- Styles:
  - Eyecup goggles cover the eye sockets completely
  - Cover goggles are worn over spectacles
- Frame types:
  - Rigid
  - Flexible
- Frame materials:
  - Plastic
  - Polyvinylchloride (PVC)
- Configurations:
  - Direct ventilation (allows direct passage of air)
  - Indirect ventilation (excludes passage of liquids and/or optical radiation)
- Lens types:
  - Impact resistant
  - Shaded for protection against non-ionizing radiation
  - Photochromic (darken when exposed to and fades when removed from ultraviolet light or sunlight)
- Bands
  - Adjustable
  - Non-adjustable

☐ Aspects of goggle design which affect sizing include the rigidity of the frame and the type of adjustment band:

- Typically, flexible goggles are designed to conform to the face when the strap is placed behind the head.
- While many flexible goggles are offered in only one size, some products are available in several sizes, usually represented as small, medium, and large.
- Rigid goggles often require a cushioning material at the edges that fit onto the face, and are more likely to have adjustable straps.

21.5 Welding Helmets and Handshields

☐ Welding helmets and handshields are special types of eye/face protection usually intended to protect the entire face, eyes, ears, and neck front from optical radiation and weld splatter.

☐ These devices may have two types of lenses:

- Stationary
- Lift-front

☐ Welding helmets include some form of headgear with sizing and fit constraints similar to those for faceshields.

☐ Handshields are simply held in front of the face by the wearer during use and have few sizing constraints except that these devices must be large enough to cover the face sizes of the general worker population.

WARNING: ***Welding helmets and handshields must be used in conjunction with safety glasses or goggles.***

Section 22

Respirator Types, Features, and Sizing

The purpose of this section is to describe the different types of protective respirators in terms of their general design and design features.

22.1 Application

☐ Respirators protect the wearer from inhalation of harmful dusts, chemicals, and other respirable substances.

☐ Respirators provides protection to the wearer by:

- Removing contaminants from the air (air-purifying)
- Supplying an independent source of respirable air (atmosphere-supplying)

22.2 Respirator Inlet Covers

☐ All respirators are equipped with respiratory inlet covers or facepieces to provide a barrier from the hazardous atmosphere and for "connecting" the wearer's respiratory system with the respirator.

☐ There are two types of respiratory inlet covers include:

- Tight-fitting
- Loose-fitting

22.2.1 Tight-fitting Respirator Inlet Covers

☐ Tight-fitting respirator inlet covers are facepieces.

☐ Facepieces provide a gas or particle-tight seal with the wearer's face.

☐ Facepieces are usually constructed of moldable elastomer material, facepieces include:

- Elastomeric face covering
- Headbands or straps
- Optional valves
- Connections to air-purifying elements or the air supply

☐ Facepieces may be of three different types:

- ***Quarter-mask facepieces*** cover the wearer's nose and mouth and have a lower sealing surface, which rests between the mouth and the chin.
- ***Half-mask facepieces*** cover the wearer's nose and mouth and have a lower sealing surface that rests under the chin.
- ***Full facepieces*** cover the wearer's nose, mouth, and eyes.

☐ Some respirators may have facepieces constructed of fabric that provides the same level of fit or integrity as elastomeric facepieces.

☐ Another type of tight-fitting respirator inlet cover is a mouthpiece with a nose clamp (generally used only for escape).

☐ Full facepiece respirators may include a nose cup to help maintain positive pressure inside the respirator and to reduce fogging of the facepiece visor.

22.2.2 Loose-fitting Respiratory Inlet Covers

☐ Loose-fitting respiratory inlet covers are clothing-like structures.

☐ Loose-fitting respiratory inlet covers may be of the following types:

- A ***loose-fitting facepiece*** covers only a portion of the head.
- A ***helmet*** covers the entire head and contains rigid headgear to protect the heat against injury.
- A ***hood*** is usually constructed of a flexible fabric and covers the wearer's head and neck or head, neck, and shoulders.
- A ***blouse*** is usually constructed of a flexible fabric and covers the wearer's head and upper torso of the body (with or without the arms).
- A ***suit*** is usually constructed of a flexible fabric and covers the wearer's entire body.

☐ A flexible tube or hose is used to supply respirable air to the covering.

WARNING: ***The supply of respirable air must be sufficient to create a positive pressure within the covering to prevent the leakage of ambient air into the wearer's breathing zone. Not all respirators create positive pressure.***

- ☐ Loose-fitting respiratory inlet covers are used only with either:
  - Air-purifying respirators
  - Atmosphere-supplying respirators

22.3 General Design

- ☐ The National Institute for Occupational Safety and Health (NIOSH) define respiratory designs and requirements in 42 CFR Part 84, *Respiratory Protective Devices, and Tests for Permissibility*.
- ☐ Respirators are generally classified according to their mode of operation:
  - Air-purifying
  - Atmosphere-supplying
- ☐ Respirators are qualified by their purpose:
  - For entry and escape
  - For escape only
- ☐ Respirators are further differentiated by the types of environments they can be used in:
  - Not for oxygen deficient atmospheres (atmospheres containing less than 19.5% oxygen)
  - Not for immediately dangerous to life or health atmospheres (hazardous atmospheres which may produce physical discomfort immediately, chronic poisoning after exposure, or acute physiological symptoms after prolonged exposure)
- ☐ Respirators which rely on finite air supplies or filtering capabilities are also classified by their service time ranging from three minutes to four hours as defined in 42 CFR Part 84 (Approval of respiratory protective devices).
- ☐ Examples of different respirators are illustrated in Figure 22-1.

22.3.1 Air-Purifying Respirators

- ☐ Air-purifying respirators (APRs) operate by removing contaminants from air as it is inhaled into the mask.

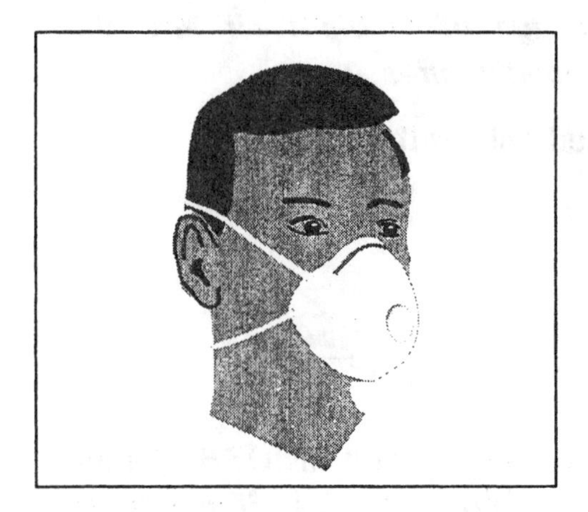

(a) Disposable, half-mask particulate respirator

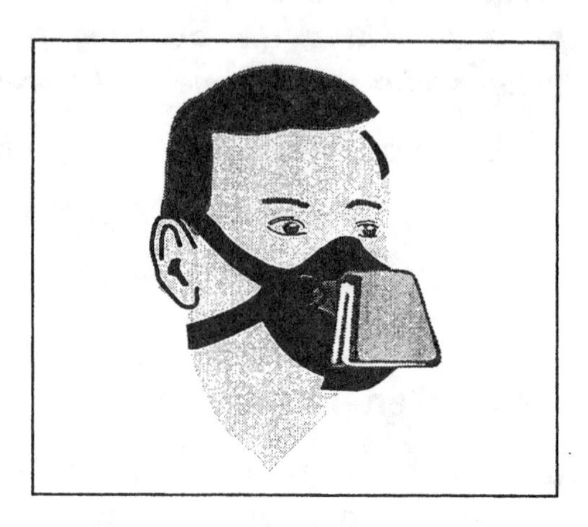

(b) Quarter-mask particulate filter respirator

(c) Half-mask, air-purifying respirator with twin cartridges

(d) Full frontpiece, chin-style gas mask

Figure 22-1. Examples of Respirators

(e) Half-mask, powered air-purifying respirator

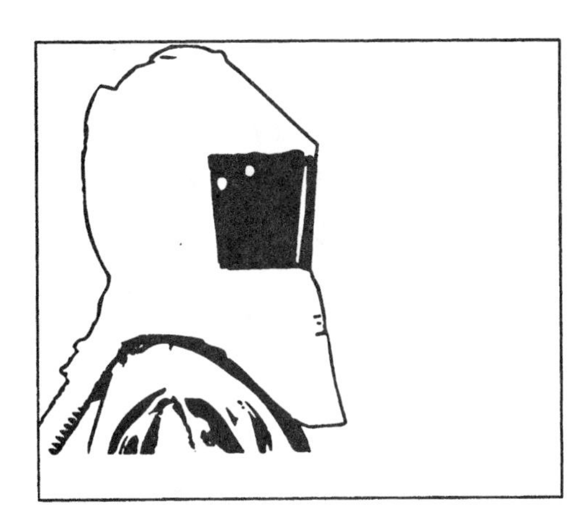

(f) Supplied-air respirator with hood

(g) Continuous flow, full facepiece supplied-air respirator

(d) Full facepiece, chin-style gas mask

Figure 22-1. Examples of Respirators *(continued)*

☐ Air-purifying respirators (APRs) are intended to provide protection to the wearer's respiratory system from inhalation of hazardous atmospheres that are:

- Contaminated with toxic particulate matter, gases, and vapors

WARNING: ***Air-purifying respirators cannot be used in ambient environments that are oxygen-deficient (less than 19.5%).***

☐ General types of air-purifying respirators include:

- ***Non-powered*** APRs which operate by breathing action of wearer:
  - When the wearer inhales, ambient air is pulled through the air-purifying material or element.
  - When the wearer exhales, the exhaled breath is expelled from the respirator through an exhalation valve or the air-purifying material or element.
- ***Powered*** APRs are equipped with a portable or fixed location blower which:
  - Blows airs through the air-purifying material or element
  - Supplies air to the respiratory inlet covering

☐ 42 CFR Part 84 recognizes eight different type of air-purifying respirators:

- Gas masks
- Non-purifying particulate respirators
- Chemical cartridge respirators
- Dust, fume, and mist respirators (particulate respirators)
- Pesticide respirators (particulate respirators)
- Paint spray respirators (particulate respirators)
- Powered air-purifying high efficiency respirators
- Combination gas masks

☐ Air-purifying respirators are provided with air-purifying materials, elements, or devices, which are:

- Particle-removing
- Gas-vapor-removing
- Combination particle-gas-vapor-removing

☐ ***Particle-removing air-purifying materials*** generally use fibrous media to remove particles from an air stream based on one or more of the following principles:

- Gravity settling (particle paths are changed)
- Impact (particles impact fibers and are held in place)
- Diffusion (random motion of particles enables entrapment on fibers)
- Sieving (large particles are trapped in spaces of media)
- Electrostatic attraction (particles are held onto fibers by static electric charges)

☐ ***Gas-vapor-removing air-purifying materials*** use solid materials that remove gases or vapors by:

- Absorption (gas, vapor, or liquid penetrates solid and is held in place)
- Adsorption (gas, vapor, or liquid is retained on surface of solid)
- Catalysis (gas, vapor, or liquid reacts with solid to become different, less harmful substance)

☐ Air-purifying materials, elements, or devices can be:

- Replaceable and must be replaced under the following conditions:
    - When their useful service life has been depleted
    - When breathing resistance becomes excessive
    - When penetration of an air contaminant is detected
- Non-replaceable, for use in disposable non-powered APRs which are constructed so that replacement is not possible

22.3.2 Atmosphere-Supplying Respirators

☐ Atmosphere-supplying respirators provide the wearer with respirable air from a source that is independent of the ambient atmosphere.

☐ Atmosphere-supplying respirators are intended to provide protection to the wearer's respiratory system from inhalation of hazardous atmospheres that are:

- Contaminated with toxic particulate matter, gases, or vapors
- Oxygen-deficient
- At temperature extremes

- ☐ Types of atmosphere-supplying respirators include:
  - Supplied-air respirators
  - Self-contained breathing apparatus
  - Combination supplied-air and self-contained breathing apparatus

- ☐ ***Supplied-air respirators (SARs)*** use a source of respirable air that is stationary and removed from the wearer. SAR use the following air sources and are available in the following types:
  - Respirable air is provided from:
    - Air pumps
    - Air compressors
    - Compressed air cylinders
  - Supplied-air respirators are of two basic types:
    - Airline
    - Hose-mask
  - ***Airline*** SAR connect the air source to the respiratory-inlet covering with a small diameter, flexible hose which is attached to the wearer by a belt or other means with a quick connect/disconnect coupling; Air line SAR may operate in one of three different modes:
    - ***Continuous-flow*** airline SAR maintain at all times a flow of respirable air which is determined primarily by the pressure of the compressed air and the length of flexible hose or by an adjustable valve or orifice located downstream of the quick connect/disconnect coupling.

NOTE: *NIOSH 42 CFR Part 84 requires a respirable air flow of at least 115 liters per minute (4 cubic feet per minute) for respirators with tight-fitting respiratory inlet coverings and 170 liters per minute (6 cubic feet per minute) for respirators with loose-fitting respiratory inlet coverings*

- ***Demand*** airline SAR are equipped with a tight-fitting facepiece and airflow regulating valves. The valve permits respirable air to flow to the facepiece only when the wearer inhales and creates a negative pressure inside the facepiece (with respect to the ambient atmosphere). The valve also stops the flow of respirable air into the facepiece when the wearer exhales and creates a positive pressure inside the facepiece (with respect to the ambient atmosphere).

  - ***Pressure-demand*** airline SAR are similar to demand airline SAR except that the regulating valves maintain positive pressure inside the facepieces during both exhalation and inhalation. Each valve contains a spring that allows it to remain slightly open at all times.

- ***Hose mask*** SAR supply respirable air from a non-contaminated area through a large diameter flexible hose to a respirator-inlet covering. Hose mask SAR are of two types:
  - Hose mask SAR without blowers have tight-fitting full facepieces connected to an area having respirable air by a hose that is limited to a maximum of 75 feet (22.8 meters) by NIOSH 42 CFR Part 84. Facepieces have separate inhalation and exhalation valves.
  - Hose mask SAR with blowers use hand- or motor-operated blowers to push air through the hose to the wearer respiratory inlet covering. The hose is limited to a maximum of 300 feet by NIOSH 42 CFR Part 84.

- NIOSH 42 CFR Part 84 designates six types of supplied air respirators:
  - Type A: Hose mask SAR with motor-driven or hand-operated blower
  - Type AE: Hose mask SAR with motor-driven or hand-operated blower, equipped with additional devices designed to protect the wearer's head and neck from rebounding abrasive material, and with shield material to protect the window of the facepiece, hood, or helmet
  - Type B: Hose mask SAR equipped with a large diameter hose from which the user draws inspired air by means of his or her lungs alone
  - Type BE: Hose mask SAR equipped with a large diameter hose from which the user draws inspired air by means of his or her lungs alone, equipped with additional devices designed to protect the wearer's head and neck from rebounding abrasive material, and with shield material to protect the window of the facepiece, hood, or helmet
  - Type C: SAR with source of respirable air, hose, detachable coupling, control valve, demand or pressure demand valve, an arrangement for attaching the hose to the wearer, and a facepiece, hood, or helmet

-- Type CE: SAR with source of respirable air, hose, detachable coupling, control valve, demand or pressure demand valve, an arrangement for attaching the hose to the wearer, and a facepiece, hood, or helmet, equipped with additional devices designed to protect the wearer's head and neck from rebounding abrasive material, and with shield material to protect the window of the facepiece, hood, or helmet.

☐ ***Self-contained breathing apparatus (SCBA)*** use a source of respirable air, oxygen, or oxygen-generating material that is carried on the body of the wearer. SCBA have the following characteristics and are available in the following types:

- Most SCBA use tight-fitting full facepieces.
- The SCBA user is completely independent of the ambient atmosphere.
- The weight of SCBA is limited to 34 pounds (15.4 kg) by NIOSH 42 CFR Part 84.
- There are two types of SCBA:
  -- Closed-circuit
  -- Open-circuit
- ***Closed-circuit SCBA*** or rebreathers remove the carbon dioxide from the exhaled air (in a canister that is part of the apparatus) and restore the oxygen content through two different means:
  -- A cylinder of gaseous or liquid air or oxygen is used to replenish the oxygen content. The canister removes carbon dioxide and moisture from the air with heat evolved in the process. Heat is absorbed as the air passes through a cooling chamber. The air then passes into a breathing bag where oxygen is introduced based on the pressure of the air bag when then flows into the facepiece.
  -- Oxygen-generating systems use a solid, granular substance that reacts chemically with the carbon dioxide and water vapor in exhaled air to produce oxygen and heat. The heat is liberated in a breathing bag.
- ***Open-circuit SCBA*** utilize a portable compressed air cylinder to provide respirable air which is supplied to a tight-fitting facepiece and exhaled through an exhalation valve; types of open-circuit SCBA include:

-- ***Continuous-flow*** open-circuit SCBA use small cylinders with air at pressures of 2000 to 3000 pounds per square inch (13,700 to 20,700 Kpa) and simple, single-stage regulators to provide respirable air at 35 to 70 liters per minute and service time periods ranging from 5 to 10 minutes.

-- ***Demand*** open-circuit SCBA use air cylinders pressurized from 2000 to 4500 pounds per square inch (psig), or 13,700 to 310,000 Kpa, and two stage regulators. The first stage of the regulator reduces the air pressure to 50-100 psig while the second stage allows air to flow to the facepiece when the wearer inhales.

-- ***Pressure-demand*** open-circuit SCBA are similar to demand open-circuit SCBA except that the second stage of the regulator uses a spring to maintain a positive pressure when the wearer inhales and exhales.

22.3.3 Combination Supplied-Air and Air-Purifying Respirators

☐ Combination supplied-air and air-purifying respirators use an auxiliary SCBA in conjunction with a supplied-air respirator for use when the airflow to the SAR is interrupted. The service time for the auxiliary SCBA usually ranges from 15 to 20 minutes.

22.3.4 Combination Atmosphere-Supplying and Air-Purifying Respirators

☐ Combination atmosphere-supplying and air-purifying respirators allow the wearer to operate in either an air-purifying or atmosphere-supplying mode.

22.4 Sizing

☐ There are no specific requirements addressing sizing of respirators, except that respirators certified by NIOSH must adapt and conform to the standard head shapes used in many of the respirator performance tests.

- The principal component of respirators which affects sizing and fit for respirators with tight-fitting respiratory inlet covers is the facepiece; facepiece sizing is affected by:

  -- Stiffness of the facepiece material (usually Neoprene, natural rubber, or silicone rubber)

    - The type and number of straps used to secure mask
    - The general contour and surface area of the sealing surface
- Loose-fitting respiratory inlet covers take on the fit characteristics of clothing or equipment design.
- The type and location of harnesses used for the respirator may also affect respirator fit.

Section 23

Hearing Protector Designs, Features, and Sizing

The purpose of this section is to describe the different types of hearing protectors in terms of their general design, materials of construction, design features, and sizing.

23.1 Application and Types

- ☐ Hearing protectors are used to reduce occupational noise levels to those below OSHA permissible exposure limits or acceptable levels.
- ☐ There are three general types of ear protectors:
  - Ear plugs fit directly into the ear.
  - Ear canal caps cover the external part of the ear canal opening.
  - Ear muffs fit over the ear.
- ☐ A new class of hearing protectors includes active hearing protection devices.
- ☐ Proper fit is necessary since hearing protectors do not keep out all noise.
- ☐ Examples of different hearing protectors are illustrated in Figure 23-1.

23.2 Ear Plugs

- ☐ Earplugs are available in three types, including:
  - Formable
  - Custom-molded
  - Pre-molded

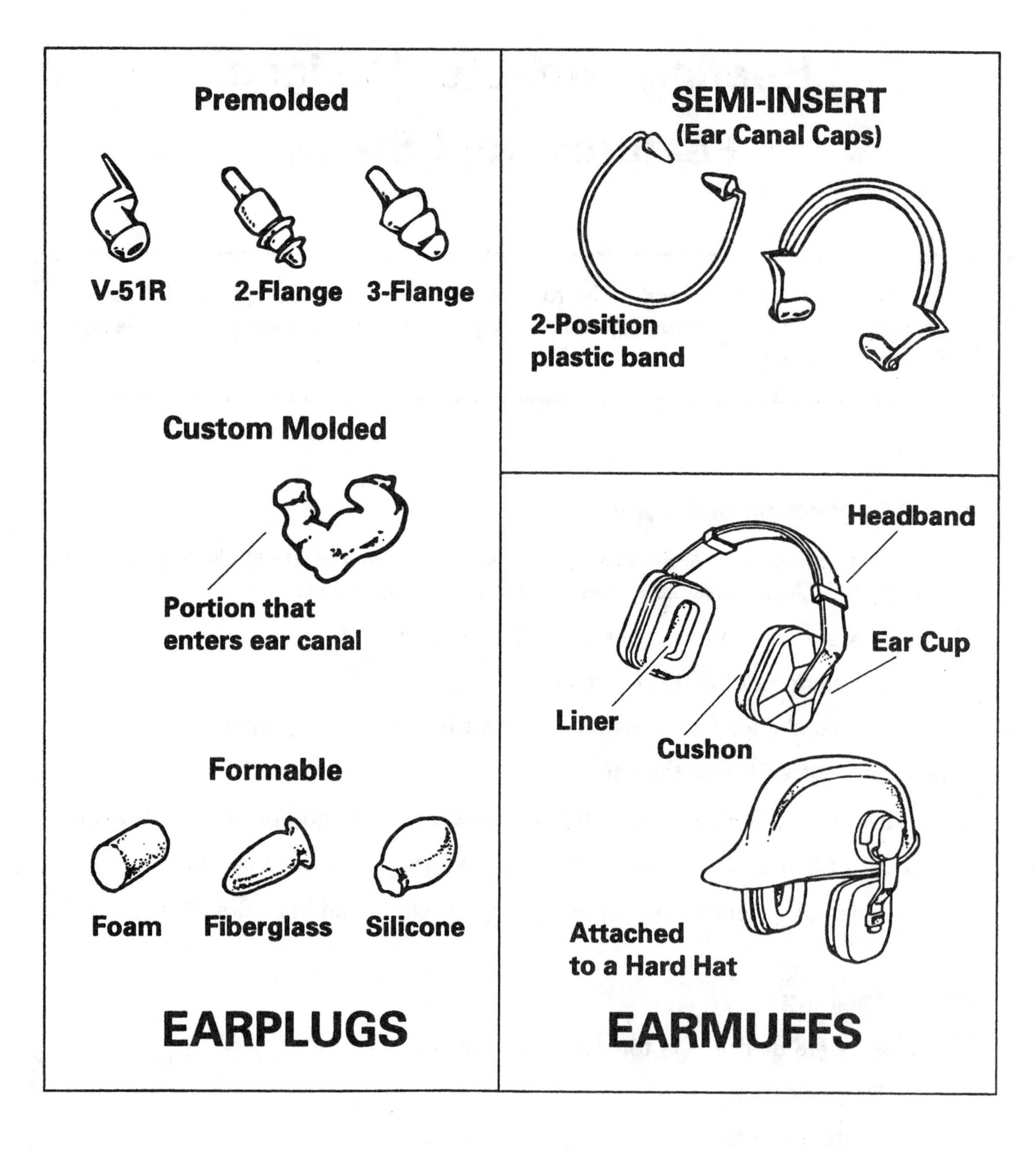

Figure 23-1. Examples of Hearing Protectors

23.2.1 Formable Ear Plugs

- ☐ Designed to adapt to the shape of the wearer's ear canal
- ☐ Typically made from fine glass fiber, wax-impregnated cotton, or expandable plastic
- ☐ Usually disposable use, and one-size-fits-all
- ☐ Require care in their insertion and wearer cleanliness of hands during insertion

23.2.2 Custom-molded Ear Plugs

- ☐ Specifically molded to the individual ear
- ☐ Use achieves a total fit for the wearer

23.2.3 Pre-molded Ear Plugs

- ☐ Composed of a soft silicone, urethane, or vinyl material
- ☐ Many styles include one or more flanges to assist in ear canal sealing
- ☐ Can be a universal size, designed to fit a wide variety of ear canal shapes and sizes
- ☐ May be available in more than one size, shape, and choice of materials

23.2.4 Ear Plug Sizing

- ☐ Ear plug fit is best determined by a trained individual because comfort for the wearer does not always translate into effective hearing protection (i.e., a loose ear plug may feel more comfortable but not provide a proper seal)

23.3 Ear Canal Caps

- ☐ Ear canal caps are a relatively new type of protection device.

23.3.1 Canal Cap Design and Construction

- ☐ Ear canal caps consist of ear plug-like caps (which are larger since they do not fit into the ear canal) and a headband that goes over or behind the head.

- ☐ Canal caps are generally made out of plastic materials.
- ☐ Headbands may be:
  - Two position (go over or behind the head)
  - Three position (go over and behind the head)
  - Metal or plastic

23.3.2 Canal Cap Features

- ☐ Canal caps may be available with or without an amplitude-sensitive, or non-linear, feature:
  - Non-linear canal caps allow conversation and announcements to reach the eardrum unimpeded, but attenuate high sound levels.
  - Linear canal caps do not preferentially affect sound levels.

23.3.3 Canal Cap Sizing

- ☐ Canal caps are generally available in only one size.

23.4 Ear Muffs

23.4.1 Ear Muff Design and Construction

- ☐ Earmuffs consist of two-cup or dome-shaped devices that fit over the entire external ear, including the lobe.
- ☐ The muffs seal against the head with a suitable cushion or pad.
- ☐ Generally, the cups are made of molded, rigid plastic and are then lined with an open cell foam material.
- ☐ The shape, size, and degree of noise attenuation of earmuffs vary from one manufacturer to another.

23.4.2 Ear Muff Features

- ☐ In nearly all industry products, the cushion material is foam, but some manufacturers are now offering liquid encased in a plastic/rubber material for improving the seal of the cup to the wearer's head.
- ☐ The headband attaching the two cups may be positioned under the chin, over the head or behind the head.
- ☐ Some earmuffs may be attached to helmets.
- ☐ The vertical distance on the head may be adjusted by changing the cup position on the headband. Depending on the specific product, the number of positions varies between one and three.
- ☐ Ear muffs may be available with or without an amplitude-sensitive, or non-linear, feature:
  - Non-linear ear muffs allow conversation and announcements to reach the eardrum unimpeded, but attenuate high sound levels.
  - Linear ear muffs do not preferentially affect sound levels.
- ☐ Some earmuffs also include communications sets with microphones, speakers, and amplifiers. The relative positioning of these accessories can also affect sizing of the hearing protector.

23.4.3 Ear Muff Sizing

- ☐ Earmuff sizing is affected by:
  - The size and shape of the ear cups
  - The choice of cushion/pad materials
  - The means for adjusting the distance between the ear cups (for differently shaped heads)
- ☐ Unlike earplugs whose size and shape is related to the ear canal, ear muffs are affected by the size, shape, and relative location of the ear lobes as well as the shape of the head.
- ☐ The majority of manufacturers offer ear muffs in one cup size, in an oval shape, sufficiently sized to cover the majority of the worker population's ear lobes, and then depend on the adjustment means to obtain proper fit over the ears on the wearer's head.

23.5 Active Hearing Protection Devices

- ☐ Active hearing protection devices are used to overcome communications difficulties when hearing protection is required.
- ☐ Active hearing protection devices resemble ear muffs but have exterior mounted microphones with electronic amplifiers for picking up ambient sounds and delivering the sounds to the wearer by earphones in the over-the-ear muff cups.
- ☐ The amplifiers in active hearing protection devices can be designed with electronic processors to selective filter or modify the level of sound within the frequency distribution reaching the hearing mechanism.

Section 24

Overview of PPE Performance Properties

The purpose of this section is to provide an overview of PPE performance properties and testing.

24.1 General Aspects of Performance Properties

- ☐ Performance properties describe specific characteristics of PPE or PPE materials or components.
- ☐ Performance properties are determined by evaluating or testing PPE.
- ☐ Performance testing may be conducted on:
  - Specific whole items of PPE, or
  - PPE materials or components
- ☐ The results of performance testing yield results which are then used to compare against:
  - Other products
  - Acceptance (or minimum performance) criteria
  - Performance classifications
- ☐ An evaluation of PPE performance test results assists in making decisions about the appropriateness of specific PPE items when:
  - The performance testing is related to the specific application of interest, and
  - The test results discriminate between acceptable and unacceptable PPE performance.

24.2 Test Methods

- ☐ Performance properties are determined by testing to specific procedures, or test methods, including:

- Standard test methods developed by consensus organizations
- U.S. or other government test methods (such as those contained in the Federal Register and Federal Test Method Standards)
- Canadian General Standard Board (CGSB) or other Canadian or provinicial standards
- European Norms (EN), European standards created by the European Committee on Standardization (CEN)
- International Standards developed by the International Standards Organization (ISO)
- Trade association standards (e.g., Association of the Nonwoven Fabrics Industry)
- Manufacturer-based test methods
- End user-defined test methods

☐ Standard test methods from consensus organizations are usually preferred because:

- Consensus organization-based test methods have typically undergone extensive review with the input of several different interests.
- Consensus organization-based test methods undergo periodic review and revision.
- The U.S. Federal government is slowly phasing out federal test methods for replacement by specification of consensus test methods.

☐ Consensus and trade organizations involved in the development of standard test methods for PPE include:

- American Association of Textile Chemists and Colorists (AATCC)
- American National Standards Institute (ANSI)
- American Society for Testing and Materials (ASTM)
- Association for the Advancement of Medical Instrumentation (AAMI)
- Association of the Nonwoven Fabrics Industry (INDA)
- Canadian General Standard Board (CGSB)
- European Committee on Standardization (CEN)
- Industrial Safety Equipment Association (ISEA)
- Institute of Environmental Sciences (IES)
- International Standards Organization (ISO)

- National Fire Protection Association (NFPA)

☐ Appropriate tests of PPE performance characteristics are those which:

- Provide detailed procedures, including
  - -- The number and type of specimens
  - -- Sampling of PPE for specimens
  - -- Specimen conditioning
  - -- Detailed descriptions of test apparatus, equipment and supplies
  - -- Methods for calibrating equipment
  - -- Complete test procedures
  - -- Requirements for reporting test results
- Are reproducible and provide accurate results
- Simulate the exposure of interest
- Provide results that assist in the selection decisions for PPE

☐ Table 24-1 provides a list of sources with addresses and phone number for obtaining test methods. Appendix D provides a complete list of standard test methods referenced in this publication.

NOTE: Consensus organization-based test methods are primarily cited in this book. There may be several other test methods which are available for measuring listed PPE performance properties that are not included.

CAUTION: Results from test methods with different designations may seem similar but may not be compared because subtle differences may existing in test methods which affect test results. Only compare test results when the same exact test method was used for generating the test results for both sets of data.

24.3 Types of Test Approaches and Test Results

☐ Tests methods differ in their:

- Application
- Type of results
- Type of specimens evaluated
- Test interpretation

Table 24-1. Sources of PPE Standard Test Methods from Consensus Organizations

Standards Organization	Address	Phone and Fax Numbers
American Association of Textile Chemists and Colorists (AATCC)	P. O. Box 12215 Research Triangle Park, NC 27709	P: (919) 549-8141 F: (919) 549-8933
American National Standards Institute (ANSI)	11 West 42nd Street New York, NY 10036	P: (212) 642-4900 F: (212) 398-0023
American Society for Testing and Materials (ASTM)	100 Barr Harbor Drive W. Conshohocken, PA 19428-2959	P: (610) 832-9500 F: (610) 832-9555
Association for the Advancement of Medical Instrumentation (AAMI)	3330 Washington Boulevard Suite 400 Arlington, VA 22201-4598	P: (800) 332-2264 F: (703) 276-0793
Association of the Nonwoven Fabrics Industry (INDA)	1001 Winstead Drive Suite 460 Cary, NC 27513	P: (919) 677-0060 F: (919) 677-0211
Canadian General Standards Board (CGSB)	CGSB, Sales Unit Ottawa, Canada K1A 1G6	P: (819) 956-0425 F: (819) 956-0426
European Committee for Standardization (CEN)	Secretariat: Deutsches Institute für Normung (DIN) D-10772 Berlin Germany	P: +49 30 26 01 24 32 F: +49 30 26 01 11 47
Industrial Safety Equipment Association (ISEA)	1901 N. Moore Street Arlington, VA 22209	P: (703) 525-1695 F: (703) 528-2148
Institute of Environmental Sciences (IES)	940 E. Northwest Highway Mount Prospect, IL 60056	P: (847) 255-1561 F: (847) 255-1699
International Standards Organization (ISO)	1, rue de Varembé Case postale 56 CH-1211 Genève 20 Switzerland	P: +41 22 749 01 11 F: +41 22 733 34 30
National Fire Protection Association (NFPA)	1 Batterymarch Park Quincy, MA 02269	P: (617) 770-3000 F: (617) 770-3500

24.3.1 Application of Test Methods

☐ Some tests are designed specifically for PPE while others may apply to broader areas (e.g., textiles) and thus can be used for PPE, when appropriate.

☐ Many test procedures are provided in standard test methods; however, some test methods may also be included as part of product specifications.

☐ Test methods are often modified by manufacturers or laboratories to meet special needs or product requirements.

24.3.2 Types of Results

☐ Testing of PPE performance properties yields either:

- Numerical test results, or
- A pass or fail determination
  - One type of pass/fail testing is "proof" testing, which involves testing a product to a certain exposure and determining if it provides the desired performance (e.g., louding a rope with a certain weight and observing if it breaks).

☐ Most PPE performance test results represent only one set of conditions; however, many test methods permit adjustment of test parameters to accommodate a broad range of conditions.

☐ Tests may be performed under

- "Worst case" conditions
- Average conditions
- A set of conditions representing an exposure of interest

24.3.3 Types of Specimens Evaluated

☐ Specimens may include:

- Sample products
- Facsimile samples representing the product
- Product components or materials

☐ Different specimen approaches are used in evaluating PPE performance properties depending on the application of the test:

- A limited number of specimens may be evaluated for determining performance for product comparison purposes.
- A statistically relevant sampling of specimens may be evaluated for determining product quality control or performance variations.

- Material or component specimens may be taken directly from items of PPE or tested individually prior to assembly in PPE.

☐ Some performance properties may be combined with different forms of specimen conditioning to ascertain how product performance is affected by the precondition; examples include:

- Measuring material strength after laundering
- Conducting thermal insulation tests when the material has been moistened
- Evaluating material liquid barrier resistance after abrasion
- Evaluating glove integrity following repeated flexing

24.3.4 Test Interpretation

☐ Results may be presented as:

- An average of all test results
- The worst result or test value
- The best result or test value
- The range of all test results

☐ Most tests, except those with pass/fail results, do not provide criteria for interpreting the results.

☐ Performance criteria are established in whole item PPE specifications or must be established by the end user or manufacturer.

24.4 Categories of Performance Properties

☐ Categories of performance properties covered in this book include:

- Physical properties
- Overall product integrity
- PPE resistance to environmental conditions
- Chemical resistance
- Biological resistance
- Heat and flame resistance
- Electrical properties

- Respirator properties
- Human factors

☐ Some performance properties apply to several types of PPE whiles others may be applied to only one PPE category.

☐ Table 24-2 shows the relationship of these properties with specific hazards.

Table 24-2. Relevant Performance Properties for Specific Hazards

Hazard Category/ Specific Hazards	Recommended PPE Items	Relevant Performance Properties (with paragraph numbers)
Physical Hazards		
Falling objects	Footwear; Headwear	Impact/compression resistance (25.4)
Flying debris	Partial body clothing Headwear Face and eyewear Respirators	Projectile and ballistic resistance (25.5) Particle-tight integrity (26.2) Particulate removal efficiency (34.3.10) Dust, fume, and mist performance (34.3.12)
Projectile/ballistic	Full body clothing Partial body clothing	Projectile and ballistic resistance (25.5)
Abrasive/rough surfaces	All protective clothing	Abrasion and scratch resistance (25.7)
Sharp edges	All protective clothing	Tear or snag resistance (25.6) Cut resistance (25.8)
Pointed objects	All protective clothing	Puncture resistance (25.9)
Slippery surfaces	Footwear; Handwear	Slip resistance (25.11)
Excessive vibration	Handwear	Shock absorption (25.12)
Environmental Hazards		
High heat/humidity	Full body clothing Partial body clothing	Thermal insulation and breathability (32.3)
Ambient cold	All protective clothing	Thermal insulation and breathability (32.3)
Environmental Hazards		
Wetness	All protective clothing	Liquid-tight integrity (26.3) Water repellency and absorption resistance

Hazard Category/ Specific Hazards	Recommended PPE Items	Relevant Performance Properties (with paragraph numbers)
Wetness *(continued)*		(27.2) Water penetration resistance (27.3)
High wind	All protective clothing Face and eyewear	Thermal insulation and breathability (32.3)
Insufficient/bright light	Face and eyewear and other PPE items covering eyes	Excess light attenuation (27.9)
Excessive noise	Ear protectors	Excess noise attenuation (27.8)
Chemical Hazards		
Inhalation	Respirators	Respirator performance requirements (34.3)
Skin contact (vapor)	All protective clothing	Gas-tight integrity (26.2) Chemical vapor penetration resistance (28.5) Chemical permeation resistance (28.6)
Skin contact (liquid)	All protective clothing	Liquid-tight integrity (26.3) Chemical degradation resistance (28.2) Chemical liquid penetration resistance (28.4)
Skin contact (particles)	All protective clothing	Particle-tight integrity (26.4) Chemical degradation resistance (28.2) Chemical particulate penetration resistance (28.3)
Ingestion	Facewear and other PPE covering face	Liquid-tight integrity (26.3) Particle-tight integrity (26.4) Chemical degradation resistance (28.2) Chemical particulate penetration resistance (28.3) Chemical liquid penetration resistance (28.4)
Injection	All protective clothing	Cut resistance (25.8) Puncture resistance (25.9)
Chemical flashover	All protective clothing Face and eyewear	Radiant heat resistance (30.4) Flame resistance (30.5)

Hazard Category/ Specific Hazards	Recommended PPE Items	Relevant Performance Properties (with paragraph numbers)
Chemical Hazards (continued)		
Chemical explosions	All protective clothing Headwear Face and eyewear	Burst strength (25.5) Projectile and ballistic resistance (25.5)
Biological Hazards		
Bloodborne pathogens	All protective clothing Face and eyewear	Biological fluid resistance (29.3) Biological fluid penetration resistance (29.4) Viral penetration resistance (29.5)
Airborne pathogens	Respirators	Microorganism filtration efficiency (29.2) Particulate removal efficiency (34.3.10) Dust, fume, and mist performance (34.3.12)
Biological toxins	All protective clothing Face and eyewear Respirators	Puncture resistance (25.9) Microorganism filtration efficiency (29.2) Biological fluid resistance (29.3) Antimicrobial resistance (29.6) Insect resistance (29.7) Particulate removal efficiency (34.3.10) Dust, fume, and mist performance (34.3.12)
Biological allergens	All protective clothing Face and eyewear Respirators	Microorganism filtration efficiency (29.2) Biological fluid resistance (29.3) Particulate removal efficiency (34.3.10) Dust, fume, and mist performance (34.3.12)
Thermal Hazards		
High heat (contact)	All protective clothing	Conductive heat resistance (30.2)
High heat (convective)	All protective clothing Headwear Face and eyewear Respirators	Convective heat resistance (30.3) Thermal protective performance (30.5)

Hazard Category/ Specific Hazards	Recommended PPE Items	Relevant Performance Properties (with paragraph numbers)
Thermal Hazards (continued)		
High heat (radiant)	All protective clothing Headwear Face and eyewear	Radiant heat resistance (30.4) Thermal protective performance (30.5)
Flame impingement	All protective clothing Headwear Face and eyewear	Flame resistance (30.6)
Hot liquids and gases	All protective clothing Headwear Face and eyewear	Water penetration resistance (27.3) Convective heat resistance (30.3)
Molten substances	All protective clothing	Resistance to molten metal contact (30.7)
Electrical Hazards		
Electric shock	All protective clothing Headwear	Electrical insulative performance (31.2)
Electrical arc flashover	All protective clothing Headwear Face and eyewear	Electric arc protective performance (31.5)
Static charge buildup	Full body garments Partial body garments Handwear	Conductivity and electrical resistivity (31.3) Static charge accumulation resistance (31.4)
Radiation Hazards		
Ionizing radiation	All protective clothing	Particle-tight integrity (26.4) Chemical degradation resistance (28.2) Chemical particulate penetration resistance (28.3)
Non-ionizing radiation	Eyewear	UV-visible light resistance (27.6) Excess light attenuation (27.9)

Hazard Category/ Specific Hazards	Recommended PPE Items	Relevant Performance Properties (with paragraph numbers)
Person-Position Hazards		
Lack of visibility	Full body garments Partial body garments Headwear	Visibility (27.10)
Drowning	Personal flotation devices	Performance area not covered in reference
Falling from elevation	Fall protection devices	Performance area not covered in reference
Person-Equipment Hazards		
Creation of particles	All protective clothing	Particle-tight integrity (26.2) Chemical particulate penetration resistance (28.3)
Creation of static electricity	Full body garments Partial body garments Handwear	Static charge accumulation resistance (31.4)
Material incompatibility	All PPE	Material biocompatibility (32.2)
Retained contamination	All PPE	No specific tests provided
Loose straps/material	All PPE	No specific tests provided
Reduced mobility	Full body garments Partial body garments Headwear Face and eyewear Respirators	Mobility and range of motion (32.4)
Reduced hand function	Handwear	Hand function (32.5)
Impaired vision	Eyewear and other PPE covering the eyes	Clarity and field of vision (32.8)
Poor communications	Ear protectors and other PPE covering the ears or mouth	Ease of communications (32.9)
Lack of ankle support	Footwear	Ankle support (32.6)

Hazard Category/ Specific Hazards	Recommended PPE Items	Relevant Performance Properties (with paragraph numbers)
Person-Equipment Hazards (continued)		
Lack of back support	Full body garments Partial body garments Respirators	Back support (32.7)
Person-Equipment Hazards		
Difficult use	All PPE	Sizing and fit (32.10) Donning and doffing ease (32.11)
Heat stress	All protective clothing Respirators	Weight (25.2) Thickness and hardness (25.3) Water repellency and absorption resistance (27.2) Stiffness and resistance to cold temperatures (27.5) Thermal insulation and breathability (32.3)
Wear through use	All PPE	Flex fatigue resistance (25.10) Salt spray or corrosion resistance (27.4) Stiffness and resistance to cold temperatures (27.5) UV-visible light resistance (27.6) Ozone resistance (27.7) Durability and serviceability (Section 33)

Section 25

PPE Physical Properties

The purpose of this section is to describe different types of PPE physical properties in terms of the measured property, applicable PPE, test principle, available standard test methods, results provided, and interpretation of test results.

25.1 Overview of Physical Properties

☐ Physical properties assess or determine:

- Weight and thickness of PPE and materials
- Strength of PPE
- Resistance to specific physical hazards
- Product durability

☐ A number of these properties apply to different types of materials which may include:

- Woven textiles
- Knit textiles
- Non-woven textiles (created by non weaving or knitting processes)
- Rubbers
- Plastics
- Coated fabrics

☐ More than one test method may be available for a performance property based on the type of material or product.

☐ Many material physical property results are reported in different material directions:

- Results that are parallel to the direction the material is fabricated or comes off the roll are referred to as:
  - -- Warp direction for woven textile fabrics
  - -- Course direction for knit textile fabrics
  - -- Machine direction for nonwoven textile, rubber, and plastic materials
- Results that are perpendicular to the direction the material is fabricated or comes off the roll are referred to as:
  - -- Fill direction for woven textile fabrics
  - -- Wales or weft direction for knit textile fabrics
  - -- Cross-machine direction for nonwoven textile, rubber, and plastic materials

☐ Performance properties included in this section include:

- Weight
- Thickness and hardness
- Breaking strength
- Burst strength
- Impact and compression resistance
- Projectile and ballistic resistance
- Tear or snag resistance
- Abrasion and scratch resistance
- Cut resistance
- Puncture resistance
- Flex fatigue resistance
- Slip resistance
- Shock absorption

☐ Representative standard test methods for different physical properties described in this section are summarized in Table 25-1.

Table 25-1. List of PPE Physical Performance Test Methods

Physical Property	Type of Test/Application	Available Test Methods
Weight	Textiles	CAN/CGSB-4.2 No. 5.1 FTMS 191A,5041
	Woven textiles	ASTM D 3776 ISO 3801
	Nonwoven textiles	ASTM D 1117 ISO 9073-1
	Coated fabrics	ASTM D 751 ISO 2286
Thickness/hardness	Thickness - textiles	CAN/CGSB-4.2 No. 37 FTMS 191A,5030
	Thickness - woven/knit textiles	ASTM D 1777 ISO 5084
	Thickness - nonwoven textiles	ASTM D 5736 ISO 9073-2
	Thickness - coated fabrics	ASTM D 751 ISO 2286
	Thickness - leather	ASTM D 1813
	Thickness - plastics, rubber, foam	ASTM D 1565 ASTM D 3767
	Compression set - rubber	ASTM D 575
	Hardness - rubber and plastic	ASTM D 2240
Breaking strength	Grab method - textiles	ASTM D 5034 CAN/CGSB-4.2 NO. 9.2 EN ISO 13934-2 FTMS 191A,5104 ISO 5082
	Grab method - leather	ASTM D 2208
	Strip method - textiles	ASTM D 5035 CAN/CGSB-4.2 NO. 9.1 EN ISO 13934-1 FTMS 191A,5102 ISO 5081
	Strip method - leather	ASTM D 2209

Physical Property	Type of Test/Application	Available Test Methods
Breaking strength *(continued)*	Nonwoven textiles	ISO 9073-3
	Rubber/coated fabrics	ASTM D 412 ASTM D 751 ISO 1421
	Plastic	ASTM D 638 ASTM D 882 ISO 1421
	Seam strength	ASTM D 751 ASTM D 1683 CAN/CGSB-4.2 No. 32.2
Burst Strength	Ball method - textiles	ASTM D 3787 ASTM D 3940 CAN/CGSB-4.2 No. 11.2 FTMS 191A,5120
	Ball method - rubber or plastic	ISO 3303
	Diaphragm method - textiles	ASTM D 3786 CAN/CGSB-4.2 No. 11.1 FTMS 191A,5122
	Diaphragm method - rubber or plastic	ISO 2960
	Coated fabrics	ASTM D 751
	Leather	ASTM D 2207
Impact and Compression Resistance	Footwear	ANSI Z41.1 S1.4.4/5 NFPA 1971 S6-18
	Face/eyewear	ANSI Z87.1 S15.1 NFPA 1971 S6-15 NFPA 1971 S6-16
	Headwear	ANSI Z879.1 S8.3
	Respirator facepieces	Milspec. GGG-M-125d
Projectile and Ballistic Resistance	Bullet resistant vests	ISO 14876 MIL-STD-662 NIJ 0101.01 NIJ 0108.01
	Eye/face wear projectiles	ANSI Z87.1 S8.4
	Helmet projectiles	ANSI Z89.1 S8.4 NFPA 1971 S6-17

Physical Property	Type of Test/Application	Available Test Methods
Tear or Snag Resistance	Elmendorf method - textiles	ASTM D 1424 CAN/CGSB-4.2 No. 12.2 FTMS 191A,5132 ISO 13937
	Elmendorf method - plastics	ASTM D 1922
	Tongue tear - textiles	ASTM D 2261 ASTM D 2262 CAN/CGSB-4.2 No. 12.1 FTMS 191A,5134 ISO 13937
	Tongue tear - nonwoven textiles	ASTM 5735
	Tongue tear - plastics	ASTM D 1938
	Tongue tear - leather	ASTM D 4704
	Trapezoidal method - textiles	CAN/CGSB-4.2 No. 12.2 NFPA 1971 S6-12
	Trapezoidal method - non woven textiles	ASTM D 5733 ISO 9073-4
	Tear resistance - plastics	ASTM D 1004 ISO 4674
	Tear resistance - coated fabrics, rubber	ASTM D 751 ISO 4676
	Puncture propagation tear resistance	ASTM D 2582 ISO 13995
	Snag resistance	ASTM D 3939 ASTM D 5362 CAN/CGSB-4.2 No. 51
Abrasion resistance	Taber method - textiles	ASTM D 3884 FTMS 191A,5306
	Taber method - coated fabrics	ASTM D 3389
	Taber method - plastics	ASTM D 1044
	Wyzenbeek method - textiles	ASTM D 4157 FTMS 191A,5304
	Martindale method - clothing materials	ASTM D 4966 EN 530 ISO 12974
	Martindale method - gloves	EN 388 S6.1

Physical Property	Type of Test/Application	Available Test Methods
Abrasion resistance *(continued)*	Flexing/abrading method - textiles	ASTM D 3885 FTMS 191A,5300
	Flexing/abrading method - coated fabrics	ISO 5981
	Inflated diaphragm method - textiles	ASTM D 3886 FTMS 191A,5302
	NBS method - rubber	ASTM D 1630
	Pico method - rubber	ASTM D 2228
	Loose abradant - plastics	ASTM D 1242
	Uniform method - textiles	ASTM D 4158
	Lens abrasion - optical properties	ASTM D 1044 NFPA 1971 S6-23 NFPA 1981 S6-9
Cut resistance	Static method - gloves and footwear	NFPA 1971 S6-22
	Dynamic method - gloves and clothing materials	ASTM F 1670 EN 388 S6.2 ISO 11393
	Impact method - gloves	EN 388 S6.5
	Chainsaw method	ASTM F 1414 ASTM F 1458 EN 381 ISO 11393
Puncture resistance	General method	ASTM F 1342 EN 863 ISO 13996
	Gloves	EN 388 S6.4
	Footwear	ANSI Z41.1 S3.5
Flex fatigue	Flat material	ASTM F 392 ISO 7854
Slip resistance	Footwear soles	ASTM F 489 ASTM F 609
	Glove grip	NFPA 1971 S6-39
Shock absorption	Cushioning materials	ASTM D 4168
	Headgear	ASTM F 429
	Respirators (SCBA)	NFPA 1981 S6-3
	Gloves	EN ISO 10819

25.2 Weight

Property Measured

☐ The weight of PPE is measured for whole PPE items, materials, or components.

Applicable PPE

☐ PPE usually subject to weight measurements includes:

- Full body garments and materials
- Partial body garments and materials
- Footwear
- Headwear
- Face or eyewear
- Respirators

Test Principles

☐ Test principles include:

- A scale used to measure the overall weight of PPE items.
- When material weight is measured, the unit area weight is determined by:
  - Sampling a specific area of material
  - Weighing the sample on a scale
  - Reporting the weight of the material in weight per unit area (e.g., ounces per square yard or grams per square meter).

Standard Test Methods

☐ Representative methods include:

- ASTM D 751 (for coated fabrics)
- ASTM D 1117 (for nonwoven textiles)
- ASTM D 3776 (for woven textiles)
- CAN/CGSB-4.2 No. 5.1 (Canadian method for textiles)
- FTMS 191A, 5041 (Federal method for textiles)
- ISO 2286 (International method for coated fabrics)
- ISO 3801 (International method for woven textiles)
- ISO 9073-1 (International method for nonwoven textiles)

Test Results Provided

☐ Test results provide the following:

- Overall "nominal" (average) PPE weight
- Unit area weight for materials

Test Results Interpretation

☐ Interpretation of test results:

- Higher weight PPE or materials can be used to compare the relative burden of the PPE on the individual wearer.
- Some standards set maximum nominal weight for PPE or materials.

25.3 Thickness and Hardness

Properties Measured

☐ The thickness and hardness of PPE is measured for whole PPE items, materials, or components.

Applicable PPE

☐ Thickness (and hardness) measurements may be applied to following PPE materials and components:

- Full body garment materials
- Partial body garment materials
- Glove materials
- Footwear materials
- Headwear
- Eye and facewear

Test Principles

☐ Thickness measurements involve using a thickness gauge to measure the thickness of a material specimen in a specific location.

- Specifications for thickness measurements include:
  - -- The type and accuracy of the thickness gauge
  - -- The area of the presser foot or area of specimen thickness measured
  - -- The force on the specimen during the thickness measurement

- Special handling is needed for materials that are easily compressed since their thickness will vary with the amount of force applied during the thickness measurement.
- The relative ease of changing a material's thickness is known as compression set and involves:
  - Measuring the starting thickness of the material
  - Placing a known force on the material for a specified period of time
  - Removing the force and allowing the specimen to recover for a specified period of time
  - Measuring the ending thickness of the material following the recovery period
- The hardness of semi-rigid and rigid materials is measured using a special apparatus that measures the force to create an indentation.

Standard Test Methods

☐ Representative methods include:

- ASTM D 575 (compression set for rubber)
- ASTM D 751 (for coated fabrics)
- ASTM D 1565 (for open cell foam)
- ASTM D 1777 (for woven textiles)
- ASTM D 1813 (for leather)
- ASTM D 2240 (hardness for rubbers and plastics)
- ASTM D 3767 (for rubber foam products
- ASTM D 5736 (for nonwoven textiles)
- CAN/CGSB-4.2 No. 37 (Canadian method for textiles)
- FTMS 191A, 5030 (Federal method for textiles)
- ISO 2286 (International method for coated fabrics)
- ISO 5084 (International method for woven or knit textiles)
- ISO 9073-2 (International method for nonwoven textiles)

Test Results Provided

☐ Test results include:

- Thickness for individual material at a given force

- Hardness of item surface for hardness test

Test Results Interpretation

☐ Interpretation of test results:

- Thicker materials generally indicate increased bulk for wearing apparel.
- Some performance properties are normalized for thickness to show relative performance per given thickness of material.
- Hardness may be used as a measured of the surface's resistance to damage from repeated wear or abuse.

25.4 Breaking Strength

Property Measured

☐ Breaking (or tensile) strength tests measure the force required to break PPE materials or components when items are pulled along one direction.

- Breaking strength tests can also be used to measure the strength of seams and closures.

Applicable PPE

☐ Breaking strength testing may be applied to following PPE materials and components:

- Full body garment materials, seams, and closures
- Partial body garment materials, seams, and closures
- Headwear retaining straps
- Face or eyewear straps
- Respirator straps
- Safety line and fall protection devices

Test Principles

☐ Most tests involve the following principal steps:

- The two ends of a specimen are placed in grips of a tensile testing machine.
- One grip is fixed while the other grip is moved at a controlled rate in terms of distance or force until the specimen breaks.
- A load cell or other measuring device indicates the force.

- Test methods vary by the type and size of the specimen and the method used for gripping the specimen in the tensile testing machine:
  - Strip methods use specimens that are the same size or narrower than the width of the grips.
  - Grab methods use specimens that are wider than the grips.
  - Some methods such as those for plastic or rubber materials use a bone shaped specimen.

Standard Test Methods

☐ Representative methods include:

- ASTM D 412 (for rubber materials)
- ASTM D 638 (for plastic materials)
- ASTM D 751 (for coated fabrics and seams)
- ASTM D 882 (for thin plastic sheeting)
- ASTM D 1683 (for textile seams)
- ASTM D 2208 (grab method for leather)
- ASTM D 2209 (strip method for leather)
- ASTM D 5034 (grab method for textiles)
- ASTM D 5035 (strip method for textiles)
- CAN/CGSB-4.2 No. 9.1 (Canadian strip method for textiles)
- CAN/CGSB-4.2 No. 9.2 (Canadian grab method for textiles)
- CAN/CGSB-4.2 No. 32.2 (Canadian method for seam strength)
- EN ISO 13934-1 (European strip method for woven fabrics)
- EN ISO 13934-2 (European grab method for woven fabrics)
- FTMS 191A, 5104 (Federal grab method for textiles)
- FTMS 191A, 5102 (Federal strip method for textiles)
- ISO 1421 (International method for rubber or plastics)
- ISO 5081 (International strip method for woven fabrics)
- ISO 5082 (International grab method for woven fabrics)
- ISO 9073-3 (International method for nonwoven textiles)

Test Results Provided

☐ Test results include:

- Breaking strength is usually reported in units of force (pounds or Newtons).
- Elongation of the specimen is reported as the percentage of specimen "stretch" before breaking as compared with its original size.
- Modulus is the force required produce a specific amount of specimen elongation and is usually reported in units of force (pounds or Newtons).
- Material results are generally reported for the two material directions for textiles, rubber, and plastic material.
- Seam and other PPE component results may also be accompanied by observations as to where the break occurs.

Test Results Interpretation

☐ Interpretation of test results:

- Higher breaking strength indicate stronger materials or components
- High elongation indicates more elastic materials
  - -- Increased elasticity may be desirable for rubber products and other PPE intended to be stretched when being donned or used.
  - -- Increased elasticity may not be desirable for materials that should stay rigid during use.

25.5 Burst Strength

Property Measured

☐ Burst strength tests measure the force or pressure required to rupture PPE materials or components when a force is directed perpendicular to the item.

- This property may be related to ability of PPE materials to prevent items from protruding through garments.

Applicable PPE

☐ Burst strength testing may be applied to PPE materials and components:

- Full body garment materials and seams
- Partial body garment materials and seams

Test Principles

☐ Two principles may be used in burst strength testing:

- A tensile testing machine is used to push a hard ball through the specimen with measurement of the required force
- A special machine uses an oil-filled diaphragm to push through the specimen

Standard Test Methods

☐ Representative methods include:

- ASTM D 751 (for coated fabrics)
- ASTM D 2207 (for leather)
- ASTM D 3786 (diaphragm method for knitted and non-woven textiles)
- ASTM D 3787 (ball method for knitted textiles)
- ASTM D 3940 (for knitted and woven stretch textiles)
- CAN/CGSB-4.2 No. 11.1 (Canadian diaphragm method for textiles)
- CAN/CGSB-4.2 No. 11.2 (Canadian ball method for textiles)
- FTMS 191A, 5120 (Federal ball method for textiles)
- FTMS 191A, 5122 (Federal diaphragm method for textiles)
- ISO 2960 (International diaphragm method for rubber or plastics)
- ISO 3303 (International ball method for rubber or plastics)

Test Results Provided

☐ Test results include:

- Burst strength results are reported as the force (in pounds or Newtons) or pressure (in pounds per square inch or Pascals) at which breakage or rupture occurs.
- Seam and other PPE component results may also be accompanied by observations as to where the break occurs.

Test Results Interpretation

☐ Interpretation of test results:

- Higher burst strengths indicate stronger materials.

25.6 Impact/Compression Resistance

Property Measured

☐ Impact and compression resistance tests evaluate the ability of PPE to resist the forces of impact from a falling heavy object, which deforms, compresses, or fractures parts of the PPE.

Applicable PPE

☐ Impact and compression resistance tests are most commonly applied to:

- Footwear toe sections
- Headwear
- Eye and facewear

Test Principles

☐ Impact and compression resistance tests involve the controlled drop of a heavy object (missile) onto part of the PPE item with measurement of the transmitted impact energy or acceleration of the PPE item or the amount of deformation, compression, or fracture of the PPE item.

Standard Test Methods

☐ Representative methods include:

- Sections 1.4.4 and 1.4.5 of ANSI Z41-1989, *Protective Footwear*
- Section 15.1 of ANSI Z87.1-1989, *Practice for Occupational and Educational Eye and Face Protection*
- Sections 9.2, 9.3, and 9.4 of ANSI Z89.1-1986, *Protective Headwear for Industrial Workers—Requirements*
- Section 6-18 of NFPA 1971 (1997), *Standard on Protective Ensemble for Structural Fire Fighting* (footwear)
- Sections 6-15 and 6-16 of NFPA 1971 (1997), *Standard on Protective Ensemble for Structural Fire Fighting* (helmets)
- Military specification GGG-M-125d (for respirator facepieces and eyepieces)

Test Results Provided

☐ Test results include:

- Impact force transmitted through the PPE item
- Maximum acceleration and duration of acceleration

- The amount of deformation or compression
- Evidence of fracture

Test Results Interpretation

☐ Better impact or compression resistance of a PPE is dependent on the method used but is generally defined by the following results:

- Lower transmission of impact forces
- Less acceleration
- Less compression or deformation of the PPE following impact
- No evidence of PPE item damage

25.7 Projectile and Ballistic Resistance

Property Measured

☐ Projectile resistance tests measure the ability of PPE items, materials, or components to deflect or prevent the damage to PPE or penetration of specific types of projectiles through PPE shot at relatively high velocities.

Applicable PPE

☐ Projectile resistance testing may be applied to PPE materials and components:

- Full body garments and materials
- Partial body garments and materials (such as vests)
- Headwear
- Eye and facewear

Test Principles

☐ Test principles include:

- A low weight, typically pointed, projectile is dropped in a controlled fashion onto the target area of the PPE item.
- Ballistic tests involve shooting a specific type of weapon onto target areas of the protective garment.

Standard Test Methods

☐ Representative standard test methods for measuring impact/compression and ballistic resistance include:

- ISO 14876 (International standard for bullet resistant vests)

- National Institute of Justice Standard 0108.01, *Ballistic Resistance Protective Material*
- National Institute of Justice Standard 0101.01, *Ballistic Resistance of Soft Body Armor*
- Military Standard MIL-STD-662, *Ballistic Test for Armor*
- Sections 15.2, 15.5, and 15.8 of ANSI Z87.1-1989, *Practice for occupational and educational eye and face protection*
- Sections 6-17 of NFPA 1971 (1997), *Standard on Protective Ensemble for Structural Fire Fighting* (helmets)

Test Results Provided

☐ Test results include:

- Projectile resistance results are reported as pass or fail depending on the level of damage to or penetration through the PPE item.
- Ballistic resistance results are reported in terms of the projectile velocity that the garment or material will prevent from penetrating.

Test Result Interpretation

☐ Interpretation of test results:

- Better projectile resistance is generally defined by low levels of damage and no penetration of the PPE item
- Better ballistic resistance is defined by higher velocities that the PPE item or material stop.

25.8 Tear or Snag Resistance

Property Measured

☐ Tear resistance tests measure the force required to continue a tear in a PPE material once initiated or the resistance of a material in preventing a tear or snag from occurring.

☐ Snagging occurs when an individual yarn of a textile is pulled away from the material

Applicable PPE

☐ Tear resistance testing may be applied to PPE materials and components:

- Full body garment materials
- Partial body garment materials

Test Principles

☐ Different approaches are used for measuring tear resistance based on the type of specimen and the way the tear is created:

- A notched material specimen is torn by a pendulum (Elmendorf method)
- The ends of a cut rectangular material specimen are pulled in opposite directions (tongue method)
- The ends of a notched, trapezoidal specimen are pulled in opposite directions (trapezoidal method)
- A material specimen is punctured and then torn by a falling sharp probe (puncture propagation tear method)
- In snag testing, a material specimen is exposed to a spiked ball (mace) or formed into a bag and rotated against a protruding pin-based surface.

Standard Test Methods

☐ Representative methods include:

- ASTM D 751 (Elmendorf and tongue methods for coated fabrics)
- ASTM D 1004 (tensile testing machine method for plastic films)
- ASTM D 1424 (Elmendorf method for woven textiles)
- ASTM D 1922 (Elmendorf method for plastics)
- ASTM D 1938 (tongue method for plastics)
- ASTM D 2261 and 2262 (tongue method for woven textiles)
- ASTM D 2582 (puncture propagation tear method for plastic films)
- ASTM D 3939 (mace snag method for textiles)
- ASTM D 4704 (tongue tear for leather)
- ASTM D 5362 (bean bag snag method for textiles)
- ASTM D 5733 (trapezoidal method for nonwoven textiles)
- ASTM D 5734 (Elmendorf method for nonwoven textiles)
- ASTM D 5735 (tongue method for nonwoven textiles)
- CAN/CGSB-4.2 No. 12.1 (Canadian tongue method for textiles)
- CAN/CGSB-4.2 No. 12.2 (Canadian trapezoid method for textiles)
- CAN/CGSB-4.2 No. 12.3 (Canadian Elmendorf method for textiles)
- CAN/CGSB-4.2 No. 51 (Canadian snag methods for textiles)

- FTMS 191A, 5132 (Federal Elmendorf method for textiles)
- FTMS 191A, 5134 (Federal tongue tear method for textiles)
- ISO 4674 (International tear method for rubber or plastics)
- ISO 9073-4 (International trapezoidal tear method for nonwoven textiles)
- ISO 13937 (International test methods for woven textiles)
- ISO 13995 (International puncture propagation tear method for clothing materials)

Test Results Provided

☐ Test results include:

- Tear resistance test results are reported as the force (in pounds or Newtons) required to create or propagate the tear for each material direction.
- Snag resistance results are reported as a visual rating.

Test Results Interpretation

☐ Interpretation of test results:

- Higher tear resistance results indicate materials which are less likely to tear; materials which show higher puncture propagation tear resistance values are also less like to snag.
- Higher ratings of snag resistance indicate materials that are less likely to snag.

25.9 Abrasion and Scratch Resistance

Property Measured

☐ Abrasion resistance tests measure the ability of PPE surfaces or materials to resist wearing away when rubbed against other surfaces.

Applicable PPE

☐ Applicable PPE includes:

- Full body garment materials and visors
- Partial body garment materials
- Glove materials
- Footwear materials and soles
- Eye and face protection device lenses

- Respirator facepiece lenses

Test Principles

☐ Test principles include:

- Abrasion testing involves a controlled technique for rubbing an abradant against the specimen under a specific pressure or tension for a given duration.
- Variables affecting how abrasion testing is conducted include:
  - -- The type of abradant
  - -- The type of abrading action or motion
  - -- The pressure behind the abradant or tension on the specimen
  - -- The number of abrasion cycles
- Common abrasion techniques include:
  - -- Use of a rotary platform with hard wheels (Taber abrasion)
  - -- Use of a oscillatory drum moving in one direction (Wyzenbeek abrasion)
  - -- Use of a constantly moving abrader test head (Martindale abrasion)
  - -- Use of loose abradant
  - -- Use of an inflated diaphragm with rotating movement
  - -- Use of combined flexing and abrasion
  - -- Use of a rotating wheel (NBS abrasion)
  - -- Use of a cam controlled specimen under abradant (uniform abrasion)
  - -- Use of a knife to scrape the specimen (Pico abrasion)

Standard Test Methods

☐ Representative methods include:

- ASTM D 1044 (plastic optical characteristics after Taber abrasion)
- ASTM D 1242 (loose abradant method for plastics)
- ASTM D 1630 (NBS method for rubber)
- ASTM D 2228 (Pico method for rubber)
- ASTM D 3884 (Taber method for textiles)
- ASTM D 3885 (flexing and abrasion method for textiles)
- ASTM D 3886 (inflated diaphragm method for textiles)

- ASTM D 3389 (Taber method for rubber coated fabrics)
- ASTM D 4157 (Wyzenbeek method for textiles)
- ASTM D 4158 (uniform method for textiles)
- ASTM D 4966 (Martindale method for textiles)
- EN 530 (European Martindale method for clothing materials)
- FTMS 191A, 5300 (Federal flexing and abrasion method for textiles)
- FTMS 191A, 5302 (Federal inflated diaphragm method for textiles)
- FTMS 191A, 5304 (Federal Wyzenbeek method for textiles)
- FTMS 191A, 5306 (Federal Taber method for textiles)
- ISO 5981 (International flexing and abrasion method for coated fabrics)
- ISO 12974 (International Martindale method for clothing materials)
- Section 6-23 of NFPA 1971 (1997), *Standard on Protective Ensemble for Structural Fire Fighting* (facewear)
- Section 6-9 of NFPA 1981 (1997), *Standard on Open-Circuit, Self-Contained Breathing Apparatus* (lens scratch resistance)
- Subclause 6.1 of EN 388 (European Martindale method for gloves)

Test Results Provided

☐ Abrasion tests can produce a number of different end points or results:

- Reporting the number of cycles of abrasion to material wear-through or appearance of fibers (for coated fabrics)
- Providing a visual or comparative rating of performance
- Measuring a different performance property following a specified number of cycles of abrasion for specific conditions (for example, a material's breaking strength can be measured before and abrasion to determine the effect of abrasion on material strength).

Test Result Interpretation

☐ Better abrasion resistance is indicated by:

- Higher numbers of cycles to wear-through or fiber appearance
- Less apparent material damage
- Lower percentage change in other measured performance properties following abrasion.

25.10 Cut Resistance

Property Measured

- ☐ Cut resistance tests measure the ability of PPE items or materials to resist cutting through by a sharp-edged objects or machinery.

Applicable PPE

- ☐ Applicable PPE includes:
  - Full body garment materials
  - Partial body garment materials (especially leggings or chaps)
  - Gloves
  - Footwear

Test Principles

- ☐ Four different principles are used for measuring cut resistance:
  - A material specimen mounted on a soft surface is slid underneath a blade which is attached to a pivoted, weighted arm; cut-through is determined by visually examining the specimen (static cut test).
  - A weighted blade is dropped in a controlled fashion onto a specimen; cut-through is determined by visually examining the specimen (impact cut test)
  - A weighted blade is moved along the surface of a specimen which is mounted on a metal mandrel until the blade makes contact with the mandrel; the distance the blade travels before cutting through the material is measured and related to the weight placed on the blade (dynamic cut test).
  - A chainsaw of specific characteristic is operated at a specific speed and contacted with the PPE until cut-through or stoppage of the chainsaw.

Standard Test Methods

- ☐ Representative methods include:
  - ASTM F 1414 (chainsaw method for leg protection)
  - ASTM F 1458 (chainsaw method for footwear)
  - ASTM F 1790 (dynamic method)
  - ISO 11393 (International chainsaw method)
  - ISO 13997 (International dynamic method)
  - EN 381 (European chainsaw method)

- Section 6-22 of NFPA 1971 (1997), *Standard on Protective Ensemble for Structural Fire Fighting* (static method for gloves and footwear)
- Subclause 6.2 of EN 388 (European dynamic method)
- Subclause 6.5 of EN 388 (European impact method)

Test Results Provided

☐ Test results include:

- Static cut tests provide a determination of whether cut-through has occurred.
- Dynamic cut tests provide length to cut-through for a specific weight (or load); results are usually presented as the force for a specific length to cut-through.
- Chainsaw cut tests provide the threshold stopping speed (in meters per second) that does not produce a cut-through of the PPE item.

Test Results Interpretation

☐ Better cut resistance is demonstrated by:

- Less damage or no cut-through under specified conditions
- Higher loads for causing a specific length to cut-through
- Higher threshold stopping speeds for chainsaw cut tests

25.11 Puncture Resistance

Property Measured

☐ Puncture resistance tests measure the ability of PPE items or materials to resist penetration through by a slow-moving, pointed object.

Applicable PPE

☐ Applicable PPE includes:

- Full body garment materials
- Partial body garment materials
- Gloves materials
- Footwear materials and soles

Test Principles

☐ In most cases, a probe of specified sharpness is pushed perpendicularly into a specimen until the specimen is punctured.

Standard Test Methods

☐ Representative methods include:

- ASTM F 1342
- ISO 13996 (International method)
- EN 863 (European method)
- Section 3.5 of ANSI Z41.1, American National Standard for Personnel Protection—Protective Footwear
- Subclause 6.4 of EN 388 (European method for gloves)

Test Results Provided

☐ Puncture resistance results are reported as the force (in pounds or Newtons) required to puncture the specimen.

Test Results Interpretation

☐ Better puncture resistance is indicated by higher puncture forces.

25.12 Flex Fatigue Resistance

Property Measured

☐ Flex fatigue resistance tests measure the ability of PPE items or materials to resist wear or other damage when repeatedly flexed.

Applicable PPE

☐ Applicable PPE includes:

- Full body garment materials
- Partial body garment materials
- Gloves
- Footwear

Test Principles

☐ Test principles include:

- A machine is used to repeatedly flex the item of PPE for a specific number of flexing cycles; specimens are then either visually examined for damage or subjected to a different performance test (e.g., material chemical resistance).

- A cut is introduced into a specimen with repeated flexing of the specimen.

Standard Test Methods

☐ Representative methods include:

- ASTM D 3629 (cut growth method for rubber)
- ASTM F 392 (for flat materials and seams)
- ISO 7854 (International method for flat materials)
- Some cleaning methods such as laundering are used as a means for introducing repeated flexing of garments

Test Results Provided

☐ Test results include:

- Findings from visual examination (e.g., number of pinholes)
- Percentage change in selected performance property as compared to unflexed material
- Increase in cut growth (percentage change)

Test Results Interpretation

☐ Better flex fatigue resistance is demonstrated by:

- Less physical damage
- Lower percentage changes in selected performance property after flexing
- Lower percentage change of cut growth

25.13 Slip Resistance

Property Measured

☐ Slip resistance tests measure the ability of PPE items to maintain grip or traction with a surface.

Applicable PPE

☐ Applicable PPE includes:

- Gloves
- Footwear sole and heel materials

Test Principles

- ☐ Test principles include:
  - For gloves, grip is measured by comparing the lifting capability of human subjects wearing gloves to the same subject's barehanded lifting capability.
  - For footwear sole materials, the coefficient of friction is measured for the material as a machine slides the material against a specific surface.

Standard Test Methods

- ☐ Representative methods include:
  - ASTM F 489 (for footwear soles and heel materials)
  - ASTM F 609 (for footwear soles and heel materials)
  - Section 6-39 of NFPA 1971 (1997), *Standard on Protective Ensemble for Structural Fire Fighting* (grip method for gloves)

Test Results Provided

- ☐ Test results include:
  - For glove grip, the percentage change in weight lifting capability
  - For footwear traction, coefficient of friction

Test Results Interpretation

- ☐ Interpretation of test results:
  - For gloves, better grip is indicated by smaller decreases or no change in weight lifting capability
  - For footwear, better traction is indicated by higher coefficients of friction.

25.14 Shock Absorption

Property Measured

- ☐ Shock absorption tests measure the ability of PPE items to provide cushioning against vibration or repeated light impact.

Applicable PPE

- ☐ Applicable PPE include:
  - Full body garment materials
  - Partial body garment materials

- Gloves
- Head protection devices
- Respirators

Test Principles

☐ Test principles include:

- Specimen PPE is subjected to repeated shock or vibration and is visually examined for evidence of damage or evaluated in a follow-on performance test.
- Specimen PPE or materials are subjected to specific vibration energy with measurement of the level and frequency of the transmitted energy.

Standard Test Methods

☐ Representative methods include:

- ASTM D 4168 (for in-place cushioning materials)
- ASTM F 429 (football protective headgear testing)
- Section 6-3 of NFPA 1981 (1997), *Standard on Open-Circuit, Self-Contained Breathing Apparatus*
- EN ISO 10819 (International/European method for gloves against mechanical vibration/shock)
- There are no current U.S. standard test methods for damping effects of vibration specifically for PPE.

Test Results Provided

☐ Test results include:

- Findings of visible damage
- Percentage change in specific performance properties compared with unexposed specimens
- Level and frequency of transmitted vibration energy

Test Results Interpretation

☐ Better resistance to shock or vibration is indicated by:

- Lower levels of damage to PPE
- Little or no change in performance as the result of vibration or shock
- Limited levels of transmitted vibration energy in "safe" frequency ranges

Section 26

PPE Overall Product Integrity

The purpose of this section is to describe different types of PPE overall product integrity in terms of the measured properties, applicable PPE, test principles, available standard test methods, results provided, and interpretation of test results.

26.1 Overview of Overall Product Integrity

- ☐ Overall product integrity performance testing provides a determination of how well PPE prevent substances from entering (or leaving) the PPE through the material, seams, closures, or any other parts of the PPE that are evaluated.
- ☐ Overall integrity tests are usually used in conjunction with barrier tests for demonstrating resistance to environmental, chemical, or biological penetration hazards.
- ☐ Three types of integrity testing included in this section are:
  - Particulate-tight integrity
  - Liquid-tight integrity
  - Gas-tight integrity
- ☐ Representative standard test methods for different types of integrity described in this subsection are summarized in Table 26-1.

26.2 Particulate-Tight Integrity

Property Measured

- ☐ Particulate-tight integrity tests determine if particles enter or leave whole items of PPE.

Applicable PPE

☐ Applicable PPE includes:

- Full body garments
- Partial body garments
- Gloves
- Respirators

Table 26-1. List of PPE Overall Product Integrity Test Methods

Type of Integrity	Type of Test/Application	Available Test Methods
Particulate-tight integrity	Cleanroom garments	IES-RP-CC003.2
	Cleanroom gloves	IES-RP-CC005.2
Liquid-tight integrity	Gloves	ASTM D 5151 EN 374-2 NFPA 1971 S6-33
	Garments - liquid spray	ASTM F 1359 EN 468 NFPA 1971 S6-48
	Garments - jet	EN 463
	Footwear	NFPA 1971 S6-34
Gas-tight integrity	Full body garments	ASTM F 1052 EN 464 29 CFR Part 1910.120 (Appendix A)
	Respirators	Fit tests in 42 CFR Part 84

Test Principles

☐ Test principles include:

- PPE is either placed on a manikin (or headform) or worn by a test subject.
- The manikin or test subject is then exposed to a particulate atmosphere, usually an aerosol formed by a non-toxic, easily detectable liquid; human subjects usually perform a series of exercises to put stress on the garment.
- The particulate-contaminated atmosphere and atmosphere inside the PPE are sampled to either:
  - -- Determine the levels of particle intrusion inside the PPE, or
  - -- Measure rates of particle release from the PPE when worn inside a "clean" chamber.
- Alternatively, the human subject indicates by taste or odor his or her detection of penetrating particulates.
- Garments and gloves may be evaluated for particle release in cleanroom applications:
  - -- Garments are tumbled in a special apparatus which are flushed with clean air from a laminar flow hood; the air is monitored for particle density.
  - -- Gloves or finger cots are immersed in water and the water is analyzed for particulate content.

Standard Test Methods

☐ Test methods include:

- No current standardized tests exist for particulate-protective garments:
  - -- Proposed U.S. methods use corn oil aerosol in a chamber to measure intrusion into the PPE by sampling air.
  - -- Proposed European methods use salt aerosol in a chamber and measure intrusion into PPE by sampling air.
- National Institute for Occupational Safety and Health (NIOSH) Title 42 Part 84 (provides guidelines for measuring protection factors or fit factors for respirators)
- Particle release methods are provided in recommended practices established by the Institute of Environmental Sciences (IES)
  - -- IES-RP-CC003.2 (for cleanroom garments)

-- IES-RP-CC005.2 (for cleanroom gloves and finger cots)

Test Results Provided

☐ Test results include:

- Garment performance is usually characterized in terms of "intrusion coefficient" which is the ratio of the outside contaminant concentration to the concentration of contaminant measured on the inside.
- Testing of respirator integrity is referred to as fit testing.
    - -- When fit testing uses quantitative means to detect penetration, a protection factor is calculated as the ratio of the outside contaminant concentration to the concentration of contaminant measured on the inside.
    - -- When qualitative fit testing is performed, results are indicated as detected or not detected.
- Results for particle release are given in particle counts by size on the basis of sampled air volume or PPE item surface area.

Test Results Interpretation

☐ Better particulate-tight integrity is indicated by:

- Higher intrusion coefficients or protection factors
- No test subject detection of penetrating particulate
- Low particle counts for particulate release methods

26.3 Liquid-Tight Integrity

Property Measured

☐ Liquid integrity tests determine if liquid enters to the interior side of the PPE or onto wearer underclothing.

Applicable PPE

☐ Applicable PPE includes:

- Full body garments
- Partial body garments
- Gloves
- Footwear
- Eye and facewear (goggles)

- Respirators

Test Principles

☐ Test principles include:

- Methods involve spraying PPE on manikin and observing penetration of water onto inner liquid absorptive garment or blotter.
- Water often treated with:
  -- Surfactant to lower surface tension and better simulate organic liquids
  -- Dyes to enhance detection of penetrating liquid
- Tests may be conducted at periods longer than expected exposure to assist in observing leakage.
- Several techniques can be used to quantify leakage such as in the use of spectophometry on dyes-based liquid challenges and electroconductivity for salt-based liquid challenges.
- Small-scale tests are possible on gloves and boots.

Standard Test Methods

☐ Standard test methods include:

- ASTM D 5151 (for gloves)
- ASTM F 1359 (for garments)
- EN 463 (European method using liquid jet)
- EN 468 (European method using liquid spray)
- EN 374-2 (European method for gloves)
- Sections 6-33, 6-34, and 6-48 NFPA 1971 (1997), *Standard on Protective Ensemble for Structural Fire Fighting* (for gloves, footwear, and garments, respectively)

Test Results Provided

☐ Test results include:

- Most results are reported as detected penetration in terms of passing or failing performance.
- Observations of areas where leakage occurred may also be reported.

Test Results Interpretation

☐ Interpretation of test results:

- Since most tests are based on pass or fail determination, the results of this test are evident.
- Observations of specific leakage areas can help determine problems in the product that limit PPE item integrity.

26.4 Gas-Tight Integrity

Property Measured

☐ Gas-tight integrity tests determine if gas can penetrate PPE.

Applicable PPE

☐ Gas-tight integrity testing can only be performed on items that can be sealed, including:

- Full body garments (totally encapsulating suit)
- Gloves (non-air impermeable materials)
- Footwear (non-air impermeable materials)
- Respirators

Test Principles

☐ Test principles include:

- The most common approach for testing gas-tight integrity of PPE is to inflate the item to a specified pressure and then observe for any change in pressure within the item after several minutes.
- The location of leaks can be identified by using soapy water solution on exterior of the PPE item.
- Alternative approaches involve placing PPE on a test subject in closed environment containing a gas (such as ammonia) and measuring concentration of challenge agent inside suit.
- Full scale testing of suits against actual hazardous materials has been done on a very limited basis in special projects.
- Tests may be modified for gloves and footwear.
- Respirator facepieces may be fit tested using gases or vapors in procedures similar to those described for particulate-tight integrity testing.

Standard Test Methods

☐ Standard test methods include:

- ASTM F 1052 (for totally encapsulating suits)
- EN 464 (European method for totally encapsulating suits)
- Appendix A of OSHA 29 CFR 1910.120 (gas-tight and ammonia leakage tests for totally encapsulating suits)
- Appendix A of OSHA 29 CFR 1910.134 (fit testing and respirators)
- National Institute for Occupational Safety and Health (NIOSH) Title 42 CFR Part 84 (provides guidelines for measuring protection factors or fit factors for respirators)

Test Results Provided

☐ Test results include:

- Testing of totally encapsulating suits is reported as the ending pressure or amount of pressure drop (in inches of water or millimeters of mercury).
- Observations of areas where leakage occurred may also be reported.
- Testing of respirator integrity is referred to as fit testing.
  - When fit testing uses quantitative means to detect penetration, a protection factor is calculated as the ratio of the outside contaminant concentration to the concentration of contaminant measured on the inside.
  - When qualitative fit testing is performed, results are indicated as detected or not detected.

Test Results Interpretation

☐ Interpretation of test results:

- Since most tests are based on pass or fail determination, the results of this test are evident.
- Observations of specific leakage areas can help determine problems in the product that limit PPE item integrity.
- For suit ammonia testing or respirator fit testing:
  - Higher intrusion coefficients or protection factors
  - No test subject detection of penetrating vapor or gas

Section 27

PPE Resistance to Environmental Conditions

The purpose of this section is to describe different types of PPE resistance to environmental conditions in terms of the measured property, applicable PPE, test principles, available standard test methods, results provided, and interpretation of test results.

27.1 Overview of Environmental Resistance

- ☐ Environmental resistance testing evaluates PPE for conditions arriving from the extremes of weather, light, or noise in occupational settings.
- ☐ Testing related to worker comfort is covered in Section 32.
- ☐ Performance properties covered in this section include:
  - Water repellency and absorption resistance
  - Water penetration resistance
  - Salt spray and corrosion resistance
  - Resistance to cold temperatures
  - UV-Light resistance
  - Ozone resistance
  - Excess light attenuation
  - Excess Noise attenuation
- ☐ Representative standard test methods for resistance to environmental conditions described in this subsection are summarized in Table 27-1.

Table 27-1. List of PPE Environmental Condition Test Methods

Physical Property	Type of Test/Application	Available Test Methods
Water repellency and absorption resistance	Surface wetting spray method - garments	AATCC 22 FTMS 191A,5522 ISO 4920
	Spray impact method - garments	AATCC 42 CAN/CGSB-4.2 No. 26.2
	Water absorption/capacity test - garments	ASTM D 5802 FTMS 191A,5504
	Rain method - garments	FTMS 191,5524
	Runoff method - garments	EN 368 ISO 6530
Water penetration resistance	Specified hydrostatic method	ASTM F 903 FTMS 191A,5512
	Increasing hydrostatic method	AATCC 127 EN 20811 FTMS 191A,5514 FTMS 191A,5516 ISO 811
	Low hydrostatic method	CAN/CGSB-4.2 No. 26.3
	High hydrostatic method	CAN/CGSB-4.2 No. 26.5
Salt spray/corrosion resistance	PPE hardware	ASTM B 117 ASTM G 27 ISO 4536 MIL-STD-810E, Method 509.3, Section II
	Face/eyewear	ANSI Z87.1 S15.6
Stiffness/resistance to cold temperatures	Stiffness - textiles	ASTM D 1388

Physical Property	Type of Test/Application	Available Test Methods
Stiffness/resistance to cold temperatures (continued)		ASTM D 4032
	Stiffness - leather	ASTM D 2821
	Stiffness - plastics	ASTM D 747 ASTM D 1043
	Stiffness - rubber	ISO 5979
	Brittleness - plastics/rubber	ASTM D 746 ASTM D 2137
	Low temperature stiffness - polymers	ASTM D 1053
	Low temperature bending - coated fabrics/rubber	ASTM D 2136 ISO 4675
	Impact - plastics/rubber	ASTM D 1790 ISO 4646
	Cold cracking - leather	ASTM D 1912
UV-light resistance	Textiles	AATCC 16 ASTM G 23 ASTM G 26 CAN/CGSB-4.2 No. 5.1 FTMS 191A,5671 ISO 105
	Rubber	ASTM D 1148
Ozone resistance	Rubber and coated fabrics	ASTM D 518 ASTM D 1149 ASTM D 1171 ISO 3011
Noise attenuation	Ear protectors	OSHA 20 CFR 1910.95 Appendix B

Physical Property	Type of Test/Application	Available Test Methods
Light attenuation	Eyewear	ANSI Z87.1 S12
Visibility	Fluorescence method	ASTM E 991 ASTM 1247 EN 471
	Retroreflective method	ASTM E 810 ASTM E 1809 EN 471
	Marking specification	ASTM E 1501 EN 471

27.2 Water Repellency and Absorption Resistance

Property Measured

☐ Property measured:

- Water repellency tests measure the ability of PPE to resist surface wetting when contacted with water from spray or impingement.
- Water absorption resistance tests measure the ability of PPE to gain weight when contacted with water from spray or impingement.

Applicable PPE

☐ Applicable PPE includes:

- Full body garment materials
- Partial body garment materials
- Glove materials
- Footwear materials
- Headwear

Test Principles

☐ Test principles include:

- Water is reproducibly sprayed or impinged onto the PPE material exterior surface that is oriented to allow water runoff.

- In some tests:
  - -- Water runoff is collected.
  - -- The weight of absorbed by the PPE item is determined (water absorption resistance).
  - -- The weight of water penetrating the PPE material and absorbed by a blotter material is determined.

Standard Test Methods

☐ Standard test methods include:

- AATCC 22 (surface wetting spray test)
- AATCC 42 (spray impact test)
- ASTM D 5802 (water absorption/capacity test)
- CAN/CGSB-4.2 No. 26.2 (Canadian spray impact test)
- EN 368 (European runoff test)
- FTMS 191A, 5504 (Federal water absorption test)
- FTMS 191A, 5522 (Federal surface wetting spray test)
- FTMS 191A, 5524 (Federal rain test)
- ISO 4920 (International surface wetting spray test)
- ISO 6530 (International runoff test)
- ISO 9865 (International rain test)

Test Results Provided

☐ Test results may include:

- A qualitative spray test rating
- Percentage by weight absorption of the PPE material specimen
- Percentage by weight absorption by underlying blotter material

Test Results Interpretation

☐ Better water repellency or water absorption resistance is indicated by

- Lower qualitative spray test ratings
- Smaller percentages of specimen water absorption
- Smaller percentages of specimen water absorption

27.3 Water Penetration Resistance

Property Measured

☐ Water penetration tests measure the ability of PPE materials the flow of liquid water onto the interior surface of PPE material surfaces when contacted with water from spray or impingement.

Applicable PPE

☐ Applicable PPE inludes:

- Full body garment materials, seams, and closures
- Partial body garment materials, seams, and closures
- Glove materials and seams
- Footwear materials
- Head protection device materials

Overall product liquid integrity testing is addressed in Subsection 26.3.

Test Principles

☐ Test principles include:

- Water is held in contact with the exterior surface of a specimen for a specified period of time under hydrostatic pressure.
  -- Pressure may be gradually increased (increasing hydrostatic pressure), or
  -- Pressure may be held at specific pressures for specific durations (specified hydrostatic pressure)
- Penetration is either observed visually or measured by the amount of water that penetrates the specimen and absorbs into an underlying blotter.

Standard Test Methods

☐ Test methods include:

- AATCC 127 (increasing hydrostatic pressure method)
- ASTM F 903 (specified hydrostatic pressure method)
- CAN/CGSB-4.2 No. 26.3 (Canadian low hydrostatic pressure method)
- CAN/CGSB-4.2 No. 26.5 (Canadian high hydrostatic pressure method)
- EN 20811 (European hydrostatic pressure method)
- FTMS 191A, 5512 (Federal specified hydrostatic pressure method)

- FTMS 191A, 5514 (Federal increasing hydrostatic pressure method)
- FTMS 191A, 5516 (Federal increasing hydrostatic pressure method)
- ISO 811 (International hydrostatic pressure method)

Test Results Provided

☐ Test results provided:

- For tests conducted under specified hydrostatic pressure conditions, results are report as pass or fail (water observed penetrating).
- For tests conducted with increasing hydrostatic pressure, the pressure at which penetration (or droplets of water appeared) is reported (in pounds per square inch or inches of water column).

Test Results Interpretation

☐ Better water penetration resistance is demonstrated by:

- Passing results
- Higher failure pressures

27.4 Salt Spray or Corrosion Resistance

Property Measured

☐ Salt spray testing evaluates the ability of PPE items to resist corrosion when exposed to salt spray.

Applicable PPE

☐ Applicable PPE includes:

- Full body garment hardware
- Partial body garment hardware
- Glove hardware
- Footwear hardware
- Headwear
- Eye and facewear
- Respirators

Test Principles

☐ The specimen is placed in a closed chamber and is subjected to salt spray or fog for a specific duration and at a specific temperature.

Standard Test Methods

- ☐ Test methods include:
  - ASTM B 117
  - ASTM G 85
  - ISO 4536 (International method)
  - Method 509.3, Section II of MIL-STD-810E
  - Section 15.6 of ANSI Z87.1, *Practice for occupational and educational eye and face protection*

Test Results Provided

- ☐ Test results may include:
  - Visual observations of specimen condition following exposure
  - Changes in PPE performance properties or functionality following the exposure.

Test Results Interpretation

- ☐ Better resistance to salt spray exposure is indicated by:
  - No evidence of corrosion or other damage
  - Little or no changes in PPE performance properties or functionality

27.5 Stiffness and Resistance to Cold Temperatures

Property Measured

- ☐ Cold temperature resistance testing evaluates changes in PPE, materials, and components when exposed to cold temperatures.

Applicable PPE

- ☐ Applicable PPE includes:
  - Full body garment materials
  - Partial body garment materials
  - Glove materials
  - Footwear materials
  - Headwear
  - Eye and facewear

- Hearing protectors
- Respirators

Test Principles

☐ Test principles include:

- Low temperature effects on materials are generally evaluated by exposing specimens to a specific temperature, subjecting the material to a bending impact force and measuring the change in stiffness or damage to the material.
  - -- Bending forces may be produced by mecahnical means or dropping objects.
  - -- Material stiffness may be measured the forces associated with bending, twisting (torsion), or suspending materials.
  - -- Testing is performance at range of temperature of interest to the user.
- Other PPE can be evaluated for cold temperature effects by placing specimens in cold temperature and conducting impact testing or other forms of testing on the cooled specimens (*see 25.4*).

Standard Test Methods

☐ Standard test methods include:

- ASTM D 746 (brittleness by impact on plastics/elastomers)
- ASTM D 747 (stiffness of plastics)
- ASTM D 1043 (stiffness of plastics)
- ASTM D 1053 (low temperature stiffness of flexible polymers)
- ASTM D 1388 (stiffness of textiles)
- ASTM D 1790 (impact on plastics)
- ASTM D 1912 (cold cracking resistance of leather)
- ASTM D 2136 (low temperature bending for coated fabrics)
- ASTM D 2137 (embrittlement point of rubber)
- ASTM D 2821 (stiffness of leather)
- ASTM D 4032 (stiffness of textile fabric)
- ISO 4646 (International low temperature impact test for rubber)
- ISO 4675 (International low temperature bend test for rubber)
- ISO 5979 (International test for rubber flexibility)

Test Results Provided

☐ Test results include:

- Bending moment or modulus (bending tests)
- Amount of deflection or angular twist (torsion tests)
- Observations of PPE or material condition following bending or impact

Test Results Interpretation

☐ Better PPE resistance to cold temperatures is demonstrated by:

- Little or no increase in bending moment with decreasing temperature
- Little or no change in deflection or angular twist for cold temperature as compared to ambient temperature
- Little or no damage to material from bending or impact forces (such as cracking, embrittlement, or breakage)

27.6 UV-Visible Light Resistance

Property Measured

☐ UV-Visible light resistance testing evaluates changes in PPE, materials, and components when exposed to ultraviolet (UV) or visible light.

Applicable PPE

☐ Applicable PPE includes:

- Full body garment materials
- Partial body garment materials
- Glove materials
- Footwear materials
- Headwear
- Eye and facewear
- Hearing protectors
- Respirators

Test Principles

☐ Test principles include:

- PPE or PPE materials are placed in open sunlight and allowed to weather over extended periods of time, or

- Exposed to artificial UV or visible light radiation using a carbon-arc or xenon-arc apparatus (weatherometer).
- Visual observations are made of the PPE condition following exposure and compared to the PPE's original condition.
- Other performance tests (such as strength) may be performed following the UV/visible light exposure to determine how the property changes with exposure.

Standard Test Methods

☐ Standard test methods include:

- AATCC 16 (colorfastness to light for textiles)
- ASTM D 1148 (for heat and ultraviolet light on rubber)
- ASTM G 23 (carbon-arc type)
- ASTM G 26 (xenon-arc type)
- CAN/CGSB-4.2 No. 18.3 (Canadian xenon-arc type)
- FTMS 191A, 5671 (Federal method for textiles)
- ISO 105 (International method)

Test Results Provided

☐ Test results include:

- Visual rating of color (colorfastness) based on change from original color.
- Percentage change in selected performance property as the result of exposure.

Test Results Interpretation

☐ Better PPE resistance to UV or visible light is indicated by:

- Less color change
- Retention of original performance following exposure

27.7 Ozone Resistance

Property Measured

☐ Ozone resistance testing evaluates changes in PPE, materials, and components when exposed to elevated concentrations of ozone.

Applicable PPE

☐ Applicable PPE includes:

- Full body garment materials
- Partial body garment materials
- Glove materials
- Footwear materials
- Headwear
- Eye and facewear
- Hearing protectors
- Respirators

Test Principles

☐ Test principles include:

- PPE or PPE materials are placed in ozone generating chamber where a specific concentration of ozone in air is generated.
- Visual observations are made of the PPE condition following exposure for signs of degradation such as cracks.
- Alternatively, samples may be placed in ozone chamber until deterioration is noted.
- Other performance tests (such as tensile strength) may be performed following the ozone exposure to determine how the property changes with exposure.

Standard Test Methods

☐ Standard test methods include:

- ASTM D 518 (for rubber)
- ASTM D 1149 (exposure followed by tensile test for rubber)
- ASTM D 1171 (for large specimens of rubber)
- ISO 3011 (International method for coated fabrics)

Test Results Provided

☐ Test results include:

- Time to deterioration under specific conditions.

- Percentage change in selected performance property as the result of exposure.

Test Results Interpretation

☐ Better PPE resistance to ozone is indicated by:

- Longer times until deterioration takes place
- Retention of original performance following exposure

27.8 Excess Noise Attenuation

Property Measured

☐ The attenuation of noise is measured to determine the effectiveness of hearing protectors in preventing unsafe levels of sound from reaching the worker's ear.

Applicable PPE

☐ This type of testing is performed on hearing protectors. Interference with communication is addressed in subsection 4.10 on Human Factors.

Test Principles

☐ Different methods evaluate the hearing levels of individuals for specific hearing protectors at different frequencies.

Standard Test Methods

☐ Appendix B to OSHA 20 CFR Section 1910.95 describes one method by the Environmental Protection Agency and three methods developed by NIOSH and described in the "List of Personal Hearing Protectors and Attenuation Data," HEW Publication No. 76-120, 1975, pages 21-37.

Test Results Provided

☐ Ear Protectors are provided a noise reduction rating (NRR).

Test Results Interpretation

☐ Interpretation of test results:

- The higher the noise reduction rating, the better the noise attenuation provided by the hearing protector.
- However, noise reduction ratings that are too high also significantly impede communications of the wearer. Noise reduction ratings often provide a measure of the change in sound level (provided in decibels), which is accomplished by the hearing protector.

- Noise reduction ratings of 5 to 10 decibels are usually recommended, but hearing protector selection should be based on the specific hazards of the situation.

27.9 Excess Light Attenuation

Property Measured

☐ Light attenuation testing measures the effectiveness of lenses or filters used in PPE for reducing the radiation produced by harmful light to acceptable levels.

Applicable PPE

☐ Light attenuation testing is applied to those portions of PPE which provide eye protection or allow vision, including:

- Full body garment visors
- Partial body garment visors
- Headwear visors
- Eye and facewear
- New methods are under development for garments for providing a rating similar to sun block tanning products.

Test Principles

☐ A spectrophotometer or other light measuring instrument is used to evaluate the transmission of UV or visible light transmission.

Standard Test Methods

☐ Test procedures are provided primarily in Section 12 of ANSI Z87.1-1989, *Practice for occupational and educational eye and face protection*.

Test Results Provided

☐ Light attenuation is characterized by the following results:

- Luminous transmittance
- Shade number
- Effective far UV transmittance
- Infrared transmittance
- Blue light transmittance

Test Results Interpretation

☐ Better visual attenuation is indicated by:

- Higher luminous transmission
- Larger shade numbers
- Lower effective far UV transmittance
- Lower infrared transmittance
- Lower blue light transmittance

27.10 Visibility

Property Measured

☐ PPE visibility is evaluated to determine how well the item of PPE can be seen as enhanced by colors or special materials during either daytime or nighttime conditions.

- Phosphorescent material are energized by sunlight and release energy as visible light for a short period of time after the light source is removed, producing a dim night time glow.
- Fluorescent materials are activated by ultraviolet radiation and provide high visual contrast in daylight.
- Retroreflective materials use light from another source and reblects light back to the source for providing high contrast during nighttime.

Applicable PPE

☐ Applicable PPE includes:

- Full body garments
- Partial body garments
- Headwear

Test Principles

☐ Test principles include:

- Retroreflectance is measured by measuring the intensity of light reflected from the material by a known light source.
- Fluorescence is measured by measuring the color intensity of material when exposed to ultraviolet light.

Standard Test Methods

☐ Standard methods include:

- ANSI/ISEA 107 (tests/requirements for high visibility clothing)
- ASTM E 810 (retroreflection test for materials)
- ASTM E 991 (fluorescence test)
- ASTM E 1247 (fluorescence test)
- ASTM E 1501 (specification of night time markings on clothing)
- ASTM E 1809 (retroreflective test for clothing)
- EN 471 (European test for high visibility clothing)

Test Results Provided

☐ Test results include:

- Observed spectral color (fluorescence)
- Coefficient of retroreflection, reported in candelas per lux per square meter or candelas per footcandle per square foot

Test Results Interpretation

☐ Better visibility is demonstrated by:

- Greater intensity color (for fluorescence)
- Higher coefficients of retroreflection

Section 28

PPE Chemical Resistance

The purpose of this section is to describe different types of PPE chemical resistance in terms of the measured property, applicable PPE, test principles, available standard test methods, results provided, and interpretation of test results.

28.1 Overview of Chemical Resistance

☐ There are three types of material-chemical interactions:

- Degradation
- Penetration
- Permeation

☐ Chemical penetration may be further distinguished by:

- Particulate penetration
- Liquid penetration
- Vapor penetration

☐ Performance properties include:

- Chemical degradation resistance
- Particulate penetration resistance
- Liquid penetration resistance
- Vapor penetration resistance
- Chemical permeation resistance

☐ Representative standard test methods for chemical resistance described in this subsection are summarized in Table 28-1.

Table 28-1. List of PPE Chemical Resistance Test Methods

Chemical Resistance Property	Type of Test/Application	Available Test Methods
Chemical degradation resistance	All PPE materials	ASTM D 471 ASTM D 543
Chemical particulate penetration resistance	Cleanroom garments	IES-RP-CC003.2
	Garments and facewear	ASTM D 2986 ASTM F 1215
	Sand and dust test for all PPE	MIL-STD-810E, Method 510.3, Section II
Chemical liquid penetration resistance	All flat PPE materials	ASTM F 903 ISO 13994
Chemical vapor penetration resistance	Respirators	42 CFR Part 84
Chemical permeation resistance	Continuous contact - gases and liquids - clothing materials	ASTM F 739
	Continuous contact - liquids - clothing materials	EN 369 ISO 6529
	Continuous contact - gloves	EN 374-3
	Intermittent contact - gases and liquids - clothing materials	ASTM F 1383
	Continuous contact - gravimetric	ASTM F 1407
	Gas transmission - gravimetric	ISO 6179
	Gas transmission - volumetric	ASTM D 1434
	Gas transmission - pressure-based	ASTM D 1434 ASTM 3985

28.2 Chemical Degradation Resistance

Property Measured

☐ Chemical degradation is defined as the deleterious change in one or more physical properties of protective clothing material as the result of chemical exposure.

Applicable PPE

☐ Applicable PPE includes:

- Full body garment materials, seams, closures, and components
- Partial body garment materials, seams, closures, and components
- Glove materials, seams, and components
- Footwear materials, seams, closures, and components
- Headwear materials
- Eye and facewear materials
- Respirator materials

Test Principles

☐ Testing involves exposing sample of PPE material to a chemical (solid, liquid, or gas) and measuring physical property before and after exposure:

- Exposure can be single sided or by total immersion
- Physical properties commonly measured include:
    - Weight change
    - Thickness change
    - Visual change
    - Hardness change
    - Change in breaking strength

Standard Test Methods

☐ Currently, there are no standardized tests specific to protective clothing.

- ASTM D 471 provides test procedures for measuring effects of chemicals on rubber materials.
- ASTM D 543 provides test procedures for measuring resistance of plastics to chemical reagents.

Test Results Provided

- ☐ Test results include:
  - Many glove manufacturers still use degradation data in their chemical resistance tables.
  - Results often reported as qualitative ratings

Test Results Interpretations

- ☐ Interpretation of test results:
  - Better qualitative ratings indicate better chemical degradation resistance.
  - Problems with degradation data
    - It is difficult to compare degradation data unless testing performed under the same conditions.
    - Degradation data can only be used to rule out candidate materials, not recommend a material.
    - Degradation resistance testing is not a true barrier test; degradation testing does not indicate whether material will provide an effective barrier.

28.3 Chemical Particulate Penetration Resistance

Property Measured

- ☐ Chemical particulate penetration is defined as the flow of chemical particles through closures, porous materials, seams, and pinholes or other opening in protective clothing.

Applicable PPE

- ☐ Applicable PPE includes:
  - Full body garment materials, seams, and closures
  - Partial body garment materials, seams, and closures
  - Glove materials and seams
  - Footwear materials, seams, and closures
  - Headwear materials
  - Eye and facewear materials
  - Respirator materials

Test Principles

☐ Test principles include:

- Specimens are subjected to a particulate environment with sampling of the specimen interior side for penetrating particulates.
- Surrogate particles are typically used in lieu of actual chemicals.
- Particulate challenge atmospheres and penetrating particulates are characterized by number and size.
- Particles used in testing generally range from 0.1 to 20 microns in size.

Standard Test Methods

☐ Standard test methods include:

- There are no specific standard test methods for protective clothing for protecting the wearing from particulate hazards.
- A material particle penetration test is provided in Institute of Environmental Sciences (IES) recommended practice IES-RP-CC003.2 (for cleanroom garments)
- ASTM D 2986 (measures penetration of DOP aerosol)
- ASTM F 1215 (measures penetration of latex spheres)
- Method 510.3, Section II of MIL-STD-810E (overall sand and dust environmental conditioning for PPE items)

Test Results Provided

☐ Test results include:

- Percent penetration efficiency is represented by particle size. The penetration efficiency is determined by the dividing of the counts for penetrating particles by the counts of the particles in the challenge atmosphere (for a specific particle size range).
- Particulate penetration may also be represented by the actual counts of particles penetrating a specimen less any background particulate contamination.

Test Results Interpretation

☐ Better particulate penetration resistance (holdout) is indicated by:

- Higher percent penetration efficiencies
- Lower particle counts

- Most particulate penetration tests require some airflow through materials. "Impermeable" materials may cause large pressure drops across the specimen, which lead to false particulate penetration results.

28.4 Chemical Liquid Penetration Resistance

Property Measured

☐ Liquid chemical penetration is defined as the flow of a liquid chemical on a non-molecular level through closures, porous materials, seams, and pinholes or other opening in protective clothing.

Applicable PPE

☐ Applicable PPE includes:

- Full body garment materials, seams, and closures
- Partial body garment materials, seams, and closures
- Glove materials and seams
- Footwear materials, seams, and closures
- Headwear materials
- Eye and facewear materials
- Respirator materials

Test Principles

☐ Liquid penetration testing involves putting chemical in contact with a material in a test cell and observing for penetration on the other side.

- Exposure periods vary and may involve pressurization; common exposure protocols include:
  - 5 minutes ambient pressure, 10 minutes pressure @ 1 psig
  - 5 minutes ambient pressure, 1 minute @ 2 psig, 54 minutes ambient pressure
- Test is qualitative - penetration is observed (FAIL) or not observed (PASS)
- Dyes and blotting techniques are used to enhance observation of penetration.

Standard Test Methods

☐ Standard test methods include:

- ASTM F 903

- ISO 13994 (International method)

Test Results Provided

☐ Test results include:

- Pass or fail determination
- Time when penetration occurs
- Observations of specimen degradation or failure

Test Results Interpretation

☐ Interpretation of test results:

- Penetration testing is appropriate for film-based or microporous fabrics in cases where the concern is to keep liquids off the skin.
- Use of the penetration test is for qualifying materials is predicated on accepting exposure to chemical gases, vapors, or by permeation of liquid chemicals.
- Penetration testing should only be used for determining acceptable material when chemical is a non-volatile or non-skin toxic chemical.

28.5 Chemical Vapor Penetration Resistance

Property Measured

☐ Chemical vapor penetration is defined as the flow of a chemical vapor on a non-molecular level through closures, porous materials, seams, and pinholes or other opening in protective clothing.

Applicable PPE

☐ Testing for vapor penetration resistance is applied to porous or microporous material combined with adsorbent materials used in:

- Full body garment materials, seams, and closures
- Partial body garment materials, seams, and closures
- Glove materials and seams
- Footwear materials, seams, closures
- Respirator canisters and cartridges

Test Principles

☐ Vapor penetration testing involves:

- Exposure of a specimen to a chemical at a specific concentration in air under specific temperature and humidity conditions
- Control of the flow rate of the contaminated air through the specimen
- Monitoring of the effluent air for contamination levels

Standard Test Methods

☐ There are no specific standard test methods for measuring vapor penetration through materials that differentiate between vapor penetration and liquid penetration.

- Methods used for measuring respirator cartridge or canister effectiveness for preventing the penetration of gaseous contaminants are specified in NIOSH 42 CFR Part 84

Test Results Provided

☐ Test results include:

- Time when penetration occurs (this time is commonly referred to as a service life for respirator cartridges and canisters.

Test Results Interpretation

☐ Interpretation of test results:

- Penetration testing is appropriate for porous or microporous fabrics which are combined with adsorbent materials to keep gaseous chemical from contacting the skin.
- Penetration time (or surface life) is significantly affected by the concentration of the chemical in air, the temperature, the relative humidity, and the flow rate through the specimen.

28.6 Chemical Permeation Resistance

Property Measured

☐ Permeation is defined as the process by which a chemical moves through a protective clothing material on a molecular level.

- Permeation involves three steps:
  - -- Adsorption of chemical onto outer surface of material
  - -- Diffusion of chemical through material
  - -- Desorption of chemical from inner surface of material

- Permeation is a function of material solubility in the material (primary chemical interaction) and diffusion through the material (primarily physical interaction).
  - -- Chemical will be soluble in material when chemical structures are similar.
  - -- Diffusion based primarily on size of permeation chemical versus structure of material polymer.

Applicable PPE

☐ Applicable PPE includes:

- Full body garment materials, seams, and closures
- Partial body garment materials, seams, and closures
- Glove materials and seams
- Footwear materials, seams, and closures
- Headwear materials
- Eye and facewear materials
- Respirator materials

Test Principles

☐ Permeation testing involves a test cell that a material specimen divides into two hemispheres; on one side (challenge side) the chemical is introduced, the other side (collection side) is monitored for permeating chemical.

- The challenge side of the test cell accommodates solids, liquids, or gases.
- Contact can be continuous or intermittent
- The collection side can use either a gas or liquid collection medium.
  - -- Dry nitrogen is a common collection medium for volatile organic chemicals.
  - -- Water is a common collection medium for inorganic liquid solutions (e.g., acids and bases).
  - -- Other collection media can be used but must be selected for efficiency in collecting permeant.
- Detection of permeating chemical is by the appropriate analytical instruments.

- -- Gas chromatographs with flame ionization, photoionization, thermal conductivity, or electron capture detector are common for many organic chemicals.
- -- Infrared spectroscopy is used for gas collection systems.
- -- Ion specific electrodes are used for inorganic liquids.

- Variations of permeation testing involve use of gravimetric, pressure-based, and volumetric techniques for measuring permeation.

Standard Test Methods

☐ Standard test methods include:

- ASTM D 1434 (volumetric and pressure-based gas transmission rate method)
- ASTM D 3985 (oxygen gas transmission rate method)
- ASTM F 739 (for continuous contact of liquids and gases)
- ASTM F 1383 (for intermittent contact of liquids and gases)
- ASTM F 1407 (simple gravimetric method for liquids)
- EN 369 (European method for liquids)
- EN 374-3 (International method for gloves against liquids and gases)
- ISO 6179 (International method for gravimetric gas transmission)
- ISO 6529 (International method for liquids)

Test Results Provided

☐ The primary test outputs are the breakthrough time and permeation rate.

- Breakthrough time is the time that elapses between contact of the chemical with the material and the time the chemical is first detected in the collection side of the test cell.
  - -- Actual breakthrough times are based on the sensitivity of the detector used.
  - -- Normalized breakthrough times are the time when the rate of permeation equals 0.1 $\mu g/cm^2 min$ (in Europe breakthrough times are normalized at 1.0 $\mu g/cm^2 min$).
  - -- Normalized breakthrough times are becoming more common and may be the only data reported by the manufacturer.
- The permeation rate is the steady state or maximum observed rate of permeation observed during the test.

-- Many chemicals will exhibit a steady state rate during the test.

-- If steady state does not occur, then the maximum rate is reported

Test Results Interpretation

☐ Uses of permeation data:

- The majority of the protective garment and glove industry provide permeation data on their products; this information is provided in various levels of detail.

- Test is appropriate when any contact (solid, liquid, or vapor) with chemical must be prevented.

- Most end users solely use breakthrough time for judging acceptability of protective clothing by using selecting material which show breakthrough times for period less than the maximum estimated exposure time.

- Permeation data is often inappropriate used, e.g., using permeation data to select an material for a non-skin toxic, non-volatile chemical when the principal exposure to the wearer will be areas which are not protected.

- Factors affecting permeation:

  -- Temperature—higher temperatures resulting in shorter breakthrough times and increased permeation rates; most manufacturers reported permeation data is conducted at room temperature (25°C).

  -- Type and length of contact—continuous exposure tests often, but not always different (more severe) than intermittent contact tests; most tests conducted for 4 or 8 hours.

  -- Mixtures—mixture permeation cannot always be predicted on the basis of permeation data for individual components of mixture against same material.

  -- Generic material classes—Similar materials of different manufacturers perform differently, i.e., one manufacturer's Neoprene glove will probably have different permeation performance as compared to a Neoprene glove from a different manufacturer; this is due to difference in polymer formulation, thickness, and substrates.

Section 29

PPE Biological Resistance

The purpose of this section is to describe different types of PPE biological resistance in terms of the measured property, applicable PPE, test principles, available standard test methods, results provided, and interpretation of test results.

29.1 Overview of Biological Resistance

- ☐ Biological resistance tests are used to assess barrier properties of PPE or PPE susceptibility for supporting microorganism growth.
- ☐ Representative standard test methods for biological resistance described in this subsection are summarized in Table 29-1.

29.2 Microorganism Filtration Efficiency

Property Measured

- ☐ The measurement of microorganism filtration efficiency evaluates the ability of PPE items and fabrics to prevent the passage of airborne microorganisms. This property is also related to particulate penetration resistance (*see 26.2*)

Applicable PPE

- ☐ Applicable PPE includes:
  - Full body garment materials and seams
  - Partial body garment materials and seams
  - Eye and facewear (facemasks)
  - Respirators

Table 29-1. List of PPE Biological Resistance Test Methods

Biological Resistance Property	Type of Test/Application	Available Test Methods
Microorganism filtration efficiency	Breathable materials	ASTM F 1608 MIL-M-36954C
Biological fluid resistance	Clothing materials	ASTM F 1819 ASTM F 1862
Biological fluid penetration resistance	Clothing materials	ASTM F 1670
Viral penetration resistance	Clothing materials	ASTM F 1671
Antimicrobial resistance	Microorganism resistance (general)	CAN/CGSB-4.2 No. 28 U.S. Pharmacopeia
	Bacterial resistance	AATCC 100 ASTM G 22
	Viral resistance	ASTM E 1053
	Mildew resistance	ASTM D 2020 FTMS 191A,5750 FTMS 191A,5760
	Fungus resistance	ASTM G 21
Insect resistance	Clothing materials	CAN/CGSB-4.2 No. 38 FTMS 191A,5764 ISO 3998

Test Principles

☐ A sample of PPE or material specimen is exposed to a challenge atmosphere containing microorganisms and the number of microorganisms penetrating the PPE or material are counted.

- Bacterial or virus microorganisms are typically used.

- Bacterial-based tests are commonly called Bacterial Filtration Efficiency (BFE).
- Viral-based tests are commonly called Bacterial Filtration Efficiency (VFE).
- Other microorganisms such as spores may be used.
- Tests generally involve using a pump or vacuum to pull air through the specimen.
- Assay or microorganisms counting techniques are specific to the type of microorganism used.

Standard Test Methods

☐ Standard test methods include:

- ASTM F 1608 (test involving spores)
- MIL-M-36954C (bacterial filtration test for surgical masks)

Test Results Provided

☐ Test results include:

- Filtration efficiency is generally represented as either a log reduction value (the number of magnitude or order of 10 that the challenge is reduced) or the percent efficiency of the PPE or material in preventing microorganism penetration.

Test Results Interpretation

☐ Interpretation of test results:

- Higher filtration efficiency indicates PPE or materials that are more likely to prevent passage of airborne microorganisms.
- Tests for one type of microorganisms and one set of conditions may not usually indicate equal performance for different microorganisms or use conditions.
- The use of efficiency results indicates that the tested material, unless 100% efficient, do not prevent the passage of all microorganisms.
- PPE or materials with less than 100% exposure will allow microorganisms to pass through.

29.3 Biological Fluid Resistance

Property Measured

- ☐ Biological fluid penetration resistance testing evaluates material resistance to penetration by blood and related body fluids.
  - This form of testing discriminates barrier characteristics of different fabrics used in apparel for preventing blood or other body fluid strike-through.
  - Fabrics with fluid resistance may still allow fluid penetration under some use conditions.

Applicable PPE

- ☐ Applicable PPE includes:
  - Full body garment materials and seams
  - Partial body garment materials and seams
  - Glove materials and seams
  - Footwear materials and seams
  - Eye and face protective devices
  - Respirators

Test Principles

- ☐ The external surface of PPE fabrics or items is contacted with a biological fluid or surrogate biological fluid under specific conditions and the interior side is examined for evidence of penetration
  - Some testing approaches substitute a biological fluid for water in tests commonly used for water repellency and penetration resistance *(see 27.2 and 27.3)*.
  - A common surrogate for blood is synthetic blood, composed of water, surfactant/dye, and thickener which simulates the surface tension, viscosity, and color of actual blood and some body fluids.
  - These tests provide some quantified result, either in terms of a "breakthrough" or "failure" pressure, or pass/fail determination at a specific pressure or exposure condition.
  - Test conditions may fluid contact which is hydrostatic or mechanical or a combination of both pressure application methods:
    - -- Hydrostatic pressure involves contact of the PPE specimen with a reservoir of liquid.

-- Mechanical pressure involves contact of the PPE specimen with a hard object (the PPE specimen usually sits on top of a pad or other soft layer which contains the biological fluid).

- One standardized procedure involves using a motor-driven press to gradually apply pressure onto a PPE material specimen lying on a pad of containing synthetic blood until synthetic blood can be observed to penetrate the fabric.

Standard Test Methods

☐ Standard test methods include:

- ASTM F 1819 (Mechanical pressure method for garments)
- ASTM F 1862 (Fluid resistance for face masks)

Test Results Provided

☐ Test methods include:

- An indication of the amount of fluid passing through the material in terms of weight gain for a blotter or visual estimate of fluid penetration stain or amount
- Pressure at which failure is noted

Test Results Interpretation

☐ Interpretation of test results:

- Any indication of penetration provides evidence that material does not provide barrier under conditions of test.
- Higher penetration pressures indicate materials with greater resistance to biological fluids under test conditions.

29.4 Biological Fluid Penetration Resistance

Property Measured

☐ Biological fluid penetration resistance testing evaluated the ability of fabrics to prevent penetration of biological fluids into the PPE. Biological fluid penetration resistance testing differs from biological fluid resistance testing in that it provides a 'proof' type determination.

Applicable PPE

☐ Applicable PPE includes:

- Full body garment materials and seams

- Partial body garment materials and seams
- Glove materials and seams
- Footwear materials and seams

Test Principles

☐ In fluid penetration resistance testing, similar principles are applied as in fluid resistance testing.

- Biological fluid or surrogate biological fluid is placed against the external surface of the PPE material or item under specific conditions with the interior side examined for evidence of penetration.
- Fluid penetration resistance tests differ from fluid resistance tests by using pass/fail criteria under specific condition without ranking of fabric performance.
- The only standardized form of fluid penetration resistance involves use of a test cell that provides a reservoir of synthetic blood against the exterior surface of the test material. The interior surface of the material is directly visible for examination during the test.
- A common surrogate for blood is synthetic blood, composed of water, surfactant/dye, and thickener which simulates the surface tension, viscosity, and color of actual blood and some body fluids.
- The conditions used in this testing involve 5 minutes of synthetic blood at ambient pressure, followed by one minute of pressurization of the synthetic blood reservoir at 2 psi (13.8 kPa), and 54 minutes at ambient pressure.
- Detection of penetration is visual.
- The test developed by ASTM (F 1670) showed the best correlation to simulated pressing against materials on synthetic blood soaked pads. Fabrics that showed penetration under simulated pressing against the synthetic blood-soaked pad also showed penetration per ASTM F 1670.

Standard Test Method

☐ The standard test method is:

- ASTM F 1670

Test Results Provided

☐ Test results include:

- Results of this testing are pass or fail for each replicate tested.

- The time of penetration may also be reported.
- The location and observation of penetration may also be reported.

Test Results Interpretation

☐ Interpretation of test results:

- Materials exhibiting passing results indicate barrier qualities for the prevention of blood and other body fluid penetration under normal use conditions.
- Materials demonstrating fluid penetration resistance generally provide a higher level of barrier performance compared to materials that only provide fluid resistance.
- PPE materials, seams, and other components should be tested under conditions representative of normal use and handling, including sterilization, laundering, and wear.

29.5 Viral Penetration Resistance

Property Measured

☐ Viral penetration resistance testing evaluates the ability of materials to prevent the passage of virus or related microorganisms.

Applicable PPE

☐ Applicable PPE includes:

- Full body garment materials and seams
- Partial body garment materials and seams
- Glove materials and seams
- Footwear materials and seams
- Eye and face protective devices
- Respirators

Test Principles

☐ A material sample is placed in a test cell where its exterior surface is contacted with a viral challenge suspension under specific conditions followed by a rinse of the sample interior surface and assay of the rinse solution for presence of virus.

- The Bacteriophage Phi-X174 is used in a suspension of nutrient broth for allowing the microorganism to grow at required titers (concentrations).
- The conditions used in this testing involve 5 minutes of bacteriophage challenge suspension at ambient pressure, followed by one minute of pressurization of the bacteriophage challenge suspension reservoir at 2 psi (13.8 kPa), and 54 minutes at ambient pressure.
- Only a portion of the rinse solution is assayed for bacteriophage.
- Assays involve microbiological techniques to measure the number of bacteriophage that are collected in the rinse solution.
  - Bacteria are used in a petri dish.
  - If bacteriophage is present, the bacteria are 'killed' leaving a clear spot in the petri dish.
  - Each clear spot represents one Bacteriophage and is called a plaque.
  - The rinse solution is successively dilutes such that a count of bacteriophage can be made.

Standard Test Method

☐ The standard test method is:

- ASTM F 1671

Test Results Provided

☐ Test results include:

- A pass or fail determination is made based on whether bacteriophage is found in the rinse solution.
- The number of plaque forming units (PFU) representative of penetrating bacteriophage is reported per milliliter of rinse solution.

Test Results Interpretation

☐ Interpretation of test results:

- Materials that do not show any viral penetration indicate potentially better barriers to virus under test conditions.
- Materials which show some viral penetration may be compared using interlaboratory test results but cannot be used to make any claims of viral penetration resistance performance.

29.6 Antimicrobial Performance

Property Measured

- ☐ Antimicrobial testing evaluates PPE and material:
  - Resistance to microbial growth (such as fungus and mildew)
  - Inactivation of microorganisms when contacted by the PPE or material

Applicable PPE

- ☐ Applicable PPE includes:
  - Full body garment materials
  - Partial body garment materials
  - Glove materials
  - Footwear materials

Test Principles

- ☐ Two general approaches are used for antimicrobial performance:
  - One approach involves placing a sample of PPE or material in a sterile environment under specified conditions and determining if microorganisms grow after a specific period.
  - A second approach entails inoculation of sample PPE or materials with specific microorganisms with either a qualitative or quantitative determination of microorganism growth on the specimen.
  - Techniques related to evaluating the efficacy of sterilization can also be used for examining PPE resistance to microorganisms.

Standard Test Methods

- ☐ Standard test methods include:
  - AATCC 100 (assessment of antibacterial finishes on textiles)
  - ASTM D 2020 (mildew resistance for paper and related products)
  - ASTM E 1053 (viricidal properties on surfaces)
  - ASTM G 21 (fungi resistance for plastics)
  - ASTM G 22 (bacterial resistance for plastics)
  - CAN/CGSB-4.2 No. 28 (Canadian methods for microorganism resistance)
  - FTMS 191A, 5750 (Federal method for single culture mildew resistance)
  - FTMS 191A, 5760 (Federal method for mixed culture mildew resistance)

- U.S. Pharmacopeia

Test Results Provided

☐ Test results include:

- An assessment of the samples ability to support microbiological growth under specific conditions
- The reduction of microorganisms as the result of PPE contact

Test Results Interpretation

☐ Interpretation of test results:

- PPE or materials that do not readily support microorganism growth may be preferred under many working conditions.
- Results that show that PPE prevents microorganism growth or intended to inactivates microorganisms are useful only for tested microorganisms under similar use conditions represented in the testing.

29.7 Insect Resistance

Property Measured

☐ Insect resistance testing evaluates the effectiveness of material treatments for repelling insects.

Applicable PPE

☐ Applicable PPE includes:

- Full body garment materials
- Partial body garment materials
- Glove materials
- Footwear materials

Test Principles

☐ Insect repellency is measured by exposing materials to specific insects with either:

- Measurement of insect activity on treated versus control specimens, or
- Determination of insect excrement weight, or
- Determination of material loss (due to insect destruction)

Standard Test Methods

- ☐ Standard test methods include:
  - CAN/CGSB-4.2 No. 38 (Canadian method for insect repellency)
  - FTMS 191A, 5764 (Federal method for insect repellency)
  - ISO 3998 (International method for insect repellency)

Test Results Provided

- ☐ Effectiveness of insect repellency is determined by:
  - Percentage of insect population killed
  - Weight gain per specimen (insect excrement weight measurement)
  - Weight loss of material specimen

Test Results Interpretation

- ☐ Interpretation of test results:
  - Superior insect repellency is indicated by high percentages of insect percentage killed, low insect excrement weight gain, and low weight loss of material specimens due to insect destruction.

Section 30

PPE Heat and Flame Resistance

The purpose of this section is to describe different types of PPE heat and flame resistance in terms of the measured property, applicable PPE, test principles, available standard test methods, results provided, and interpretation of test results.

30.1 Overview of Heat and Flame Resistance

☐ Heat and flame resistance test methods evaluate PPE, materials, and components for:

- The effects of either heat or flame exposure, or
- The relative protection offered when exposed to either heat or flame

☐ Performance properties covered in this section include:

- Convective heat resistance
- Conductive heat resistance
- Radiant heat resistance
- Thermal protective performance
- Flame resistance
- Resistance to molten metal contact

☐ Representative standard test methods for heat and flame resistance described in this section are summarized in Table 30-1.

Table 30-1. List of PPE Heat and Flame Resistance Test Methods

Heat or Flame Property	Type of Test/Application	Available Test Methods
Convective heat resistance	Heat effects/thermal shrinkage - textiles, other PPE	NFPA 1971, S6-6 EN 469, Annex A FTMS 191A,5870 ISO 1161, Annex A
	Heat aging - textiles	FTMS 191A,5870
	Heat aging - rubbers	ASTM D 573
	Heat aging - plastics	ASTM D 3045
	Blocking - textiles	FTMS 191A,5872
	Blocking - plastics	ASTM D 3345 ISO 5978
	Blocking - coated fabrics	ASTM D 751 ISO 5978
Conductive heat resistance	Clothing materials	ASTM D 1518 ASTM F 1060 EN 702 ISO 12127
Radiant heat resistance	Clothing materials	EN 366 ISO 6942 NFPA 1976, S5-3
Thermal protective performance	Clothing and glove materials	ASTM D 4108 CAN/CGSB-4.1 No, 70.1 ISO 9151 NFPA 1971, S6-10
	Whole clothing items	ISO 13506
	Respirators (SCBA)	NFPA 1981, S6-11
Flame resistance	Flame spread, 45 degree, surface exposure, butane	16 CFR Part 1610 NFPA 702

Heat or Flame Property	Type of Test/Application	Available Test Methods
Flame resistance *(continued)*		ASTM D 1230 CAN/CGSB-4.2, No. 27.5
	Flame spread, 90 degree, edge, butane	ISO 6941
	Ease of ignition, 90 degree, surface, butane	ISO 6940
	Burn time, 90 degree, edge, methane	ASTM D 3659 EN 532 FTMS 191A,5903.1 CAN/CGSB-4.2, No. 27.1 CAN/CGSB-4.2, No. 27.2 ISO 15025 NFPA 701 NFPA 1971, S6-2
	Burn time, 90 degree, edge, propane	FTMS 191A,5905.1
	Burn time, 90 degree, folded edge, methane	ASTM F 1358
	Burn time, 180 degree, edge, propane	ASTM D 635
	Limiting oxygen index - plastics	ASTM D 2863
	Smoke generation - plastics	ASTM D 4100 ASTM E 906
	Helmets	NFPA 1971, S6-3
	Gloves	NFPA 1971, S6-4
	Footwear	NFPA 1971, S6-5
	Respirators (SCBA)	NFPA 1981, S6-11
Molten metal resistance	Splash method	ASTM F 955 EN 373 ISO 9185
	Small splash (droplet) method	EN 348 ISO 9150

30.2 Convective Heat Resistance

Property Measured

☐ Convective heat resistance testing measures the effects of convective heat on different PPE items, materials, or components.

- Blocking resistance refers to prevention of material adhering to itself under high heat conditions.

Applicable PPE

☐ Appliable PPE includes:

- Full body garment materials
- Partial body garment materials
- Glove materials
- Footwear materials
- Headwear
- Eye and facewear
- Respirators

Test Principles

☐ A sample item, material, or component is placed in a forced air-circulating oven at a specified temperature for a specified period of time. The condition of the sample is then examined for effects of heat exposure.

- Tests may specify oven dimensions, techniques for exposing or suspending samples, and recovery time for oven.
- Small items of PPE such as gloves and boots may be filled with vermiculite or other items to simulate the heat sink effects of the body and to limit heat exposure to interior layers.
- Physical dimensions of specimens may be measured before and after exposure to determine shrinkage or deformation.
- Oven exposures may also be combined with other tests as a precondition to determine how the heat exposure affects a PPE item, material, or component performance property.
- Convective heat resistance testing may also be referred to as blocking resistance when materials are folded, placed in the oven with a weight on top of the specimen, and examined for evidence of self-adhesion following the heat exposure.

- Some tests may be performed on very small items such as thread by using a heated stage on which a sample of the item is placed and observed under a magnifying glass for the onset of melting.

Standard Test Methods

☐ Standard test methods include:

- Section 6-6 of NFPA 1971 (Structural fire fighting protective clothing)
- Annex A of EN 469 (European standard on fire fighting protective clothing)
- Annex A of ISO 11613 (International standard on fire fighting protective clothing)
- ASTM D 573 (heat aging of rubbers)
- ASTM D 751 (blocking of coated fabrics)
- ASTM D 3045 (heat aging of plastics)
- ASTM D 3354 (blocking of plastics)
- FTMS 191A, 1534 (melting temperature of synthetic fibers)
- FTMS 191A, 5850 (heat aging of textiles)
- FTMS 191A, 5870 (heat effects on textiles)
- FTMS 191A, 5872 (blocking of textiles)
- ISO 5978 (International method for blocking)
- ISO 11645 (International method for glove leather heat stability)

Test Results Provided

☐ Test results include:

- Observations of physical condition followed by exposure, including:
  - -- Ignition
  - -- Melting
  - -- Dripping
  - -- Separation
  - -- Charring
- Actual dimensional distortion or percentage shrinkage
- Indication of functional performance following heat exposure (compared to unexposed samples).

- Percentage change in performance property compared with unexposed samples
- Qualitative rating of blocking resistance or pass/fail determination of blocking

Test Results Interpretation

☐ Better heat resistance is demonstrated by:

- Materials which show fewer signs of heat degradation (under test conditions)
- None or small distortion/shrinkage of specimens
- PPE Materials which retain original performance or functionality following heat exposure
- The absence of blocking

30.3 Conductive Heat Resistance

Property Measured

☐ Conductive heat resistance testing measures the effects of conductive heat on PPE items or material and the amount of insulation provided by these items or materials when in contact with hot surfaces.

Applicable PPE

☐ Applicable PPE includes:

- Full body garment materials
- Partial body garment materials
- Glove materials
- Footwear materials

Test Principles

☐ The specimen of PPE or material is placed on a hot plate at a controlled temperature with measurement of heat transfer through the material.

- The hot plate temperature is typically set at a specific temperature.
- The pressure on the specimen must be specified since large pressures can compress material layers and increase rate of heat transfer.

- A calorimeter is used to measure heat transfer or temperature rise on the interior material side. The calorimeter is designed to simulate skin response to heat.
- Tests may be conducted with moisture in the system (the effect of moisture will depend on the amount and location).
- Tests may also be conducted as a precondition to other tests (e.g., breaking strength).

Standard Test Methods

☐ Standard test methods include:

- ASTM D 1518 (steady state heat transfer at low temperatures)
- ASTM F 1060 (human threshold times at high temperatures)
- EN 702 (European method)
- ISO 12127 (International method)

Test Results Provided

☐ Test methods:

- Heat transfer may be measured as:
  - -- The time for a specific temperature rise to occur (Heat Transfer Index or HTI); for contact heat, a time to 10°C is usually reported
  - -- The total or cumulative amount of heat transferred
  - -- The time required to cause pain or second degree burn as determined by comparison of heat transferred to human tolerance levels
- Observations of material condition following exposure:
  - -- Ignition
  - -- Charring
  - -- Embrittlement
  - -- Break open
  - -- Melting
  - -- Dripping
  - -- Shrinkage
- Percentage change in specific performance property as compared to unexposed samples

Test Results Interpretation

☐ Interpretation of test results:

- In low heat conditions (hot ambient temperatures), high rates of heat transfer are preferred as an indication of body heat losses.
- Under high contact conditions, resistance to conductive heat transfer is indicated by:
  - -- High HTIs
  - -- Low cumulative heat transfer
  - -- Long time-to-pain or time-to-burn times
- The difference between time-to-pain and time-to-burn is known as the alarm time and indicates the relative amount of warning between detection of high heat (pain) and actual injury (burns). Longer alarm time is another consideration in evaluating material systems for conductive heat transfer.
- Absence of heat degradation from contact also demonstrates integrity of material to contact heat exposure (under test conditions).
- Retention of other performance properties also indicates conductive heat transfer resistance.

30.4 Radiant Heat Resistance

Property Measured

☐ Radiant heat resistance measures the effects of radiant heat on PPE items or material and the amount of insulation provided by these items or materials when exposed to different levels of radiant heat.

Applicable PPE

☐ Applicable PPE includes:

- Full body garment materials
- Partial body garment materials
- Glove materials
- Footwear materials
- Eye and face protective devices
- Respirators

Test Principles

☐ A gas-fired or electrically controlled radiant panel is used to create radiant energy to expose PPE or PPE materials with measurement of heat effects or transfer through the sample.

- The samples are oriented vertically to minimize convective heat transfer from circulating air.
- Tension may be applied on specimens in some test methods.
- A shutter between the radiant heat source and the specimen is used to control exposure.
- The exposure heat flux can be varied, however, the majority of testing is conducted at a heat fluxes ranging between 0.5 and 2.0 cal/cm^{2}s (21 to 84 kW/m^2).
- A calorimeter is used to measure heat transfer or temperature rise on the interior material side. The calorimeter is designed to simulate skin response to heat.
- Tests may be conducted with moisture in the system (the effect of moisture will depend on the amount and location)
- Tests may also be conducted as a precondition to other tests (e.g., breaking strength).

Standard Test Methods

☐ Standard test methods include:

- Section 5-3 of NFPA 1976 (Proximity fire fighting protective clothing)
- EN 366 (European method)
- ISO 6942 (International method)

Test Results Provided

☐ Test methods include:

- Heat transfer may be measured as:
  - -- The time for a specific temperature rise to occur (Heat Transfer Index); times to a 12 or 24 degree Celsius rise are typically reported, (HTI_{12} and HTI_{24}, respectively)
  - -- The total or cumulative amount of heat transferred (or percent of incident heat transferred)
  - -- The time required to cause pain or second degree burn as determined by comparison of heat transferred to human tolerance levels

- -- The Radiant Protective Performance (RPP) rating which is the product of the time-to-pain and the incident heat flux in calories per square centimeter per second

- Observations of material condition following exposure:
  - -- Ignition
  - -- Charring
  - -- Embrittlement
  - -- Break open
  - -- Melting
  - -- Dripping
  - -- Shrinkage

- Percentage change in specific performance property as compared to unexposed samples

Test Results Interpretation

☐ Interpretation of test results:

- Resistance to radiative heat transfer is indicated by:
  - -- High HTIs or RPP ratings
  - -- Low cumulative heat transfer (or low percent heat transmission)
  - -- Long time-to-pain or time-to-burn times

- The difference between time-to-pain and time-to-burn is known as the alarm time and indicates the relative amount of warning between detection of high heat (pain) and actual injury (burns). Longer alarm time is another consideration in evaluating material systems for conductive heat transfer.

- Absence of heat degradation also demonstrates integrity of material to radiative heat exposure (under test conditions).

- Retention of other performance properties also indicates radiative heat transfer resistance.

30.5 Thermal Protective Performance

Property Measured

- ☐ Thermal protective performance testing measures the insulative and barrier properties of PPE materials when exposed to heat (usually convective or convective and radiant heat).

Applicable PPE

- ☐ Applicable PPE includes:
    - Full body garment materials
    - Partial body garment materials
    - Glove materials
    - Footwear materials

Test Principles

- ☐ A PPE material specimen is placed in a horizontal specimen holder and is exposed to heat source for a specified period of time. A calorimeter is used to measure heat transfer through the material and is related to a threshold value for either temperature rise or burn injury.
    - Some methods using a single burner as the heat exposure source.
    - Other methods use burners combined with a radiant heat panel (quartz tubes) for a combined convective and radiant heat exposure.
    - The exposure heat flux can be varied, however, the majority of testing is conducted at a heat flux of 2.0 cal/cm^{2}s (84 kW/m^2).
    - A shutter between the heat source and the specimen is used to control exposure.
    - A calorimeter is used to measure heat transfer or temperature rise on the interior material side. The calorimeter is designed to simulate skin response to heat.
    - Test methods are being developed which allow exposure of overall PPE items; these tests involve:
        - -- A manikin with multiple sensors which provide measurements over the manikin's body.
        - -- Manikin exposure to a series of propane fueled jets for enveloping flame over the manikin body.

Standard Test Methods

- ☐ Standard test methods include:
  - ASTM D 4108 (one burner method discontinued in 1995)
  - Section 6-10 in NFPA 1971 (combined radiant panel and two-burner method for structural fire fighting protective clothing)
  - CAN/CGSB-4.2 No. 70.1 (Canadian method for thermal insulation)
  - ISO 9151 (International method involving single burner)
  - ISO 13506 (International method for whole garments on manikin)
  - Section 6-11 of NFPA 1981 (Overall heat and flame test for self-contained breathing apparatus).

Test Results Provided

- ☐ Test results include:
  - Heat transfer may be measured as:
    - The time for a specific temperature rise to occur (Heat Transfer Index); times to a 12 or 24 degree Celsius rise are typically reported, (HTI_{12} and HTI_{24}, respectively)
    - The total or cumulative amount of heat transferred
    - The time required to cause pain or second degree burn as determined by comparison of heat transferred to human tolerance levels
    - The Thermal Protective Performance (TPP) rating which is the product of the time-to-pain and the incident heat flux in calories per square centimeter per second
  - Observations of material condition following exposure:
    - Ignition
    - Charring
    - Embrittlement
    - Break open
    - Melting
    - Dripping
    - Shrinkage
  - Percentage change in specific performance property as compared to unexposed samples

- Percentages of manikin body sustaining 2nd or 3rd degree burns (for whole garment tests)

Test Results Interpretation

☐ Interpretation of test results:

- Resistance to heat transfer is indicated by:
  - High HTIs or TPP ratings
  - Low cumulative heat transfer
  - Long time-to-pain or time-to-burn times
- The difference between time-to-pain and time-to-burn is known as the alarm time and indicates the relative amount of warning between detection of high heat (pain) and actual injury (burns). Longer alarm time is another consideration in evaluating material systems for conductive heat transfer.
- Absence of heat degradation also demonstrates integrity of material to heat and flame exposure (under test conditions).
- Retention of other performance properties also indicates heat transfer resistance.
- Lower body burn percentages indicate better overall thermal protective performance of full garments.

30.6 Flame Resistance

Property Measured

☐ Flame resistance testing measures the effects of flame contact on an item of PPE or PPE material. Different effects may be measured depending on the nature of the PPE or material and its intended application.

Applicable PPE

☐ Applicable PPE includes:

- Full body garments and materials
- Partial body garments and materials
- Glove materials
- Footwear materials
- Headwear
- Eye and facewear

- Respirators

Test Principles

☐ A flame of specified fuel, heat, and height contacts a material specimen with subsequent observation or measurement of flame impingement effects:

- Methane, propane, butane, or synthetic gases are typical flame fuels.
- Specimen orientation may be vertical, horizontal, or at a specified angle.
- Specimens may be loosely held relative to flame or restrained in a specimen holder.
- The length of flame contact may be for a specified period of time or until the specimen ignites.
- The surface, edge, or a folded edge of the specimen may be contacted with the flame.
- Most tests evaluate different burning characteristics of materials related to ease of ignition, self-extinguishing characteristics, or rate of burning.
- Some tests involve determination of ignition conditions such as the amount of oxygen required to sustain combustion (oxygen limiting index).
- Some tests involve measurement of the amount of smoke generated from flaming materials.

Standard Test Methods

☐ Materials are characterized in terms of (fuel, specimen orientation, sample restraint, flame application, and primary determination):

- 16 CFR Part 1610 (butane, 45°, restrained, surface, flame spread rate)
- ASTM D 635 (propane, 180°, semi-restrained, edge, burn time for plastics)
- ASTM D 1230 (butane, 45°, restrained, surface, flame spread rate)
- ASTM D 2863 (limiting oxygen index for plastics)
- ASTM D 3659 (methane, 90°, semi-restrained, edge, flame spread rate)
- ASTM D 4100 (smoke particulates from plastics)
- ASTM E 906 (heat and visible smoke release rates)
- ASTM F 1358 (methane, 90°, restrained, folded edge, burn time)
- CAN/CGSB-4.2 No. 27.1 (Canadian, methane, 90°, restrainted, burn time)
- CAN/CGSB-4.2 No. 27.2 (Canadian, methane, 90°, surface, burn time)

- CAN/CGSB-4.2 No. 27.5 (Canadian, butane, 45°, surface, flame spread rate)
- EN 532 (European, methane, 90°, restrained, surface, burn time)
- FTMS 191A, 5903.1 (methane, 90°, restrained, edge, burn time)
- FTMS 191A, 5905.1 (propane, 90°, restrained, edge, burn time)
- FTMS 191A, 5908 (butane, 45°, restrained, surface, flame spread rate)
- ISO 6940 (International, butane, 90°, restrained, surface, ease of ignition)
- ISO 6941 (International, butane, 90°, restrained, surface, flame spread)
- ISO 15025 (International, methane, 90°, restrained, edge, burn time)
- NFPA 701 (methane, 90°, restrained, edge, burn time)
- NFPA 702 (butane, 45 °, restrained, edge, flame spread rate)
- Section 6-3 of NFPA 1971, *Standard on Protective Ensemble for Structural Fire Fighting* (tests on helmets with methane flame)
- Section 6-4 of NFPA 1971, *Standard on Protective Ensemble for Structural Fire Fighting* (tests on gloves with propane flame)
- Section 6-5 of NFPA 1971, *Standard on Protective Ensemble for Structural Fire Fighting* (tests on footwear with methane flame)
- Section 6-11 of NFPA 1981, *Standard on Open-Circuit Self-Contained Breathing Apparatus* (Overall heat and flame test).

Test Results Provided

☐ Test results include:

- Occurrence of ignition
- Rate of flame spread (the time for flame to burn the distance of the specimen)
- After flame time (time material continues to show flame after removal of flame)
- After glow time (time material continues to show flame after material after flame stops)
- Burn time (similar to after flame)
- Char length (length of specimen destruction)
- Burn distance (similar to char length)
- Percentage consumption of specimen mass

- Observations of melting, dripping, shrinkage or other burning behavior.

Test Results Interpretation

☐ Interpretation of test results:

- Flame resistance results are generally intended to answer the following questions:
  - -- How easily will ignition?
  - -- Once ignited, how long will burning continue?
  - -- Will the PPE or material self-extinguished when removed from flame contact?
  - -- Will the burning of the material create additional hazards to the wearer?
- Materials that demonstrate high levels of ignition resistance require long periods of flame contact for ignition to occur or do not ignite in the specified flame contact period.
- Rate of flame spread indicates how quickly materials will burn once ignited.
  - -- Fabrics for apparel in the U.S must demonstrate flame spread times of 4 second or greater when tested in accordance with 16 CFR 1610.
  - -- 16 CFR 1610 and other related standards specify other classes of material performance.
- After flame time and after glow time are used to assess material self-extinguishing characteristics.
  - -- Longer after flame times (or burn times) indicate material do not extinguish quickly when removed from contact with the flame.
  - -- Longer after glow times indicate the possibility that a material could be a continued ignition source or be reignited given certain conditions.
  - -- Some standards do not permit the use of the term "self-extinguishing" because of its dependency on test conditions and concerns over misuse for this product characterization.
- Char length, burn distance, and percent consumption of specimen mass are indications of how much material is destroyed by flame as the result of flame impingement (under test conditions):
  - -- Longer char length or burn distance indicates less flame resistant materials or PPE.

-- Higher percent consumption also indicates less flame resistant materials or PPE.

- Observations of melting or dripping indicate additional hazards to the wearer since melted or dripping residue from flaming may worsen burn injury and cause additional heat transfer.

- Observations of shrinkage indicate a material or PPE that retreats away from the flame.

  -- Shrinkage is a phenomenon caused by melting and can cause injury when shrunken items contact the wearer's skin.

- Flame resistant materials may use flame-retardant treatments or be constructed of inherently flame resistant fibers.

  -- Flame retardants may be removed over time when PPE is used, cleaned, or maintained.

  -- Given sufficient heat, inherently flame resistant fibers will melt or be destroyed.

30.7 Resistance to Molten Metal Contact

Property Measured

☐ Molten metal contact resistance testing evaluates the effects of and heat transfer through for PPE contacted by molten metal.

Applicable PPE

☐ Applicable PPE includes:

- Full body garment materials
- Partial body garment materials
- Glove materials
- Footwear materials

Test Principles

☐ Specified volumes of molten metal are allowed to contact a material in a specific orientation over a sensor board.

- The sensor board uses:

  -- Thermocouples to measure temperature rise, or

  -- Calorimeters to measure heat transfer, or

-- A PVC skin simulant

- The volume of molten metal may vary between repeated droplets (small splash method) and relatively larger volumes (splash method).
- The splash method may use different types of molten metals (e.g., iron, aluminum, and brass).
- In the small splash method, the heating of a welding rod produces droplets of molten metal.
- Specimen orientation is at an angle for splash methods and vertical for the small splash method.

Standard Test Methods

☐ Standard test methods include:

- ASTM F 955 (splash method)
- EN 348 (European small splash method)
- EN 373 (European splash method)
- ISO 9150 (International small splash method)
- ISO 9185 (International splash method)

Test Results Provided

☐ Test results include:

- The number of droplets required to produce a 40°C temperature rise on the thermocouples (small splash method)
- Time-to-second degree burn as determined by comparing the heat transfer to human skin threshold data (ASTM method)
- Level of damage to skin simulant (EN and ISO methods)
- Observations of material condition following exposure:

-- Ignition

-- Charring

-- Embrittlement

-- Break open

-- Melting

-- Dripping

-- Shrinkage

-- Adhesion of molten metal

Test Results Interpretation

☐ Interpretation of test results:

- In small splash tests, larger numbers of drops to the specified temperature rise indicate materials more resistant to heat transfer from small splashes of molten metals.
- In splash tests, longer times to second degree burn or less damage to skin simulants represents greater resistance to heat transfer from splashes of molten metal.
- Less observed damage or effects on the material also denote greater material resistance to molten metal contact.

Section 31

PPE Electrical Properties

The purpose of this section is to describe different types of PPE electrical properties in terms of the measured property, applicable PPE, test principles, available standard test methods, results provided, and interpretation of test results.

31.1 Overview of Electrical Properties

☐ Performance properties covered in this section include:

- Electric insulative performance
- Conductivity
- Static charge accumulation resistance
- Electric arc protective performance

☐ Representative standard test methods for electrical resistance described in this subsection are summarized in Table 31-1.

31.2 Electrical Insulative Performance

Property Measured

☐ Electrical insulative performance evaluates PPE for their ability to reduce contact with electrically energized parts.

Applicable PPE

☐ Applicable PPE includes:

- Partial body garments
- Gloves
- Footwear

Table 31-1. List of Electrical Property Test Methods

Electrical Property	Type of Test/Application	Available Test Methods
Electrical insulation	Gloves	ASTM D 120
	Sleeves	ASTM F 1051
	Footwear	ASTM F 1116 ANSI Z41.1, S4.4.1
	Headwear	ANSI Z89.1, S9.7
Conductivity and electrical resistivity	Textiles	AATCC 76 FTMS 191A,5903
	Footwear	ANSI Z41.1, S3.5
Static charge accumulation resistance	Clothing materials	FTMS 101C,4046 NFPA 1991, S9-29 INDA IST 40.2
Electrical arc protective performance	Clothing materials	ASTM PS 58
	Overall clothing	ASTM PS 57

Test Principles

- ☐ Different performance tests are applied to different items of PPE for measuring electrical insulative performance:
  - Gloves and sleeves are separately subjected to increasing A-C and D-C voltage to determine the breakdown voltage (in which electrical arcing occurs) and the amount of current leakage.
    - Tests are done under dry and wet conditions.
    - Different voltage levels are established for proof (no occurrence of arcing) and breakdown testing.
  - The principal technique for measuring electrical insulative performance of footwear and headwear involves determining current leakage through the PPE item.

- Footwear tests involve
  - -- Placement of saline-soaked blotter paper inside the footwear
  - -- Placement of the footwear on a metal base
  - -- Insertion of a weighted electrode in the footwear
  - -- Application of 14,000 volts
  - -- Measurement of current leakage through the footwear
  - -- Variations of the test include using footwear submerged in water and filling the footwear with metal shot so that the inner sole is covered
- Headwear tests involve:
  - -- Inversion of the helmet
  - -- Filling the helmet to a specified level with tap water
  - -- Placement of the helmet in a water bath
  - -- Application of voltage (either 2,200 or 20,000 volts depending on the class of the helmet)

Standard Test Methods

☐ Standard test methods include:

- Section 18 of ASTM D 120 (AC and DC proof/breakdown tests for insulating gloves)
- Section 18 of ASTM F 1051 (AC and DC proof/breakdown tests for insulating sleeves)
- ASTM F 1116 (dielectric strength of footwear)
- Section 4.4.1 of ANSI Z41.1, American National Standard for Personnel Protection—Protective Footwear
- Section 9.7 of ANSI Z89.1-1997, American National Standard for Personnel Protection—Protective headwear for industrial workers—Requirements

Test Results Provided

☐ Test results include:

- Voltage at which breakdown occurs
- Current leakage in amperes

Test Results Interpretation

- ☐ Interpretation of test results:
  - Higher breakdown voltages indicate greater electrical insulative performance.
  - Lower current leakages indicate greater electrical insulative performance.

31.3 Conductivity and Electrical Resistivity

Property Measured

- ☐ Conductivity testing evaluates PPE or material for protection against hazards of static charge build-up (*see 31.5*) and for the equalization of electrical potential of personnel and energized high voltage sources.

Applicable PPE

- ☐ Applicable PPE includes:
  - Full body garment materials
  - Partial body garment materials
  - Gloves
  - Footwear

Test Principles

- ☐ Techniques are different with the type of item evaluated:
  - For garment and glove materials, electrodes are placed on the material surface and the resistance for the distance separating the electrodes is measured.
  - For footwear, a weighted electrode is placed inside the heel section of the footwear and an electrical resistance measurement is made.

Standard Test Methods

- ☐ Standard test methods include:
  - AATCC 76 (for textile materials)
  - FTMS 191A, 5930 (for textile materials)
  - Section 3.5 of ANSI Z41.1, American National Standard for Personnel Protection—Protective Footwear

Test Results Provided

☐ Test results include:

- Resistance measurement in ohms

Test Results Interpretation

☐ Interpretation of test results:

- The lower the resistance measurement, the more conductive the material or footwear.
- NFPA 99 (Health Care Facilities) sets a requirement for materials used in operating room environments that the surface resistivity be less than 1×10^{11} ohms.

31.4 Static Charge Accumulation Resistance

Property Measured

☐ Static charge accumulation resistance testing evaluate the static charge generated and the rate of its discharge.

Applicable PPE

☐ Applicable PPE includes:

- Full body garment materials
- Partial body garment materials
- Glove materials
- Footwear materials

Test Principles

☐ Two techniques are used for measuring static charge accumulation resistance on materials:

- In the first method, a specified voltage is applied to a material with subsequent measurement of the voltage at specific time intervals following the applied voltage (electrostatic decay method).
- In the second approach, static charge on the material is generated by rubbing a Teflon-felt wheel against the material for several seconds and measuring the resulting charge and its decay (triboelectric method).
- The triboelectric method is considered to be more representative of actual static charge generation and can be modified to include different rubbing surfaces and contact times.

- Some materials may not be charged by either method because of conductive or other properties.
- Test results are heavily dependent on the test chamber temperature and relative humidity. Lower humidity results in greater static charge accumulation.

Standard Test Methods

☐ Standard test methods include:

- FTMS 101C, 4046 (electrostatic decay method)
- Section 9-29 of NFPA 1991, *Standard on Vapor-Protective Suits for Hazardous Chemical Emergencies* (triboelectric test method)
- INDA IST 40.2 (electrostatic discharge method)

Test Results Provided

☐ Test results include:

- Peak voltage and voltage at prescribed interval following material charging (usually 0 to 5 seconds)
- Decay time to a specific percentage of the original applied voltage.

Test Results Interpretation

☐ Interpretation of test results:

- For triboelectric testing:
  - -- Higher peak voltages indicate ready accumulation of static charge.
  - -- Little change in voltage following material charging indicates materials that do not readily dissipate static charge.
  - -- NASA uses criteria of 350 volts as the potential required to cause a spark and requires that materials not exceed this voltage 5 seconds after charging.
- For electrostatic decay testing:
  - -- Long decay charges indicate materials that do not readily dissipate static charge.
  - -- NFPA 99 (Health Care Facilities) sets a requirement for materials used in operating room environments that the voltage decay to 500 volts within 0.5 second of the charge application.

31.5 Electrical Arc Protective Performance

Property Measured

☐ Electrical arc protective performance testing measures the heat transfer from exposure to high-energy electrical arcs.

Applicable PPE

☐ Applicable PPE includes:

- Full body garment materials
- Partial body garment materials
- Glove materials
- Footwear materials

Test Principles

☐ New tests define whole product performance for ignitability and material thermal insulation from arc exposure.

- Whole product tests involve placement of clothing items on a manikin.
  - -- The manikin with test clothing is placed between the electrodes and in a partial faraday cage to balance magnetic and electric field surrounding the arc.
  - -- Discharge of the electrical arc with subsequent measurement of of afterflame on clothing.
- Material tests use specimen panels, which are instrumented with calorimeters for measuring heat transfer.
  - -- Discharge of the electrical arc with subsequent measurement of heat transferred through the clothing.
  - -- Calorimeter output is related to human skin threshold data for second-degree burn times.

Standard Test Methods

☐ Standard test methods include:

- ASTM PS 57 (ignitability of clothing by electric arc)
- ASTM PS 58 (arc thermal performance of materials)

Test Results Provided

☐ Test results include:

- Arc power over time

- Incident energy over time
- Afterflame times for whole garment exposures
- Arc thermal performance exposure value (AVTPV)—the incident energy that caused the onset of a second degree burn
- Observations of specimen condition following exposure

Test Results Interpretation

☐ Interpretation of test results:

- Shorter afterflame times for whole garments indicate better performance for resisting ignition from electrical arc exposure.
- Higher ATPV values indicate greater protective qualities of clothing from electrical arc exposure (under test conditions).
- Electrical arc energies may vary significantly with the operation and system.

Section 32

PPE Human Factors

The purpose of this section is to describe different types of PPE human factors in terms of the measured property, applicable PPE, test principles, available standard test methods, results provided, and interpretation of test results.

32.1 Overview of Human Factors

☐ Human factors describe how PPE affects the wearer in terms of functionality, fit, comfort, and overall well being.

☐ Most human factor properties represent tradeoffs with protection, for example, barriers to chemicals versus thermal comfort.

☐ Performance properties covered in this section include:

- Material biocompatibility
- Thermal insulation and breathability
- Mobility and range of motion
- Hand function
- Clarity and field of Vision
- Ease of communications
- Sizing and fit (adjustment)
- Donning and doffing ease
- Visibility

☐ Representative standard test methods for human factors described in this subsection are summarized in Table 32-1.

Table 32-1. List of PPE Human Factors Test Methods

Human Factor	Type of Test/Application	Available Test Methods
Material biocompatibility	Medical device (gloves, gowns) biocompatibility tests	ISO 10993
	Natural rubber latex proteins	ASTM D 5712
	pH level in gloves	ISO 4045
	Chromium content in leather	EN 420, S6.1
Thermal insulation and breathability	Air permeability - materials	ASTM D 737 CAN/CGSB-4.2 No. 36
	Moisture vapor transmission - materials	ASTM E 96 CAN/CGSB-4.2 No. 49
	Thermal resistance - materials	ASTM D 1518 ISO 5085
	Thermal resistance - garments	ASTM F 1291
	Total heat loss - materials	ISO 11092 NFPA 1977, S6-5
	Overall product evaluation	Field studies
Mobility and range of motion	Overall product evaluation	ASTM F 1154 NFPA 1991, S9-4 EN 943-1, S7.2 Field studies
Hand function	Gross dexterity	NFPA 1971, S6-38
	Fine dexterity	EN 420, S6.3 NFPA 1999, S6-14
	Grip	NFPA 1971, S6-39
Ankle support	Footwear	Field studies
Back support	Garments or accessories	Field studies
Clarity and field of vision	Clarity/field of vision	NFPA 1991, S9-13 EN 943-1, S7.2

Human Factor	Type of Test/Application	Available Test Methods
Clarity and field of vision *(continued)*	General optical properties	ISO 4854
	Prismatic power	ANSI Z87.1, S15.4
	Refractive power, resolving power, astigmatism	ANSI Z87.1, S15.4
	Haze and transmittance	ANSI Z87.1, S15.4 ASTM D 1003
Ease of communications	Hearing protectors	ANSI S3.2
	Respirators	NFPA 1981, S6-10
Sizing and fit (adjustment)	Body measurements	ASTM D 5585 ASTM D 5586 CAN/CGSB-49
	Garment sizing/dynamic fit test	ASTM F 1731 ANSI/ISEA 101
	Glove fit test	NFPA 1977, S6-32
	Size designation	ISO 3635
Donning and doffing ease	Full body suits	NFPA 1991, S9-1 DOE Los Alamos National Laboratory test

32.2 Material Biocompatibility

Property Measured

- ☐ Material biocompatibility evaluates the potential for skin irritation or advserse reactions due to contact with certain substances that may be present in or on PPE.

Applicable PPE

- ☐ Applicable PPE includes:
  - Full body garment materials
  - Partial body garment materials

- Glove materials
- Footwear materials
- Headwear
- Eye and facewear
- Hearing protectors
- Respirators

Test Principles

☐ A battery of tests is used to assess the potential incompatibility of PPE materials with skin or other forms of human contact. Other related tests exist for evaluating the presence of certain substances suspected of or considered to be hazards when part of PPE.

- Tripartite guidelines have been established by Canada, the United Kingdom, and the United States for evaluating the potential toxicity of medical devices.
    - These guidelines define nine different categories of medical devices based on their expected contact with the patient's body.
    - A total of 14 biological tests are specified based on the medical device category.
- Medical device categories include:
    - Non-contact devices which do not contact the body directly
    - External devices for intact surfaces that contact external body surfaces only
    - External devices for breached or compromised surfaces
    - Externally communicating devices for intact natural channels
    - Externally communicating devices with indirect blood path contact
    - Externally communicating devices with direct blood path contact
    - Internal devices contacting the bone
    - Internal devices contacting the tissue and tissue fluid
    - Internal devices contacting the blood
- Most PPE in the medical industry such as examination gloves and surgical gowns are categorized as external devices for breached or compromised surfaces.
- Biological tests include:

-- Sensitization assay (estimates the potential for sentization of a material or its extracts with an animal)

-- Irritation tests (estimates the irritation and sensitization of a material or its extracts using implant tissue on an animal)

-- Cytotoxicity (uses cell culture techniques to determine cell death, inhibition of cell growth, or other toxic effects on cells by a material or its extracts)

-- Acute systemic toxicity (estimates the harmful effects of single and multiple exposures to a material or its extracts using animals during a period less than 24 hours)

-- Hemocompatibility (evaluates any effects of blood contacting material on blood properties using animals)

-- Pyrogenicity (evaluates effect of the material or its extracts in causing fever in animals)

-- Hemolysis (determines the degree of red blood cell destruction and the separation of hemoglobin cause by the material or its extracts in an artificial organism model)

-- Implantation tests (evaluates the local toxic effects on living tissue, at both gross and microscopic levels, surgically implanted into animal tissue, muscle, or bone)

-- Mutagenicity (determines the likelihood of gene mutations or other gene toxicity caused by the material or its extracts)

-- Sub-chronic toxicity (determines harmful effects from multiple exposures to the material or its extracts during a period representing between one day and 10% of the total life of the test animal)

-- Chronic toxicity (determines harmful effects from multiple exposures to the material or its extracts during a period representing 10% of the total life of the test animal)

-- Carcinogenesis bioassay (determines the tumor causing potential of the test material or its extracts from multiple exposures over the total life of the test animal)

-- Pharmacokinetics (determines the metabolic processes involved in the absorption, distribution, biotransformation, and elimination of toxic leachable and degradation products of the test material or its extracts)

-- Reproductive and developmental toxicity (evaluates the potential effects of the material or its extracts on fertility, reproductive function, and prenatal/early postnatal development)

- Most PPE subject to biocompatibility tests is evaluated in sensitization assay, irritation tests, and cytotoxicity.
- PPE that is expected to come in contact with breached skin is also evaluated for acute systemic toxicity, hemocompatibility, and hemolysis.
- Levels of latex proteins in natural rubber-containing PPE, suspected as a cause for latex allergies, are measured using a colorimetric technique with spectrophotometric determinations.
- Other substances may be evaluated using specific analytical procedures; for example, chromium content and pH levels in leather gloves can be measured.

Standard Test Methods

☐ Standard test methods include:

- ASTM D 5712 (colorimetric test for latex protiens)
- ISO 4045 (test for pH level in leather)
- ISO 10993-1 (selection of tests)
- ISO 10993-2 (animal welfare requirements)
- ISO 10993-3 (tests for genotoxicity, carcinogenicity, and reproductive toxicity)
- ISO 10993-4 (tests for interactions with blood)
- ISO 10993-5 (tests for cytotoxicity)
- Section 6.1 of EN 420 (chromium content in leather gloves)

Test Results Provided

☐ Test results include:

- Each of the biocompatibility tests provides a pass or fail determination with respect to the specific measurements of the test.

Test Results Interpretation

☐ Interpretation of test result:

- Materials used in PPE classified as medical devices must meet the requirements of recommended biocompatibility tests.

32.3 Thermal Insulation and Breathability

Property Measured

☐ Thermal insulation and breathability tests evaluate the ability of protective clothing materials to allow the passage of air, moisture, and the heat associated with body evaporative cooling and environmental conditions.

Applicable PPE

☐ Applicable PPE includes:

- Full body garment materials
- Partial body garment materials
- Glove materials
- Footwear materials

Test Principles

☐ Different test method approaches are used for evaluating breathability, including:

- Measurement of amount of air which can pass through a material (air permeability)
- Quantification of the amount of water vapor which passes through a material (water vapor transmission rate)
- Measurement of heat transfer through the material associated with dry and wet heat loss (total heat loss)
- Water vapor transmission rate is typically measured by sealing a material specimen onto a cup:
    - -- Dessicant is placed in the cup before sealing the material, the cup is placed in a controlled humid atmosphere, and the amount of weight gain of the cup, material, and dessicant system is related to the amount of water vapor passing through the material specimen.
    - -- Alternatively, water is placed in the cup for sealing the material, the cup is place in a controlled atmopshere, and the amount of weight loss of the cup, material, and dessicant system is related to the amount of water vapor passing through the material specimen.
    - -- One technique involves sandwiching the materials between two layers of specific microporous membrane and comparing water loss through the system to a control system without a specimen.

- Total heat loss is measured using a apparatus called a sweating hot plate which measures the energy required to keep the hot plate at a specific temperature when the plate and overlaying material specimen are placed in a controlled environment:
  - One part of the test is conducted under dry conditions to provide the dry thermal resistance of the material (for some test methods, this is the only determination made).
  - A second part of the test is conducted under wet conditions, with the hot plate covered with a thin layer of water to represent sweating, to determine the heat loss associated with moisture passage through the material (evaporative heat transfer).
  - Testing may be conducted isothermally (with the plate and the test environment at the same temperature) or non-isothermally (with the plate and the environment at different temperatures).
- Thermal insulation is similarly measured using a hot plate where the surface temperature is maintained and the energy used to maintain the plate temperature indicates the amount of heat loss or thermal insulaiton of the test material (thermal insulation).
- Thermal insulation may also be measured for whole garments using a manniquin designed to maintain a constant surface temperature.
  - Sample clothing is placed on the manikin and the manikin is placed in a controlled environmental chamber.
  - The energy required to keep the manikin surface temperature constant is measured for the selected clothing and environmental conditions.
- PPE effects on individual wearing comfort can also be assessed using human subject-based tests where:
  - Human subjects wear PPE under controlled circumstances (such as in an environmental chamber set at specific conditions).
  - Human subjects are connected to instrumentation for measuring specific physiological performance.
  - Human subjects walk on a treadmill to achieve certain work rates or do a series of specified tasks.
  - Physiological responses are used to compare different types of PPE.
  - Test protocols need to be reviewed by a human subjects committee; testing should be performed in the presence of a physician or other qualified medical personnel.

WARNING - Thermal comfort testing involving human subjects must be designed and supervised to allow termination of testing when physiological limits established for the individual are reached, on the judgement of the attending physician or other trained supervisor, or at the request of the test subject.

Standard Test Methods

☐ Standard test methods include:

- ASTM D 737 (air permeability test)
- ASTM D 1518 (thermal resistance test)
- ASTM E 96 (moisture vapor transmission rate test)
- ASTM F 1291 (manikin thermal resistance test)
- CAN/CGSB-4.2 No. 36 (Canadian air permability test)
- CAN/CGSB-4.2 No. 49 (Canadian moisture vapor transmission test)
- ISO 5085 (International thermal resistance test)
- ISO 11092 (International total heat loss test)
- Section 6-5 of NFPA 1977 (total heat loss test)
- No specific consensus standards exist for physiological evaluations of PPE.

Test Results Provided

☐ Test results include:

- Air permeability, usually expressed in cubic feet per square foot of material per minute (ft^3/ft^2 min) or cubic centimeters per square centimeter of material per second (cm^3/cm^2 sec)
- Water vapor transmission rate, expressed as grams of water per square meter of material per 24 hour period; alternatively, the resistance to moisture transfer is represented as height of still air (in millimeters)
- Total insulation, expressed as the total resistance to dry heat loss from body and air layer surrounding body
- Total heat loss, expressed as Watts of heat energy released per square meter of material, accounts for both dry and wet heat transfer
- Permeability index (i_m), a dimensionless number, from 0 to 1, where 0 represents no evaporative heat transfer for the material and 1.0 indicates that the clothing has achieved the theoretical maximum amount of heat transport

- Clo, a unit used to express the resistance to dry heat transfer provided by clothing; 1 clo is the equivalent of a summer business suit with a value of 0.155°K m^2/W
- Physiological responses in human subject tests including:
  - -- Core temperature
  - -- Skin temperature
  - -- Heart rate
  - -- Rate of oxygen consumption
  - -- Sweat (weight) loss
- Human subject tests may also involve human subject ratings of comfort conducted during and after testing.

Test Results Interpretation

☐ Interpretation of test results:

- More breathable materials are indicated by:
  - -- Higher air permeability
  - -- Larger moisture vapor transmission rates
  - -- Higher values of total heat loss
- Greater thermal insulation is demonstrated by:
  - -- Higher thermal insulation values
  - -- Lower total heat loss values
- Better thermal comfort in human subject tests is generally indicated by:
  - -- Lower ending core and skin temperatures
  - -- Lower heart rates
  - -- Less sweat (weight) loss
  - -- Less oxygen consumption
  - -- Better subject ratings of PPE comfort
- PPE evaluated in human subject tests cannot be compared unless tested under the same study for the specific set of test subjects, work tasks, and environmental conditions.
- Greater breathability is usually achieved at the expense of barrier-type properties for clothing materials.

- PPE factors besides breathability must be considered in determining potential wearer thermal comfort, such as material water absorption and stiffness.
- Work comfort is primarily affected by the individual's physical conditions, work environment conditions, the type of activity and the length and duration of breaks in work activity.

32.4 Mobility and Range of Motion

Property Measured

☐ Tests for PPE mobility and range of motion evaluate the effects of PPE on wearer function in performing work tasks.

Applicable PPE

☐ Applicable PPE includes:

- Full body garments
- Partial body garments
- Gloves
- Footwear
- Headwear
- Eye and facewear
- Ear protectors
- Respirators

Test Principles

☐ Wearer mobility and range of motion must be evaluated using human subject-based tests, which simulate work tasks or operations of interest to the user organization.

- Work tasks can be either:
  - -- Common reaching exercises such deep knee bend or arm reaches, or
  - -- Specific tasks representative of the work environment, such as moving drums or boxes, coiling hose, climbing a ladder, etc.
- In many cases, mobility and range of motion may be a function of PPE sizing and fit.
- Comfort and integrity tests can be incorporated into human subject tests.

Standard Test Methods

☐ Standard test methods include:

- ASTM F 1154 (for chemical protective suit ensembles)
- Section 9-4, Overall Suit Function and Integrity Test, of NFPA 1991, *Standard on Vapor-Protective Suits for Hazardous Chemical Emergencies*
- Subclause 7.2, Practical Performance Test, in EN 943, Part 1 (ventilated and non-ventilated chemical protective clothing)

Test Results Provided

☐ Test results include:

- Test measurements can be quantitative or subjective:
  - Anthropometric measurements can be made to evaluate the extent of reach or movement for specific PPE and individuals.
  - Test subjects can be asked to rate the effects of different PPE based on their own experience and preferences.

Test Results Interpretation

☐ Interpretation of test result:

- Interpretation of test results depends on the critical needs of the application as to what degree of function impairment is acceptable.
- PPE should not cause any additional wearing hazards that put workers at risk.
- Subjective ratings are appropriately used to rank and compare PPE that is tested in the same study; results from one study with different conditions and test subjects cannot be compared with results from other studies.

32.5 Hand Function

Property Measured

☐ Hand function tests assess the effects of handwear on specific functions of the hand used to perform tasks or manipulations such as:

- Dexterity, the ability for the individual glove wearer to manipulate small or large objects with their hands
- Tactility, the ability of the individual glove wearer to sense by touch differences in surfaces

- Grip, the ability of the individual glove wearer to grasp objects under both dry and wet conditions

Applicable PPE

☐ Applicable PPE includes:

- Gloves and other handwear

Test Principles

☐ Test principles include:

- Dexterity tests for handwear involve comparing the ability of test subjects to perform tasks with and without gloves to determine how wearing gloves diminishes hand function.
    -- In one type of dexterity testing, the ability of the person to perform a series of hand-based tasks is evaluated on a pass/fail basis with a specific period of time allowed for performing these tasks.
    -- Another type of dexterity testing entails timing the test subject for completing a set of tasks and comparing that test time with similar tests without using gloves.
    -- The majority of tests used for evaluating glove dexterity are principally tests for human coordination that involve repetitive, but simple, functions.
    -- Dexterity tests usually involve gross motor skills (for manipulating large objects) or fine motor skills (for manipulating small objects).
- Tactility testing involves determining a test subject's ability to detect changes in object surfaces usually by having the test subject manipulate successively more intricate or difficult to sense surfaces, until no discernment between surfaces can be made.
    -- Some tactility tests involve test subject discernment between the size and number of indentations in test object surfaces.
    -- Most tactility tests are done while the test subject is blindfolded or is unable to see the test objects being manipulated.
- Grip testing involves measuring the ability of a test subject to maintain his or her grasp on a object or his or her weight lift capacity on wet and dry objects.
    -- Most grip tests use a rope or lanyard attached to a set of weights to determine lifting capacity of the test subject with or without gloves.

-- Testing may be performed with different diameter lanyards, or lanyards which are either wet or dry, or gloves which are either wet or dry.

Standard Test Methods

☐ Standard test methods include:

- Section 6-38 of NFPA 1971, *Standard on Protective Ensemble for Structural Fire Fighting* (gross motor dexterity)
- Section 6-14 of NFPA 1999, *Standard on Protective Clothing for Emergency Medical Operations* (fine motor dexterity)
- Section 6-39 of NFPA 1971, *Standard on Protective Ensemble for Structural Fire Fighting* (grip)
- Subclause 6.3 of EN 420 (European test for fine motor dexterity and tactility)

Test Results Provided

☐ Test methods include:

- Pass/fail (on some dexterity performance tests)
- Percent of barehanded control (dexterity timed comparisons)
- Smallest item discernable (some tactility tests)
- Percent of weight lifting capacity (grip tests)

Test Results Interpretation

☐ Interpretation of test results:

- Gloves or other handwear which permits the wearer the same hand function when not wearing gloves are preferred over gloves which have a significant effect on hand function.
- Percentages of barehanded control for dexterity tests which are closer to 100% indicate improved handwear dexterity; percentages well over 100% indicate problems with dexterity for the tested handwear.
- The smaller the item than can manipulated or sensed by touch, the better the handwear in permitting fine motor dexterity or tactility.
- Percentages of weight lifting capacity for grip tests which are closer to 100% indicate improved handwear dexterity; percentages well under 100% indicate problems with dexterity for the tested handwear.
- In many instances, gloves with special surface finishes may provide better grip with gloves than without gloves.

32.6 Ankle Support

Property Measured

☐ Ankle support testing evaluates the ability of footwear to maintain support for the ankle under conditions of use.

Applicable PPE

☐ Applicable PPE includes:

- Footwear

Test Principles

☐ Test principles include:

- Ankle support is assessed by end user tests by walking on different surfaces or under different conditions with measurement of the test subject's energy expenditure.
- In many cases ankle support is a function of footwear sizing.

Standard Test Methods

☐ Standard test methods include:

- There are no standardized test methods for measuring footwear ankle support at this time.

Test Results Provided

☐ Test results include:

- Comparison of test subjects' energy expenditure for different footwear types
- Subjective ratings from test subjects

Test Results Interpretation

☐ Interpretation of test results:

- Ankle support is important in reducing ankle injuries.
- Footwear which fits snugly around the ankle while still provide flexibility in movement is likely to provide better ankle support than footwear which does not offer those properties.
- Footwear which produces less energy expenditures for test subject offers better ankle support when other factors related to energy expenditure are accounted for.

- Better ankle support is indicated by comparably better subjective ratings from test subjects.

32.7 Back Support

Property Measured

☐ Back support testing evaluates the ability of garment PPE and accessories to provide support for the lower back during lifting and other strenuous activity under conditions of use.

Applicable PPE

☐ Applicable PPE includes:

- Full body garments
- Partial body garments

Test Principles

☐ Test principles include:

- Back support can only be assessed by end user tests where test subjects determine relative back comfort from the device. PPE must be evaluated using large test subject populations over a long period of time (months) to determine if the device provides any improvement.
- Interface effects of back support devices incorporated into PPE must be isolated from other aspects of PPE design.

Standard Test Methods

☐ Standard test methods include:

- There are no standardized test methods for measuring the effectiveness of back support PPE.

Test Results Provided

☐ Test results include:

- Comparison of test subjects' relative back support and comfort with and without use of back support-based PPE based on indications of strain or injury.
- Subjective ratings from test subjects

Test Results Interpretation

☐ Interpretation of test results:

- Back support is important in reducing lower back injury and strain.
- Better back support is indicated by comparably better subjective ratings from test subjects.

32.8 Clarity and Field of Vision

Property Measured

☐ Clarity testing evaluates the ability of an individual to see through a visor or a faceshield; field of vision testing evaluates peripheral vision for an individual wearing the visor or faceshield.

Applicable PPE

☐ Applicable PPE includes:

- Full body garments with visors
- Partial body garments with visors
- Headwear faceshields
- Eye and facewear
- Respirators which cover the eyes

Test Principles

☐ Several different tests are used to evaluate lens, eyepiece, facepiece, or visor clarity and field of vision:

- Prismatic power tests evaluate how lenses, eyepieces, facepieces, or visor cause deviation of light. A telescope and optical system are used to determine the amount of difference in viewing the left eye as compared to the right eye and vice versa.
- Refractive power, resolving power, and astigmatism tests evaluate the ability of lenses, eyepieces, facepieces, or visors to focus light by using an optical device to measure distortion of images.
- Haze tests evaluate the clarity of the material used in lenses, eyepieces, facepieces, or visors in terms of refracting or scattering light away from the user using a spectrophotometer to measure the relative amount of scattered light.
- Transmittance tests evaluate the amount of incident light passing through the lenses, eyepieces, facepieces, or visors onto a light sensor for a spectrophotometer; Transmittance tests can be conducted for light which is:

-- Ultraviolet

-- Luminous

-- Infrared

-- Blue-light

- Simple clarity tests require test subjects to look through the lenses, eyepieces, facepieces, or visors at a standard eyechart to determine their visual acuity while wearing PPE.
- Field of vision testing assesses the range of sight for an individual wearing an item of PPE by either:

  -- Having a test subject determine the extent of his or her peripheral vision while wearing sample PPE, or

  -- Using a lamp inside the PPE item inside a darkened, circular room with measurement of the angles of light transmission from the normal center line of the PPE wear vision

Standard Test Methods

☐ Test methods include:

- Section 15.4 of ANSI Z87.1-1989, *Practice for occupational and educational eye and face protection* apply to clarity assessments for eye and facewear
- Section 9-13, Clarity Test, of NFPA 1991, *Standard on Vapor-Protective Suits for Hazardous Chemical Emergencies*
- Subclause 7.2, Practical Performance Test, in EN 943, Part 1 (European test for ventilated and non-ventilated chemical protective clothing)
- ISO 4854 (International collection of optical test methods)

Test Results Provided

☐ Test results include:

- Prismatic power and refractive power test results are expressed in terms of diopters (1 diopter is equal to angular deviation of 1 centimeter over 1 meter).
- Haze and transmittance test results are expressed in the percentage of light scattered (haze) or passing straight through (transmittance) for a specific wavelength of light, or range of light wavelength.
- Simple clarity test results are reported in the measured visual acuity (e.g., 20/20 or 20/100) for the individual wearer.

- Field of vision tests report the overall angular range of vision for the individual PPE or PPE wearer.

Test Results Interpretation

☐ Interpretation of test results:

- Different types of eye and facewear, and other PPE, have different requirements for acceptable optical properties.
- In general, better clarity of lenses, eyepieces, facepieces, or visors is indicated by:
  -- Low diopter measurements for prismatic power or refractive power
  -- Low percentage haze
  -- High percent light transmittance for luminous light, but low light transmittance for other forms of light.
  -- Better visual acuity of test subjects
- Lower switching indexes are preferred; switching index depends on the type of shade provided by the lenses (e.g., longer switching indexes are provided for lenses which offer a greater or darker shade).
- Better field of vision is demonstrated by PPE with a high range of angular view for the wearer.

32.9 Ease of Communications

Property Measured

☐ Communications tests evaluate the ability of PPE to allow intelligible (understood) communications of the wearer to other persons.

Applicable PPE

☐ Applicable PPE includes:

- Full body garments covering the wearer's head
- Partial body garments covering the wearer's head
- Headwear
- Eye and facewear
- Hearing protectors
- Respirators

Test Principles

- ☐ Communications testing involves a number of human test subjects. One test subject wears sample PPE and reads a specified list of words while the other test subjects write down what they hear.
  - Test subjects must be classified as audiometrically normal to participate in testing.
  - Testing is conducted in a soundproof or audiometric chamber.
  - Background noise of a certain frequency range and loudness may be added to the test chamber.
  - Tests are conducted with and without PPE to compare results.

Standard Test Methods

- ☐ Standard test methods include:
  - ANSI S3.2, *Method for Measuring the Intelligibility of Speech over Communication Systems*
  - Section 6-10, Communication Test, of NFPA 1981, *Standard on Open-Circuit, Self-Contained Breathing Apparatus*

Test Results Provided

- ☐ Test results include:
  - The percentage of words understood correctly for the 'listening' test subjects

Test Results Interpretation

- ☐ Interpretation of test results:
  - Higher percentages of understood words indicate potentially less interference of PPE with communications.

32.10 Sizing and Fit (Adjustment)

Property Measured

- ☐ Sizing of PPE relative to the individual body determines how well PPE fits the individual wearer

Applicable PPE

- ☐ Applicable PPE includes:
  - Full body garments

- Partial body garments
- Gloves
- Footwear
- Headwear
- Eye and facewear
- Ear protectors
- Respirators

Test Principles

☐ As described in Sections 15-22, different sizing practices are used for the various type of PPE.

- The relative fit of the PPE item pertains to its functional use and comfort and is determined subjectively by the wearer in a series of motions or exercises intended to evaluate fit.
- Sizing standards use a combination of:
    -- Specified size dimension for design of the PPE, and
    -- Range of motion and functional determinations

Standard Test Methods

☐ Standard test methods include:

- ASTM D 5585 (body measurements for female misses sizes 2-20)
- ASTM D 5586 (body measurements for women aged 55 and older)
- ASTM F 1731 (sizing and fit of fire service and other rescue uniforms)
- ANSI/ISEA 101 (sizing and dynamic fit for disposable and limited use coveralls)
- CAN/CGSB-49 (Canadian standards on garment sizing)
- ISO 3635 (International standard on size designation of clothes)
- Section 6-32, Glove Fit Test, in NFPA 1977, *Standard on Protective Clothing and Equipment for Wildland Fire Fighting*

Test Results Provided

☐ Test results include:

- No quantitative results are provided
- Some tests are pass/fail

Test Results Interpretation

☐ Interpretation of test results:

- Fit is determined by the individual or organization as acceptable or not acceptable.

Warning: Proper fit is important for providing appropriate protection to the wearer. Poorly fit PPE may provide diminished protection to the wearer.

32.11 Donning and Doffing Ease

Property Measured

☐ Donning and doffing tests evaluate how easily or how quickly individuals can don or doff PPE.

Applicable PPE

☐ Applicable PPE includes:

- Full body garments
- Partial body garments
- Gloves
- Footwear
- Headwear
- Eye and facewear
- Ear protectors
- Respirators

Test Principles

☐ PPE is given to trained individuals with assessments made on the ease of proper donning/doffing or the time required for either donning or doffing.

Standard Test Methods

☐ Standard test methods include:

- Section 9-1, Donning and Doffing Test, of NFPA 1991, *Standard on Vapor-Protective Suits for Hazardous Chemical Emergencies*
- Escape Test specified in Los Alamos National Laboratory requirements for supplied-air suits.

Test Results Provided

☐ Test results include:

- Subjective rating of donning or doffing ease
- Time for either donning, doffing, or both by trained individual

Test Result Interpretation

☐ Interpretation of test results:

- Easier donning (doffing) or shorter donning (doffing) times indicate PPE that is easier for the wearer to put on (off).
- Some types of PPE may require relatively rapid donning or doffing.
- Speed of donning should not prevent proper wearing of PPE; improperly donned PPE can create exposure hazards for the wearer. Likewise, improper clothing may expose the wearer to contaminants on the outside of the PPE.

Section 33

PPE Durability and Serviceability

The purpose of this section is to describe how PPE durability or reservicing can be addressed for applicable performance assessments.

33.1 Overview of Durability and Serviceability

- ☐ Durability refers to how PPE maintain its performance with use.
- ☐ Serviceability refers to the user's ability to care for, maintain, and repair PPE so that it remains functional for further use.
- ☐ Representative standard test methods for durability and serviceability described in this subsection are summarized in Table 33-1.

33.2 Assessment of PPE Durability

- ☐ Conventional approaches for measuring durability include measurement of PPE performance following different conditioning techniques intended to simulate PPE use or expected effects on PPE.
- ☐ Specific effects on PPE include:
  - Repeated wearing
  - Repeated cleaning, decontamination, or sterilization
  - Different storage conditions
- ☐ Types of conditioning techniques include:
  - Repeated laundering or dry cleaning
  - Repeated sterilization
  - Abrasion (*see 25.7*)
  - Repeated flexing (*see 25.10*)
  - Thermal exposures, low or high heat (*see 30.2, 30.3, 30.4, and 30.5*)

- Simulated wearing (*see 32.4*)

☐ Additional information on many conditions is contained in other sections; conditioning procedures for laundering, dry cleaning, and sterilization follow.

Table 33-1. List of Durability Test Methods

Durability Property	Type of Test/Application	Available Test Methods
Wear through laundering	Home laundering - clothing	AATCC 135 CAN/CGSB-4.2 No. 58 ISO 6330
	Commercial laundering- clothing	AATCC 96 CAN/CGSB-4.2 No. 24 ISO 5077
	Dry cleaning- clothing	AATCC 85 ISO 3175
	Elevated temperature laundering- clothing	ISO 675
	Effects of steam- clothing	ISO 3005
	Effects of cold water- clothing	ISO 7771
	Cleanability - face/eyewear	ANSI Z87.1, ?15.5
Wear through sterilization	Steam sterilization	AAMI ST46 ISO 11138 U.S. Pharmacopeia
	Liquid sterilization	ISO 14160
	Ethylene oxide sterilization	AAMI ST41
Serviceability	All PPE	Field studies

33.2.1 Laundering or Dry Cleaning Procedures

- ☐ PPE garments and gloves intended for reuse are subject to repeated cleaning by either laundering or dry cleaning.
- ☐ Garment, glove, and footwear material may be subject to shrinkage as well as changes in performance properties.
- ☐ Repeated cleaning can affect other types of PPE.
- ☐ Standard laundering and dry cleaning procedures include:
  - AATCC 96 (commercial laundering)
  - AATCC 132 (dry cleaning)
  - AATCC 135 (home laundering)
  - CAN/CGSB-4.2 No. 24 (Canadian test for commercial laundering)
  - CAN/CGSB-4.2 No. 30 (Canadian test for dry cleaning)
  - CAN/CGSB-4.2 No. 58 (Canadian test for home laundering)
  - ISO 695 (International test for elevated temperature laundering)
  - ISO 3005 (International test for effects of steam)
  - ISO 3175 (International test for dry cleaning)
  - ISO 5077 (International test for industrial laundering)
  - ISO 6330 (International test for home laundering)
  - ISO 7771 (International test for cold water effects)
  - Section 15.7, Cleanability Test, of ANSI Z87.1-1989, *Practice for occupational and educational eye and face protection*
- ☐ Laundering and dry cleaning procedures differ in:
  - Washing and rinsing temperatures
  - Type of washing machine and machine action
    - Top loading
    - Front loading
  - Type of detergent
  - Type of drying and drying temperature
  - Specification of material samples
    - Material pieces

-- Whole garments

☐ The number of laundering and drying cycles can be varied to simulate different amounts of wearing or cleaning time over the expected life of the PPE.

☐ Measurement of material or garment dimensions before and after laundering and shrinkage can provide an estimate of overall shrinkage.

- Shrinkage is reported in percent change from original dimensions.
  - -- For material specimens, shrinkage is reported by material direction (e.g., for coated textiles, machine and cross-machine directions).
  - -- For garment specimens, shrinkage is reported by orientation on garment (e.g., girth and length).
- High levels of shrinkage can affect clothing performance for thermal insulation by reducing the air layer.
- Shrinkage of whole garments in excess of 5% is considered to represent a change in garment size but is dependent on the clothing's design.

33.2.2 Sterilization Procedures

☐ Sterilization procedures are used to ensure that PPE is free of biological contamination before reuse or disposal.

☐ Common sterilization procedures include:

- Liquid sterilization by using
  - -- Alcohols
  - -- Formalin
  - -- Glutaraldehyde
  - -- Halogens
- Vapor and gas sterilization by using:
  - -- Paraformaldehyde
  - -- Ethylene Oxide
- Steam Sterilization
- Ionizing Radiation

- ☐ Standard sterilization procedures include:
  - AAMI ST41 (Ethylene oxide sterilization and sterility assurance)
  - AAMI ST46 (Steam sterilization and sterility assurance)
  - ISO 11138 (International steam sterilization and sterility assurance)
  - ISO 14160 (International liquid sterilization and sterility assurance)
  - U.S. Pharmacopeia
- ☐ An example test method (test method) for addressing the effects of sterilization agents on PPE is:
  - Section 6-11, Ultimate Tensile Strength, Elongation, and Modulus Test, in NFPA 1999, *Standard on Protective Clothing for Emergency Medical Operations* (effects of alcohol exposure on medical glove tensile properties)
- ☐ Sterilization of PPE can affect PPE materials primarily due to chemical (ethylene oxide or other sterilant liquids/gases), thermal (steam), and radiation effects.
  - Some materials or types of PPE may be degraded by sterilization.
  - Sterilization may also cause future breakdown or weakening of PPE; this may affect other performance areas.
- ☐ The number of sterilization cycles can be varied to simulate different amounts of sterilization over the expected life of the PPE.
- ☐ In addition to effects on the PPE, sterilization testing is used to assure that sterilization can be achieved with the selected sterilization technique and conditions for the PPE.
  - Sterilization assurance testing involves placing selected microorganisms on sample items of PPE, subjecting the PPE to the sterilization technique, and then determining remaining counts of microorganisms.
  - The level of sterilization assurance will vary with the application; some level of remaining microorganism growth may be acceptable for some applications but not for others.

33.3 Assessment of Serviceability

- ☐ Serviceability generally applies to equipment items where special care and maintenance other than routine cleaning are required.

- ☐ Service of PPE includes the following operations:
  - Cleaning
  - Decontamination
  - Sterilization
  - Inspection
  - Maintenance
  - Non-destructive testing
  - Repair
  - Storage
- ☐ Serviceability is assessed by putting PPE through a trial period of use, care, and maintenance to determine how long the PPE can continue to give acceptable performance.
- ☐ Evaluation of serviceability should account for the ease of understanding manufacturer instructions for providing service to some items.
- ☐ Manufacturers of PPE may prohibit or warn against certain types of service (such as repairs) or require servicing of PPE at their own or other authorized facilities.

Section 34

Respirator Design and Performance Properties

The purpose of this section is to describe different types of respirator performance properties in terms of the measured property, applicable PPE, test principles, available standard test methods, results provided, and interpretation of test results.

34.1 Overview of Respirator Design and Testing Standards

☐ Evaluation of respirator design and testing of respirator performance properties is performed in accordance with Public Health Service Title 42 Code of Federal Regulations Part 84, Approval of Respiratory Protective Devices. (42 CFR Part 84).

- 42 CFR Part 84 provides comprehensive design and performance requirements for respirators in the United States.
- All approvals and testing are carried out by the National Institute for Occupational Safety and Health (NIOSH) in accordance with 42 CFR Part 84.
- 42 CFR Part 84 contains provisions for:
    - Application for approval
    - Testing fees
    - Procedures for approval and disapproval
    - Quality control
    - Classification of approval respirators
    - General construction and performance requirements
    - Specific construction (design) and performance requirements for each class of respirator

- 42 CFR Part 84 establishes different classes of respirators for approval:
  - -- Self-contained breathing apparatus (SCBA); Subpart H
  - -- Gas masks; Subpart I
  - -- Supplied-air respirators; Subpart J
  - -- Non-powered air-purifying particulate respirators; Subpart K
  - -- Chemical cartridge respirators; Subpart L
  - -- Special use respirators; Subpart N
  - -- Dust, fume, and mist; pesticide; paint spray; powered air-purifying high efficiency respirators; and combination gas masks; Subpart KK
- 42 CFR Part 84 (in Subparts H through L) further distinguishes between respirators which are classified as:
  - -- Use for entry into and escape from a hazardous atmosphere; or
  - -- Use only for escape from a hazardous atmosphere
- 42 CFR Part 84 also provides for approval of respirators for any or all of the following respiratory hazards:
  - -- Oxygen deficiency
  - -- Gases and vapors
  - -- Particles, including dusts, fumes, and mists
- 42 CFR Part 84 establishes minimum rated service times for respirators (where applicable) at 3, 5, 10, 15, 30, or 45 minutes, and 1, 2, 3, or 4 hours.

☐ Fit factor testing is a common method for evaluating the fit of the facepiece to the specific wearer or the overall effectiveness of the respirator.

☐ Specific tests for open-circuit, self-contained breathing apparatus for the fire service are not covered by 42 CFR Part 84 but are included in NFPA 1981, *Standard on Open-Circuit, Self-Contained Breathing Apparatus* (SCBA).

☐ Specific tests for supplied-air suit are not covered in 42 CFR Part 84 but have been established by the Department of Energy and the certification program for these respirator/protective clothing products is administered by Los Alamos National Laboratory.

34.2 Respirator Design, Construction, and Component Requirements

- ☐ Under 42 CFR Part 84, there are minimum requirements for design, construction, and components for each class of respirator.

- ☐ Table 34-1 summarizes the design requirements specified by 42 CFR Part 84 for the different classes of respirators.

34.2.1 Common Design, Construction, and Component Requirements

- ☐ Common design minimum requirements for all classes of include:
  - A description of the respirator design
  - Types of components for the specific respirator class
  - Breathing gas requirements
  - Breathing tube requirements
  - Harness installation and construction requirements
  - Container requirements for the respirator
  - Half mask or full facepiece, mouthpiece, hood, and helmet fit requirements
  - Facepiece and eyepiece requirements
  - Inhalation, exhalation, and check valve requirements
  - Head harness requirements

- ☐ Breathing gas requirements include:
  - Breathing gas containing no less than 19.5% volume percent of oxygen
  - Oxygen being not less than 99.0% oxygen and nor more than 0.03% carbon dioxide or 0.001% carbon monoxide
  - Compressed breathing air meeting minimum requirements for Type I gaseous air in Compressed Gas Association Commodity Specification for Air, G-7.1 (Grade D or higher quality)
  - Liquefied breathing air meeting minimum requirements for Type II liquid air in Compressed Gas Association *Commodity Specification for Air*, G-7.1 (Grade B or higher quality)

Table 34-1. Summary of Respirator Design Requirements Specified in 42 CFR Part 84

Design Requirement	Type of Respirator (Subpart in Regulations)*					
	Self-Contained Breathing Apparatus (Subpart H)	Gas Masks (Subpart I)	Supplied-Air Respirators (Subpart J)	Non-Powered Air-Purifying Particulate Respirators (Subpart K)	Chemical Cartridge Respirators (Subpart L)	Dust, Fume, and Mist; Pesticide; Paint Spray? (Subpart KK)
Required components	84.71	84.111	84.131	84.171	84.191	84.1131
Canisters/cartridges in parallel		84.112			84.192	
Canister/cartridge color marking		84.113			84.193	84.1154
Canister/cartridge filters		84.114			84.194	84.1155
Breathing tube design	84.72	84.115	84.132	84.172	84.195	84.1132
Harness installation/construction	84.73	84.116	84.133	84.173	84.196	84.1133
Respirator container	84.74	84.117	84.134	84.174	84.197	84.1134
Facepiece/mouthpiece fit	84.75	84.118	84.135	84.175	84.198	84.1135
Facepiece/eyepiece design	84.76	84.119	84.136	84.176	84.199	84.1136
Inhalation/exhalation valves	84.77	84.120	84.137	84.177	84.200	84.1137
Head harness design	84.78	84.121	84.138	84.178	84.201	84.1138
Head and neck protection			84.139			
Air velocity and noise levels			84.140		84.202	84.1139

* Numbers refer to section numbers in 42 CFR Part 84.

H Also includes Powered air-purifying high efficiency respirators and combination gas masks

Table 34-1. (continued)

Design Requirement	Type of Respirator (Subpart in Regulations)*					
	Self-Contained Breathing Apparatus (Subpart H)	Gas Masks (Subpart I)	Supplied-Air Respirators (Subpart J)	Non-Powered Air-Purifying Particulate Respirators (Subpart K)	Chemical Cartridge Respirators (Subpart L)	Dust, Fume, and Mist; Pesticide; Paint Spray? (Subpart KK)
Breathing gas quality	84.79		84.141			
Air supply source - blowers			84.142			
Chamber fitting - terminal ends			84.143			
O_2/air interchange prohibition	84.80					
Breathing gas cylinder design	84.81					
Gas pressure gauge design	84.82					
Timer/service life indicator design	84.83					
Hand-operated valve design	84.84					
Breathing bag design	84.85					
Exposed part O_2 compatibility	84.86					
Compressed gas filters	84.87					
Weight	84.89					
Filter identification				84.179		

* Numbers refer to section numbers in 42 CFR Part 84.

H Also includes Powered air-purifying high efficiency respirators and combination gas masks

- ☐ Requirements for breathing tubes address design that prevents:
  - Restriction of head movement
  - Disturbance of facepiece or mouthpiece fit
  - Interference with wearer activities
  - Accidental shutoff of airflow from wearing pressure or kinking
- ☐ Harness requirements address:
  - Adequate construction to allow respirator components to be held close to wearer's body
  - Ease of removal or replacement of parts
- ☐ Apparatus containers are required to be constructed of durable materials, include appropriate labelling, and permit easy removal.
- ☐ Fit requirements for half-mask and full facepieces, mouthpieces, hoods, and helmets address:
  - Provision of more than one size or use of a one size design that accommodates a wide range of facial features
  - Provision of adjustment to permit sizing wide range of head sizes (hoods and helmets)
  - Optional use of corrective lenses (for full facepieces)
  - Use of attached nose clips for mouthpiece designs that provide an air-tight seal
  - Prevention of eyepiece, spectacle, or lens fogging
- ☐ Facepiece and eyepiece requirements include provision of adequate vision without distortion by the eyepiece and eyepiece resistance to impact and penetration (*see 25.4*)
- ☐ Requirements for inhalation, exhalation, and check valves involve:
  - Protection from damage or distortion
  - Prevention of adverse effects of excess exhaled air on cartridges or canisters
  - Protection from external influences (exhalation valves)
  - Construction of exhalation valves to prevent inward leakage
  - Check valve design permitting flow towards facepiece only

☐ Head harness requirements entail providing adjustable and replaceable head harnesses for facepieces or mouthpieces; the harness must provide adequate tension and even distribution of pressure of the face contact area.

34.2.2 Specific Design Requirements for Self-Contained Breathing Apparatus (SCBA)

☐ 42 CFR Part 84 classifies SCBA into two general and several subcategories:

- Closed-circuit apparatus
  - Compressed oxygen source
  - Chemical oxygen source
  - Liquid oxygen source
- Open-circuit apparatus
  - Demand-type apparatus
  - Pressure-demand-type apparatus
- Combination respirator of SCBA with Type C supplied-air respirator (SAR):
  - SAR has rated service time with 3, 5, or 10 minutes
  - SAR with 15 minutes or longer service time

☐ Required components for self-contained breathing apparatus include:

- Facepiece or mouthpiece and nose clip
- Respirable breathing gas container
- Supply of respirable breathing gas
- Gas pressure or liquid level gauges
- Timer (if applicable)
- Remaining service life indicator or warning device
- Hand-operated valve
- Breathing bags (if applicable)
- Safety relief valve or relief system
- Harness

- ☐ Specific minimum requirements in 42 CFR Part 84 that apply to SCBA include:
  - Prohibition of interchangeability of oxygen and air
  - Compressed gas and liquefied breathing gas container requirements
  - Gas pressure gauge requirements
  - Timer, elapsed time, or service life indicator requirements
  - Hand-operated valve requirements
  - Breathing bag requirements
  - Compatibility of exposed component parts to oxygen pressure above atmospheric pressure
  - Downstream filter to remove particulates from gas stream
  - Overall SCBA weight
- ☐ Breathing gas or liquefied breathing gas container (cylinder) requirements include:
  - Meeting requirements of the U.S. Department of Transportation for interstate shipment
  - Permanent and legible marking of container contents
  - Provision of a gauge that shows the internal pressure of the container.
  - Use of threads on valves and charging system consistent with American Standards Association, *Compressed Gas Cylinder Valve Outlet and Inlet Connections*, B57.1-1965
- ☐ SCBA gauge requirements include:
  - Calibration in pounds per square inch (for gases) or fractions of liquid-level for liquid-level gauges
  - Reliability to ±5% of scale
  - Minimum graduations and a full scale graduation not exceeding 150 percent of the maximum rated cylinder pressure
  - Failure of gauge not impairing SCBA operation
  - Visibility of gauge to wearer
  - Special labelling of oxygen gauges
- ☐ Requirements for timers, elapsed time indicators and service life indicators include:

- Calibration in minutes of remaining service life
- Readability by the wearer
- Preset alarms that last 7 seconds indicating that the remaining service life of SCBA is reduced to 20 to 25% of the rated service time
- Automatic operation

☐ Hand-operated valve requirements include designs for:

- Preventing removal of valve stem from valve body
- Preventing accidental opening or closing
- Ready adjustment by wearer
- Conservation of air supply in event of failure
- Easy distinguishment between valve types by wearer
- Use of red color for bypass valve
- Pressure release on closed-circuit systems

☐ Breathing bags are required to

- Have sufficient volume to prevent waste
- Be flexible and resist gasoline vapors
- Located in an area to prevent damage or collapse (except escape only apparatus)

☐ The overall, fully charged SCBA cannot exceed 35 pounds (16 kg) except when the SCBA weight decreases by more than 25% of its initial weight during its rated service life, in which it is permitted to weigh no more than 40 lbs (18 kg).

34.2.3 Specific Design Requirements for Gas Masks

☐ 42 CFR Part 84 provide three design classifications of gas masks:

- Front-mounted or back-mounted gas mask
- Chin-style gas mask
- Escape gas mask

☐ 42 CFR Part 84 further classifies gas masks according to the type of gases or vapors they designed to provide respirator protection against:

- Acid gas

- Ammonia
- Carbon monoxide
- Organic vapor
- Combination of two or more gases or vapors

☐ Required components for gas masks include:

- Facepiece or mouthpiece and nose clip
- Canister or cartridge
- Canister harness (if applicable)
- External check valve
- Breathing tube (if applicable)

☐ Specific minimum requirements in 42 CFR Part 84 which apply to gas masks include:

- Canister or cartridge used in parallel having equal resistance to airflow
- Marking and color coding of canisters and cartridges in accordance with ANSI K13.1, *American National Standard for Identification of Air-Purifying Respirator Canisters and Cartridges*.
- Location of filters on inlet side of canister or cartridge which permit easy removal and replacement

34.2.4 Specific Design Requirements for Supplied-Air Respirators (SAR)

☐ 42 CFR Part 84 classifies SARs into three types:

- Type A, which uses a hand-operated or motor-driven blower to provide air from a remote source
- Type B, which uses a large diameter hose for the transport of air from the wearer
- Type C, which uses a source of respirable air such as from a compressor or air cylinders

☐ 42 CFR Part 84 further designates respirators as AE, BE, and CE for SARs that include additional devices for protection against rebounding abrasive material (such as from sand blasting).

☐ Required components for supplied air respirators include:

- Facepiece, hood, or helmet
- Air-supply valve, orifice, or demand or pressure-demand regulator
- Hand-operated or motor-driven air blower (if applicable)
- Air supply hose
- Detachable couplings
- Flexible breathing tube
- Respirator harness

☐ Specific minimum requirements in 42 CFR Part 84 that apply to SAR include:

- Head and neck protection from impact and abrasion resistance for abrasive application supplied-air respirators (Types AE, BE, and CE) (*see 25.5*)
- Limitation of noise generated in respirator during maximum airflow to 80 decibels on both the inside and outside of respirator (*see 27.7*)
- Air supply source for Type A supplied air respirators
- Terminal fittings for Type B supplied air respirators
- Requirements for Type C continuous-flow supplied air respirators
- Requirements for Type C demand and pressure-demand supplied air respirators

☐ Blowers for Type A SAR are required to deliver an adequate amount of air to the wearer in either blower rotation direction and to allow the passage of air when not operating.

☐ Terminal fittings for Type B SAR are required to:

- Install in the inlet of the hose
- Provide for drawing air through a hose that removes particulate matter greater than 0.149 mm in diameter
- Provide a means for fastening or anchoring the hose

☐ Continuous-flow Type C SAR requirements include:

- Maximum pressure at the inlet no greater than 125 pounds per inch (863 kN/m^2)
- Use of a pressure-release mechanism to prevent hose pressures greater than 125 pounds per inch (863 kN/m^2)

- ☐ Demand and pressure-demand Type C SAR requirements include:
  - Manufacturer specification of air pressure range for the point of hose attachment and the range of hose length
  - Maximum pressure at the point of attachment no greater than 125 pounds per inch (863 kN/m^2)
  - Use of a pressure-release mechanism to prevent hose pressures at the point of attachment greater than 125 pounds per inch (863 kN/m^2)
  - Use of a pressure-release mechanism to prevent hose pressures 20% above the manufacturer's highest specified pressure

34.2.5 Specific Design Requirements for Non-Powered Air-Purifying Particulate Respirators (SAR)

- ☐ Non-powered air-purifying particulate respirators are classified into three series:
  - N-series filters are restricted to use in workplaces free of oil aerosols.
  - R- and P-series filters are intended for removal of any particulate that including oil-based liquid particulates.
- ☐ Non-powered air-purifying particulate respirators are also classified by the efficiency level of the filter as determined in test requirements:
  - N100, R100, and P100 filters have a minimum efficiency of 99.97%.
  - N99, R99, and P99 filters have a minimum efficiency of 99%.
  - N95, R95, and P95 filters have a minimum efficiency of 95%.
- ☐ Required components for non-powered air-purifying particulate respirators include:
  - Facepiece, mouthpiece with nose clip, hood, or helmet
  - Filter unit
  - Harness
  - Attached blower
  - Breathing tube
- ☐ Specific minimum requirements in 42 CFR Part 84 that apply to non-powered air-purifying particulate respirators include:
  - Filter identification for type and efficiency

-- N100, R100, and P100 filters are labelled as particulate filters with the appropriate designation (e.g., "N100 Particulate Filter") and use a magenta color.

-- N99, R99, and P99 filters are labelled as particulate filters with the appropriate designation (e.g., "N99 Particulate Filter") and use a color other than magenta.

-- N95, R95, and P95 filters are labelled as particulate filters with the appropriate designation (e.g., "N95 Particulate Filter") and use a color other than magenta.

34.2.6 Specific Design Requirements for Chemical Cartridge Respirators

☐ Chemical cartridge respirators are classified by the chemical they are intended to protect against, up to a specified maximum use concentration:

- Ammonia
- Chlorine
- Hydrogen chloride
- Methyl amine
- Organic vapor
- Sulfur dioxide
- Vinyl chloride

☐ Required components for chemical cartridge respirators include:

- Facepiece, mouthpiece with nose clip, hood, or helmet
- Cartridge
- Cartridge with filter
- Harness
- Attached blower
- Breathing tube

☐ Specific minimum requirements in 42 CFR Part 84 that apply to chemical cartridge respirators include:

- Canister or cartridge use in parallel having equal resistance to airflow

- Marking and color coding of canisters and cartridges in accordance with ANSI K13.1, *American National Standard for Identification of Air-Purifying Respirator Canisters and Cartridges*.
- Location of filters on inlet side of canister or cartridge which permit easy removal and replacement
- Limitation of noise generated in respirator during maximum airflow to 80 decibels on both the inside and outside of respirator (*see 27.7*)

34.2.7 Specific Design Requirements for Special Purpose Respirators

☐ Vinyl chloride respirators include all respirators designed for respiratory protection against vinyl chloride:

- Front-mounted or back-mounted gas masks
- Chin-style gas masks
- Chemical-cartridge respirators
- Powered air-purifying respirators
- Combination respirators

☐ Required components for chemical cartridge respirators include:

- Facepiece
- Canister with end-of-service life indicator
- Cartridge with end-of-service life indicator
- Harness
- Attached blower
- Breathing tube

☐ Vinyl chloride respirators must meet the respective requirements for the type of respirator class that apply in 42 CFR Part 84.

☐ Vinyl chloride respirators are required to have end-of-service life indicators which provide an indicator change before 1 part per million (ppm) vinyl chloride penetration occurs.

34.2.8 Specific Design Requirements for Dust, Fume, and Mist Respirators; Pesticide Respirators; Paint Spray; Powered Air-Purifying High Efficiency Respirators; and Combination Gas Masks

☐ 42 CFR Part 84 distinguishes between different type of specialized respirators, including:

- Dust, fume, and mist respirators
- Pesticide respirators
- Paint spray respirators
- Powered air-purifying high efficiency respirators
- Combination gas masks

☐ Required components for chemical cartridge respirators include:

- Facepiece, mouthpiece with nose clip, hood, or helmet
- Filter unit, canister with filter, or cartridge with filter
- Harness
- Attached blower
- Breathing tube

☐ Specific minimum requirements in 42 CFR Part 84 that apply to chemical cartridge respirators include:

- Canister or cartridge used in parallel having equal resistance to airflow
- Marking and color coding of canisters and cartridges in accordance with ANSI K13.1, *American National Standard for Identification of Air-Purifying Respirator Canisters and Cartridges*.
- Location of filters on inlet side of canister or cartridge which permit easy removal and replacement
- Limitation of noise generated in respirator during maximum airflow at 80 decibels on both the inside and outside of respirator (*see 27.7*)

34.3 Respirator Performance Requirements

☐ Respirators are evaluated for different performance or functional properties under 42 CFR Part 84, depending on the class of respirator.

☐ Some tests are common for each class of respirator but may involve a different minimum requirement.

☐ Categories of performance requirements for respirators include:

- Breathing and air flow resistance (inhalation and exhalation)
- Exhalation valve leakage
- Gas-flow
- Service time
- Carbon dioxide in inspired gas
- Low temperature operation
- Breathing bag gasoline permeation
- Manned tests
- Gas-tightness
- Particulate protection
- Canister performance
- Blower performance
- Air-supply line performance
- Harness performance
- Breathing tube performance

☐ Table 34-2 shows the general performance tests specified by 42 CFR Part 84 for each class of respirator.

34.3.1 Breathing and Air Flow Resistance

Property Measured

☐ Breathing resistance tests evaluate the resistance to breathing or airflow in the facepiece during inhalation and exhalation by measuring the pressure during either inhalation or exhalation.

Applicable Respirators

☐ Applicable respirators include:

- Self-contained breathing apparatus (SCBA)
- Gas masks
- Supplied-air respirators (SAR)

Table 34-2. Summary of Respirator Performance Requirements Specified in 42 CFR Part 84

Performance Requirement	Type of Respirator (Subpart in Regulations)*					
	Self-Contained Breathing Apparatus (Subpart H)	Gas Masks (Subpart I)	Supplied-Air Respirators (Subpart J)	Non-Powered Air-Purifying Particulate Respirators (Subpart K)	Chemical Cartridge Respirators (Subpart L)	Dust, Fume, and Mist; Pesticide; Paint Spray? (Subpart KK)
Breathing bag performance	84.88		84.153-157			
Breathing/air flow resistance	84.90-91	84.122		84.180	84.203	84.1149
Exhalation valve leakage	84.92	84.123	84.158	84.182	84.204	84.1150
Gas flow	84.93-94					
Service time	84.95-.96					
CO_2 level	84.97					
Low temperature operation	84.98					
Manned performance	84.99-103		84.159-163			
Gas tightness/facepiece performance	84.104	84.124			84.205	84.1141-2

* Numbers refer to section numbers in 42 CFR Part 84.
H Also includes Powered air-purifying high efficiency respirators and combination gas masks

Table 34-2. (continued)

Performance Requirement	Type of Respirator (Subpart in Regulations)*					
	Self-Contained Breathing Apparatus (Subpart H)	Gas Masks (Subpart I)	Supplied-Air Respirators (Subpart J)	Non-Powered Air-Purifying Particulate Respirators (Subpart K)	Chemical Cartridge Respirators (Subpart L)	Dust, Fume, and Mist; Pesticide; Paint Spray? (Subpart KK)
Particulate removal efficiency		84.125		84.181	84.206	84.1148
Chemical removal efficiency		84.126			84.207	84.1157
Blower performance			84.144-145			
Air supply line performance			84.150			
Harness performance			84.151			
Breathing tube performance			84.152			
Silica dust removal efficiency						84.1147
Lead fume removal efficiency						84.1146
DOP filter efficiency						84.1141
Silica dust loading						84.1152

* Numbers refer to section numbers in 42 CFR Part 84.

H Also includes Powered air-purifying high efficiency respirators and combination gas masks

- Non-powered, air-purifying particulate respirators
- Chemical cartridge respirators
- Dust, fume, and mist respirators
- Pesticide respirators
- Paint spray respirators
- Powered air-purifying high efficiency respirators
- Combination gas masks

Test Principles

☐ Test principles include:

- Breathing or air flow resistance tests are performed with the respirator placed on a breathing machine operating (breathing resistance) or attached to a air supply (air flow resistance) at rate specified for the type of the respirator:
  - -- SCBA use 24 respirations per minute and 40 Liters per minute
  - -- Gas masks, non-powered, air-purifying particulate respirators, and chemical cartridge, dust, fume, and mist respirators, pesticide respirators, paint spray respirators, powered air-purifying high efficiency respirators, and combination gas masks use continuous flow at 85 Liters per minute.
  - -- Supplied-air respirators use continuous flow through the air-supply line at 85 or 115 Liters per minute depending on the type of SAR being tested.
- Pressure is measured inside the respirator facepiece at inhalation and exhalation.

  -- For gas masks, both initial and final inhalation breathing resistance is measured (at the start and the end of the respirator service time).

Test Results Provided

☐ Breathing resistance test results are reported as a pressure in millimeters or inches of water column or as a difference between the inhalation and exhalation pressure.

Test Results Interpretation

☐ Interpretation of test results:

- While some breathing resistance is expected, lower inhalation or exhalation pressures indicate favorable breathing resistance.
- 42 CFR Part 84 establishes specific inhalation and exhalation breathing resistance maxima for different classes of respirators.

34.3.2 Exhalation Valve Leakage

Property Measured

☐ Exhalation valve leakage tests evaluate the amount leakage that occurs when the valve is subjected to a vacuum.

Applicable Respirators

☐ Applicable respirators include:

- Self-contained breathing apparatus (SCBA)
- Gas masks
- Supplied-air respirators (SAR)
- Non-powered, air-purifying particulate respirators
- Chemical cartridge respirators
- Dust, fume, and mist respirators
- Pesticide respirators
- Paint spray respirators
- Powered air-purifying high efficiency respirators
- Combination gas masks
- Exhaust valves on totally encapsulating suits are also evaluated using this test.

Test Principle

☐ Exhalation valves are placed in a closed cell apparatus and subjected to a 25 mm (1 inch) water-column height vacuum with measurement of the airflow through the valve under these conditions.

Test Results Provided

☐ Test results include:

- Flow rate in milliliters per minute.

Test Results Interpretation

☐ Interpretation of test results:

- Lower or no measurable leakage is an indication that the exhalation valve will prevent the inward leakage of outside environmental contaminants.
- 42 CFR Part 84 establishes a maximum leakage of 30 mL/min for all respirator exhalation valves.

34.3.3 Gas Flow

Property Measured

☐ Gas flow tests evaluate the how much breathing gas flows under a condition representing failure of the facepiece seal

Applicable Respirators

☐ Applicable respirators include:

- Self-contained breathing apparatus (SCBA), open-circuit and closed-circuit

Test Principles

☐ Test principles include:

- Testing is performed statically with the respirator on a headform or manikin.
- For open-circuit SCBA, the pressure inside the facepiece is lowered 51 mm (2 in.) water column height with measurement of flow rate from the breathing air supply.
- For closed circuit SCBA, the flow rate is measured over the rated service time or when the respirator is operated in a fully open position.
- Flow rate into the facepiece is measured.

Test Results Provided

☐ Test results include:

- The flow rate in Liters per minute.

Test Results Interpretation

☐ Interpretation of test results:

- 42 CFR Part 84 establishes specific requirements for acceptable minimum gas flow rates for different types of respirators.

34.3.4 Service Time

Property Measured

☐ Service time tests evaluate the how long breathable air is provided by the respirator under simulated use conditions.

Applicable Respirators

☐ Applicable respirators include:

- Self-contained breathing apparatus (SCBA), open-circuit and closed-circuit

Test Principles

☐ Test principles include:

- A respirator is either placed on a breathing machine or evaluated in a specific human subject tests to determine how long the respirator will continue to provide breathing air of acceptable quality.

Test Results Provided

☐ Test results include:

- The rated service time of the respirator in minutes or hours according the classification scheme established in 42 CFR Part 84.

Test Results Interpretation

☐ Interpretation of test results:

- 42 CFR Part 84 establishes specific rated service times, but no minimum requirements are set other than the minimum rated service life of 3 minutes.
- Some rated service times are not appropriate for different types of respirators.
- The respirator may not provide acceptable breathing air for the full rated service time under actual use conditions because of the work activity and use conditions faced by the wearer.

34.3.5 Carbon Dioxide Levels in Inspired Gas

Property Measured

- ☐ These tests evaluate the how much carbon dioxide is present or accumulates in the breathing gas under simulated use conditions.

Applicable Respirators

- ☐ Applicable respirators include:
  - Self-contained breathing apparatus (SCBA), open-circuit and closed-circuit

Test Principles

- ☐ Test principles include:
  - The concentration of carbon dioxide is measured inside the facepiece in the mouth area.
  - The respirator is placed on a breathing machine that is operated at 14.5 respirations per minute and 10.5 Liters per minute.

Test Results Provided

- ☐ Test results include:
  - The concentration of carbon dioxide provided in percent (%) or parts per million (ppm) measured in the facepiece versus time.

Test Results Interpretation

- ☐ Interpretation of test results:
  - 42 CFR Part 84 sets maximum carbon dioxide levels for different closed-circuit SCBA and the acceptable percentage of carbon dioxide in exhaled gas for open-circuit SCBA.

34.3.6 Low Temperature Operation

Property Measured

- ☐ Respirators are evaluated in low temperature conditions to determine acceptable respirator performance at test conditions.

Applicable Respirators

- ☐ Applicable respirators include:
  - Self-contained breathing apparatus (SCBA), open-circuit and closed-circuit

Test Principles

☐ Test principles include:

- Testing is performed inside a climate controled chamber at the minimum safe temperature specified by the manufacturer or the respirator.
- Test subjects wear the respirator according to manufacturer instruction after the respirator has been precooled to the test temperature for 4 hours.
- The respirator is worn inside the chamber while the test subject performs a series of exercises.
- Assessments are made of respirator function, wearer vision, and wear comfort.

Test Results Provided

☐ Test results include:

- Subjective evaluation of respirator function, vision through the facepiece, and ease of breathing.

Test Results Interpretation

☐ Interpretation of test results:

- 42 CFR Part 84 establishes qualitative requirements for acceptable respirator function, wearer vision, and wear comfort.

34.3.7 Breathing Bag Gasoline Resistance

Property Measured

☐ Breathing bag gasoline resistance tests evaluate the ability of the breathing bag to prevent the penetration or permeation (*see Section 28*) of gasoline vapors under simulated use conditions.

Applicable Respirators

☐ Applicable respirators include:

- Self-contained breathing apparatus (SCBA), open-circuit and closed-circuit

Test Principles

☐ Test principles include:

- Breathing bags are tested in at atmosphere saturated with gasoline vapor at room temperature for twice the rated service time (except escape devices, which are tested only at the rated service time).
- The breathing bag is operated on a breathing machine at 24 respirations per minute and 40 Liters per minute.
- The concentration of gasoline inside the breathing bag is measured at the end of the test period.

Test Results Provided

☐ Test results include:

- Gasoline concentration in parts per million (ppm)

Test Results Interpretation

☐ Interpretation of test results:

- 42 CFR Part 84 prohibits use of breathing bags which permit gasoline concentrations of 100 ppm or more.

34.3.8 Human Subject or Man Tests

Property Measured

☐ Human subject testing (or man testing) is intended to evaluate the performance of respirators under simulated use conditions for a number of performance areas depending on the type of respirator being evaluated.

Applicable Respirators

☐ Applicable respirators include:

- Self-contained breathing apparatus (SCBA), open-circuit and closed-circuit
- Supplied-air respirators (SAR), all types

Test Principles

☐ Test principles include:

- A number of human subject tests are conducted to:
    - -- Familiarize the wearer with the operation of the respirator
    - -- Provide for a gradual increase in work activity
    - -- Evaluate the respirator under different types of work and physical orientation

-- Provide information on the breathing characteristics of the respirator during actual use

- During human subject tests, respirators are evaluated for:
  - -- Provision of adequate breathing air for the indicated service time
  - -- Breathing gas temperature inside the respirator facepiece
  - -- Fogging of the facepiece or eyepiece
  - -- Subjective complaints of discomfort by the wearer
- Exercises during human subject tests include:
  - -- Walking at various speeds
  - -- Running at a specific speed
  - -- Crawling
  - -- Carrying or pulling weight
  - -- Lying on side or back
  - -- Climbing a vertical treadmill
- The specific sequence and number of exercises is dependent on the type of respirator, the rated service time, and the specific test requirements.
- Samples and readings are taken periodically during the human subject testing.
- Additional human subject tests are conducted to:
  - -- Determine the maximum length of time the respirator will supply breathing air while the wearer is sitting at rest with measurement of oxygen and carbon dioxide levels
  - -- Evaluate respirator function specifically for liquefied air respirators when the wearer is in other than vertical orientations

Test Results Provided

☐ Test results include:

- The flow rate in Liters per minute.

Test Results Interpretation

☐ Interpretation of test result:

- 42 CFR Part 84 establishes specific requirements for acceptable minimum gas flow rates for different types of respirators.

34.3.9 Gas-Tightness

Property Measured

☐ Gas-tightness testing evaluates the adequacy of the facepiece seal on the wearer's face and/or the overall function of the respirator in preventing penetration of a test agent under simulated use conditions.

Applicable Respirators

☐ Applicable respirators include:

- Self-contained breathing apparatus (SCBA), open-circuit and closed-circuit
- Gas masks
- Supplied-air respirators (SAR)
- Chemical cartridge respirators
- Dust, fume, and mist respirators
- Pesticide respirators
- Paint spray respirators
- Powered air-purifying high efficiency respirators
- Combination gas masks

Test Principles

☐ Test principles include:

- Testing is performed by a number of human subjects who wear the respirator for a specified period of time.
- Human subjects with respirators are exposed to a specified concentration of isoamyl acetate, perform some limited exercises, and are asked if they can detect this chemical by odor or taste:
  - 1,000 parts per million of isoamyl acetate is used for SCBA and full facepiece gas masks
  - 100 ppm of isoamyl acetate is used for half-mask gas masks, dust, fume, and mist respirators; pesticide respirators; and paint spray respirators
  - 0.1% of isoamyl acetate is used for SAR (described in 42 CFR Part 84 as manned tests)

Test Results Provided

☐ Test results include:

- Pass or fail determination of gas-tightness as determined by the human subject's detecting or not detecting isoamyl acetate.

Test Results Interpretation

☐ Interpretation of test results:

- 42 CFR Part 84 requires that there be no detection of isoamyl acetate under these test conditions.

34.3.10 Particulate Removal Efficiency

Property Measured

☐ Particulate removal efficiency testing evaluates the ability of filter elements in the respirator to prevent the penetration of particulates.

Applicable Respirators

☐ Applicable respirators include:

- Gas masks
- Non-powered, air-purifying particulate respirators
- Chemical cartridge respirators

Test Principles

☐ Test principles include:

- Filter elements are first conditioned in a standard atmosphere at high temperature and humidity for 25 hours.
- Different test atmospheres are used depending on the type of filter series:
    - N-series filters are tested against a solid sodium chloride particulate aerosol with a median diameter of 0.075 microns at a concentration of 200 milligrams per cubic meter.
    - R- and P-series filters are tested against a dioctyl phthalate (DOP) aerosol with a median particulate diameter of 0.185 microns at a concentration of 200 milligrams per cubic meter.
- An air flow rate of 85 Liters per minute is used to challenge the filters with a challenge mass of no more than 200 milligrams per cubic meter.

- Particulate efficiency is measured during the test using a light-scattering photometer.

Test Results Provided

☐ Test results include:

  - Overall particulate removal efficiency

Test Results Interpretation

☐ Interpretation of test result:

  - 42 CFR Part 84 provides classification of filters by their overall particulate removal efficiency.

34.3.11 Canister Performance

Property Measured

☐ Canister performance testing evaluates the effectiveness of a canister in removing a contaminant from an air stream under specific simulated use conditions.

  - Special purpose tests are provided for vinyl chloride respirators.

Applicable Respirators

☐ Applicable respirators include:

  - Gas masks
  - Chemical cartridge respirators
  - Vinyl chloride gas masks, chemical cartridge respirators, and powered, air-purifying respirators
  - Combination gas masks

Test Principles

☐ Test principles include:

  - Testing is performed under different specific temperature, humidity and flow conditions:
    -- Gas masks and chemical cartridge respirators are tested both at room temperature and 32 or 64 Liters per minute at both 50% and 25% relative humidity.
  - Different canister types are evaluated against specific test atmospheres:

-- Sulfur dioxide and chlorine gas (acid gas canisters)

-- Carbon tetrachloride (organic vapor canisters)

-- Ammonia (ammonia canisters)

-- Carbon monoxide (carbon monoxide canisters)

-- Vinyl chloride (vinyl chloride special purpose canisters)

- Different cartridge types are evaluated against specific test atmospheres:

-- Ammonia

-- Chlorine

-- Hydrogen chloride

-- Methylamine

-- Carbon tetrachloride (organic vapor cartridges)

-- Sulfur dioxide

-- Vinyl chloride (special purpose cartridge)

- The concentration of the test atmosphere depends on the respirator type.

Test Results Provided

☐ Test results include:

- Chemical penetration in part per million (ppm) during at end of test

- Service time in which level of test agent(s) remain below the specified levels

Test Results Interpretation

☐ Interpretation of test results:

- 42 CFR Part 84 requires minimum service times for different types of canisters and respirators.

- Longer service times indicate lower penetration of contaminants.

WARNING: The limited number of tests and test conditions do not support wearing of gas masks or use of cartridges under all conditions.

34.3.12 Dust, Fume, and Mist Filtration Performance

Property Measured

☐ Dust, fume, and mist filtration performance tests evaluate respirator effectiveness in preventing the penetration of different types of dust, fumes, or mists.

Applicable Respirators

☐ Applicable respirators include:

- Dust, fume, and mist respirators
- Pesticide respirators
- Paint spray respirators
- Powered air-purifying high efficiency respirators
- Combination gas masks

Test Principles

☐ Different test approaches are used for evaluating respirator effectiveness against various dusts, fumes, and mists.

- Test agents include:
  - Silica dust of 0.4 to 0.6 micron median diameter at 50 to 60 milligrams per cubic meter
  - Lead fumes at no more than 20 milligrams per cubic meter
  - Lacquer mist at 95-125 milligrams per cubic meter
  - Enamel mist at 95-125 milligrams per cubic meter
- Flow rates and the conduct of the test depend on the type of respirator

Test Results Provided

☐ Test results include:

- Specific mass of penetrating test agent for test conditions

Test Results Interpretation

☐ Interpretation of test results:

- 42 CFR Part 84 specifies the maximum amount of test agent that is allowed to penetrate for specific classes and types of respirators.

- Better respirator effectiveness is demonstrated by lower penetrating masses of test agent.

34.3.13 Blower Performance

Property Measured

☐ Blower performance tests evaluate the ability of hand-operated or motor-driven blowers to provide a sufficient amount of air to the SAR wearer.

Applicable Respirators

☐ Applicable respirators include:

- Type A Supplied Air Respirators (SAR)

Test Principles

☐ Test principles include:

- Hand-operated blowers are evaluated by attaching the blower to a mechanical drive and operating the blower 6 to 8 hours daily for a total of 100 hours at a speed necessary to deliver 50 Liters per minute using the maximum length of hose permitted.
- Motor-driven blowers are evaluated by running the blower at the specified running speed and operating the blower 6 to 8 hours daily for a total of 100 hours.

Test Results Provided

☐ Test results include:

- In hand-operated blower tests, the following assessments are made:
  -- Required crank speed
  -- Required torque
  -- Determination of operation effectiveness (without failure)
- In motor-driven blower tests, the following assessments are made:
  -- Required power
  -- Rate of air delivery

Test Results Interpretation

☐ 42 CFR Part 84 establishes minimum and maximum performance requirements for each aspect of blower performance.

34.3.14 Air-Supply Line Performance

Property Measured

☐ Air-supply lines are tested for:

- Maximum length
- Air flow
- Air flow control (air-regulating valve)
- Noncollapsibility
- Nonkinkability
- Hose and coupling strength
- Gas-tightness in couplings
- Gasoline permeation resistance
- Ease of detaching coupling

Applicable Respirators

☐ Applicable respirators include:

- Supplied air respirators (SAR), all types

Test Principles

☐ Test principles include:

- Each hose performance area is evaluated using a specific test procedure.
- Hose length is measured and cannot exceed 300 feet (91 meters).
- The minimum and maximum airflows permitted by the air-supply line are measured.
- Damage to the air regulating value is determined after a specific number of actuations during simulated use.
- A force is applied to the hose to test ease of collapse.
- A hose is suspended in a loop and must maintain a circular shape to demonstrate resistance to kinking.
- Hose sections and couplings are subjected to a tension force with examination of hose and couplings for damage.
- Rates of leakage through the hose are checked for assessing air-supply line gas-tightness.

- 25 feet of hose are immersed in gasoline for a specified period of time with measurement of gasoline concentration in the exit stream.
- Ease of coupling detachment is assessed qualitatively.

Test Results Interpretation

☐ Interpretation of test results:

- Pass/fail criteria are provided by 42 CFR Part 84 for each of the air-supply line requirements.

34.3.15 Harness Performance

Property Measured

☐ Harness performance tests involve evaluation of:

- Shoulder strap strength
- Belt, ring, and life-line attachment strength
- Interference of harness with wearer movement or comfort.
- Strength of hose attachment to harness
- Drag provided by hose on facepiece or respirator

Applicable Respirators

☐ Applicable respirators include:

- Supplied air respirators, all types

Test Principles

☐ Test principles include:

- Strength tests involve pulling SAR components under specified tension to determine breaking strength.
- Hose-to-harness strength tests involve use of human subjects to determine effects of hose drag

Test Results Provided

☐ Test results include:

- Breaking strength
- Assessment of effects of pulling or dragging over rough surfaces

Test Results Interpretation

☐ Interpretation of test result:

- 42 CFR Part 84 sets specific criteria in each of the performance areas.

34.3.16 Breathing Tube Performance

Property Measured

- ☐ Breathing tube performance tests evaluate the ability of the breathing tube to continue supply of breathing gas to the facepiece, hood, or helmet in different wearer orientations.

Applicable Respirators

- ☐ Applicable respirators include:
  - Supplied air respirators

Test Principles

- ☐ The resistance of the breathing tube to kinking, collapsing, or interfering with facepiece sealing is evaluated by a test subject undergoing various head movements.

Test Results Provided

- ☐ Test results include:
  - Qualitative assessment of hood interference and resistance to kinking and collapse during simulated use.

Test Results Interpretation

- ☐ Interpretation of test result:
  - 42 CFR Part 84 requires that the breathing tube not interfere with head movement and show no evidence of kinking or collapse during simulated use.

34.4 Fit and Protection Factor Testing

- ☐ Fit and protection factors evaluate the effectiveness of the facepiece seal on the wearer or the effectiveness of the overall respirator in preventing the inward leakage of outside contaminants.
- ☐ The gas-tightness tests described in 34.3.9 are a form of fit testing specified in 42 CFR Part 84.
- ☐ Fit testing involves:

- Placement of the respirator on a headform or wearing of the respirator by an individual
- Exposure of the respirator or individual to some external test agent
- Determination of test agent penetration into the breathing zone of the respirator

☐ There are two types of fit testing:

- Qualitative fit testing involves having a test subject wear the respirator and determine whether he or she detects the test agent inside the facepiece by sensing odor.
- Quantitative fit testing uses instrumentation to measure the ratio of the concentration of the airborne test agent in the air of the ambient atmosphere and in the air inside the facepiece in the breathing zone of the wearer.

☐ Protection factors are the ratio of the measured average concentration of an air contaminant present in the workplace to the measured concentration of the contaminant in the air inside the facepiece of a respirator on the wearer:

- Protection factors are rarely determined in the field but rather determined by NIOSH or other organizations.
- Protection factors differ from fit factors in that:
  -- The fit factor is specific to a particular respirator on a specific wearer.
  -- The protection factor is a more general assessment of particular types of respirator for workplace applications, which takes into account different work activity and different face, head, and body movements.
- Both NIOSH and ANSI have assigned protection factors for specific classes of respirators.

34.4.1 Qualitative Fit Testing

☐ Qualitative fit tests include:

- Isoamyl acetate method
  -- A piece of cotton or cloth is saturated with isoamyl acetate liquid (banana oil) and is passed close to the respirator-faceseal surface.
  -- Isoamyl acetate is easily detected by its pleasant odor and taste.

-- Alternatively, a closed chamber is used for evaporating isoamyl acetate to a concentration of approximately 100 ppm (now specified in 42 CFR Part 84).

- Irritant smoke method
    - -- Irritating smoke is created by tubes containing stannic chloride or titanium tetrachloride impregnated pumice stone.
    - -- Smoke is directed at the facepiece seal and leakage is detected by the wearer's involuntary coughing or sneezing due to irritation.
    - -- Prescribed in OSHA lead standard, 29 CFR §1910.1025
- Sodium saccharin method
    - -- Involves use of sodium saccharin which creates irritant fume
    - -- Alternative test in OSHA lead standard, 29 CFR §1910.1025
- Other methods involve other test agents including:
    - -- Talcum powder followed by removal of the facepiece and inspection of wearer's face for power or dust streaks
    - -- Uranin, a fluorescent dye, followed by removal of the facepiece and examination of the wearer's face under ultraviolet light

34.4.2 Quantitative Fit Testing

☐ Quantitative fit tests include methods that use both:

- Gaseous challenge agents
- Aerosol challenge agents

☐ Quantitative fit tests with gaseous challenge agents include:

- Argon (100%) with mass spectroscopy detection
- Ethylene (2%) with length-of-stain tube detection
- Dichlorodifluoromethane (1,000 ppm) with halide meter detection
- Helium (10%) with mass spectroscopy detection
- Pentane (120 ppm) with gas chromagraph/flame ionization detection
- Sulfur hexafluoride (50 ppm) with hydrogen-flame detection

☐ Quantitative fit tests which use aerosol challenge agents include:

- Biological agent fit testing systems using a bacterium challenge agent with standard bacteriological plating procecures for detection

- Uranin fit testing systems using uranin, a fluorescent powder, with detection using a fluorescence meter
- Coat dust fit testing systems using di-octyl phthalate (DOP), di-2-ethylhexyl phthalate (DEHP), or corn oil with a detection by light scattering photometry
- Sodium chloride fit testing systems using sodium chloride aerosols with detection by light scattering photometry
- Ambient aersol systems using submicron particles found commonly in ambient air by detection from a continuous flow condensation nuclei counter

34.5 Specific Requirements for Fire Service Open-Circuit Self-Contained Breathing Apparatus

☐ Specific requirements for fire service respirators are not covered in 42 CFR Part 84.

☐ Specific Requirements for fire service open-circuit SCBA in NFPA 1981, *Standard on Open-Circuit Self-Contained Breathing Apparatus (SCBA)*, include tests for:

- Air flow performance
- Environmental temperature effects
- Vibration effects
- Accelerated corrosion effects
- Particulate effects
- Heat and flame effects
- Textile fabric flame and heat resistance (*see Section 30*)
- Thread heat resistance (*see Section 30*)
- Facepiece abrasion (*see 25.7*)
- Communications (*see 32.9*)

☐ SCBA performance is primarily evaluated using the air flow performance test with is a breathing resistance test (*see 34.3.1*) conducted with a specific type of breathing machine which operates at 30 respirations per minute and 100 Liters per minute:

- The air flow performance test is conducted as part of other tests for evaluating the effects of different environmental temperatures, vibration, accelerated corrosion, particulates, and overall heat and flame exposure.
- The pressure inside the facepiece must remain positive and not exceed 3½ inches (89 millimeters) water column pressure at anytime during the test.

34.6 Specific Requirements for Air-line Suits

☐ Specific requirements for airline suits are not covered in 42 CFR Part 84.

☐ Air-line suits are suits to which breathing air is supplied via an air-line:

- Air-line suits do not have a tight-fitting facepiece
- The suit serves as the loose-ftting enclosure around the wearer's breathing zone.

☐ Specific requirements for testing air-line suits are established by the Respiratory Advisory Committee of Los Alamos National Laboratory (LANL) through the U.S. Department of Energy and include tests for:

- Air flow performance
- Quantitative protection factor
- Aerosol penetration
- Ease of escape
- Supplied-air hose performance
- Level of noise (*see 27.7*)

☐ Air flow performance (*similar to air flow resistance described in 34.3.1*) is evaluated at 170 and 196 Liters per minute (6 and 6 cubic feet per minute).

☐ Quantitative protection factor testing is conducted using:

- A polydisperse aerosol challenge with a mean diameter of 0.6 microns and a concentration of 25 milligrams per cubic meter
- A test subject that performs a series of exercises

☐ Aerosol penetration testing is conducted after the quantitative protection factor testing with the flow of breathing air stopped, the test subject standing still, and the measurement of the time required for the aerosol concentration inside the suit to reach 0.05, 0.1, and 1%.

☐ Ease of escape testing (*see 32.11*) is evaluated by timing how long the wearer takes to get out the suit (LANL requires escape times of 6 seconds or less).

Section 35

PPE Service Life and Life Cycle Cost

The purpose of this section is to describe the factors affecting the service life of PPE and how the service can be determined.

35.1 Overview of PPE Service Life and Cost

☐ The selection of PPE must involve consideration of product service life and cost.

- Service life affects the longevity of the product and must meet user expectations for durability and serviceability.
- Cost is an important factor in most PPE selection decisions particularly in comparing products.

☐ PPE service life generally fits into one of three categories:

- PPE may be considered disposable when these products are:
  - Inexpensive and cannot be adequately, cleaned, reserviced, or maintained, or
  - Easier to be disposed of and replaced than to be cared for and maintained
- PPE may be considered to limited use where:
  - Some cleaning, care, and maintenance is possible, but the products may not be reusable under hard physical conditions, or
  - Cleaning, care, and maintenance processes eventually degrade the PPE.
- PPE may be considered to be reusable when:
  - Products can be readily cleaned and maintained, and still continue to provide acceptable performance, or

-- Products can be reserviced or have portions or components replaced to provide acceptable performance.

☐ Manufacturers may not specify the service life of their products and may also not indicate conditions for retiring of PPE.

☐ Some PPE may have a limited shelf life (i.e., time in storage before use) because of material degradation that can take place due to heat, ozone, or material self-degradation.

☐ In comparing products, compare the cost of the product based on life cycle cost, not purchase cost, particularly if the products have different service lives.

35.2 Factors Affecting PPE Service Life

☐ PPE service life is a function of several factors including:

- PPE durability or ease of reservicing
- Life cycle cost

☐ PPE durability is demonstrated by the length of time that PPE provides acceptable performance given the range of use conditions, care, and maintenance (*see Section 33 on Durability and Serviceability*).

- Unacceptable performance may be evident through physical changes in the PPE such as:
  -- Rips, tears, or separation of textile materials
  -- Thin spots in coated materials or protruding fibers
  -- Cracks in rubber or plastic components of materials
  -- Material discoloration
  -- Diminished functionality of PPE component parts
- Unacceptable performance may also not be readily evident unless products are carefully examined by product manufacturers or subjected to destructive testing.
- Estimates of product durability are derived both through product testing for simulating product wear and from field experience involving extended product actual use.
- Some products are expected to lose some performance in certain propertiy areas; the acceptability of any drops in performance related to protection must be examined by the end user.

- ☐ Many types of PPE can be serviced or repaired to extend service life; servicing and repair is considered part of a regular care and maintenance progam to allow PPE meet its expected service life.
- ☐ PPE has a specific life cycle that encompasses the following stages:
  - Initial selection and purchase
  - Use
  - Decontamination (if required)
  - Cleaning (and sterilization, if required)
  - Routine maintenance and testing
  - Repairs
  - Replacement of parts
  - Storage
  - Disposal

35.3 Determination of Life Cycle Cost

- ☐ Life cycle cost is the overall cost of the product over its actual service life.
- ☐ Life cycle cost (LCC$) has the following individual costs:
  - Costs associated with the selection process (S$):
    - -- Committee or individual review and comparison of products
    - -- Field trials of PPE
    - -- Time spent with manufacturer representatives, looking through catalogs or at trade shows
  - Purchase cost (P$): actual price paid for product plus shipping and taxes
  - Costs associated with inspecting product (I$):
    - -- Upon receipt
    - -- Periodic inspections as requried
  - Costs associated with caring for product (C$):
    - -- Cleaning after use
    - -- Decontamination
    - -- Sterilization

    -- Storage (time and space)

  - Costs associated with maintaining product (M$):

    -- Periodic testing (time and supplies)

    -- Service or repairs

  - Costs associated with retirement and disposal (D$)

☐ Life cycle cost should include the labor costs, supplies, and services associated with each stage of the PPE life cycle.

☐ Life cycle cost is calculated by adding each cost and dividing the total cost by the expected number of uses:

$$LCC\$ = \frac{S\$ + P\$ + I\$ + C\$ + M\$ + D\$}{\text{Number of Uses}}$$

☐ Table 35-1 provides a form which can be used calculate and compare the life cycle cost of several products.

☐ When the cost of several products are compared, the product with the lowest life cycle is the least expensive product.

Table 35-1. Determination of Life Cycle Cost (Sample Worksheet for Comparing Products)

Life Cycle Category	Type of Cost	Product 1			Product 2			Product 3		
		Qty. or Hours	Price or Rate	Total	Qty or Hours	Price or Rate	Total	Qty or Hours	Price or Rate	Total
Selection (S$)	Labor									
Purchase (P$)	Unit cost									
Inspection (I$)	Labor									
	Supplies									
Care (C$)	Labor									
	Supplies									
	Storage cost									
Maintenance (M$)	Labor									
	Equipment									
	Supplies									
Disposal ($D)	Labor									
	Shipping									
Subtotal (total of all items = S$ + P$ + I$ + C$ + M$ + D$)										
Expected service life or number of uses (n)										
TOTAL LIFE CYCLE COST (subtotal ? n)										

Section 36

PPE Product Labeling and Information

The purpose of this section is to describe the type of information that may be present on PPE labels and other user information that may be provided with PPE.

36.1 Product Labeling

☐ PPE is labeled to provide different kinds of information to the user about the specific product.

☐ The purposes of PPE product labels are to:

- Provide identification of the product
- Describe the uses of the product
- Warn about product limitations
- Provide care information and recommendations
- Indicate certification to specific standards

☐ Labels are also important in providing product traceability, particularly in the case of a product recall or identifying potential defects.

36.1.1 Types of Labeling Information and Symbols

☐ Product labels may contain any one or more of the following product identification information items:

- Product name
- Manufacturer name, address, and phone number
- Classification
- Composition
- Model number

- Serial number
- Lot number

☐ The PPE product label may also contain information for the intended use of the product (such as "for splash protection only")

☐ Many labels will contain warning or other language that indicates limitations of PPE; examples include:

- "Do not use around open flame, sparks, or high heat."
- "This (PPE item) will not protect you from all hazards associated with (activity)."

☐ Specific signal words may be used with labels; signal words include:

DANGER: indicates an imminently hazardous situation which, if not avoided, will result in death or serious injury (this signal word is to be limited to the most extreme situations)

WARNING: indicates a potentially hazardous situation which, if not avoided, could result in death or serious injury

CAUTION: indicates a potentially hazardous situation which, if not avoided, may result in minor or moderate injury. It may also be used to alert against unsafe practices.

- Each signal word should be proceded by the ⚠ symbol; Certain color schemes are used for each signal word:
  - -- DANGER uses a red background with white lettering.
  - -- WARNING uses an orange background with black lettering
  - -- CAUTION uses a yellow background with black lettering

☐ The warnings portion of labels may also provide indications of consequences if the warnings are not heeded and understood, and if the product is not properly used; for example,

- "Failure to understand these warnings may result in serious injury or death."

☐ Labels may refer the wearer to more complete information provided in the instructions for the product:

- "See instructions for additional information"
- "Refer to instructions before use"

- ☐ Labels may also provide care information:
  - "Dry clean only"
  - "Do not use bleach"
  - "Clean after using"
- ☐ Symbols or "pictograms" may be used on labels to identify product classification, care information, or certification to specific standards.
  - The European Community provides a number of "pictograms" for PPE to represent intend applications (these pictograms are shown in Figure 36-1).
  - Several standards provide symbols for care and maintenance of PPE (primarily textiles) including:
    - ASTM D 5489 (permanent care labels on consumer textile products)
    - ISO 3758 (International care labelling code using symbols)
    - Federal Trade Commision Trade Regulation Rule, *Care Labeling of Textile Wearing Apparel and Certain Piece Goods*, January 2, 1974.
    - *The National Standard of Canada—Care Labelling of Textiles* (CAN/CGSB-86.1-M87)
- ☐ Standards that apply to the general labelling of PPE include:
  - ASTM F 1301 (labelling of chemical protective clothing)
  - ANSI Z535.4-1991, *American National Standard for Product Safety Signs and Labels* (requirements for use of signal words and the design of labels)
  - ISO 13688 (International standard for general requirements on protective clothing)
  - EN 340 (European standard for general requirements on protective clothing)

36.1.2 Types of Labels and Label Location

- ☐ The amount of the label information may be limited by the size of the product or materials used in construction.
- ☐ Label information may be distributed over several individual labels.

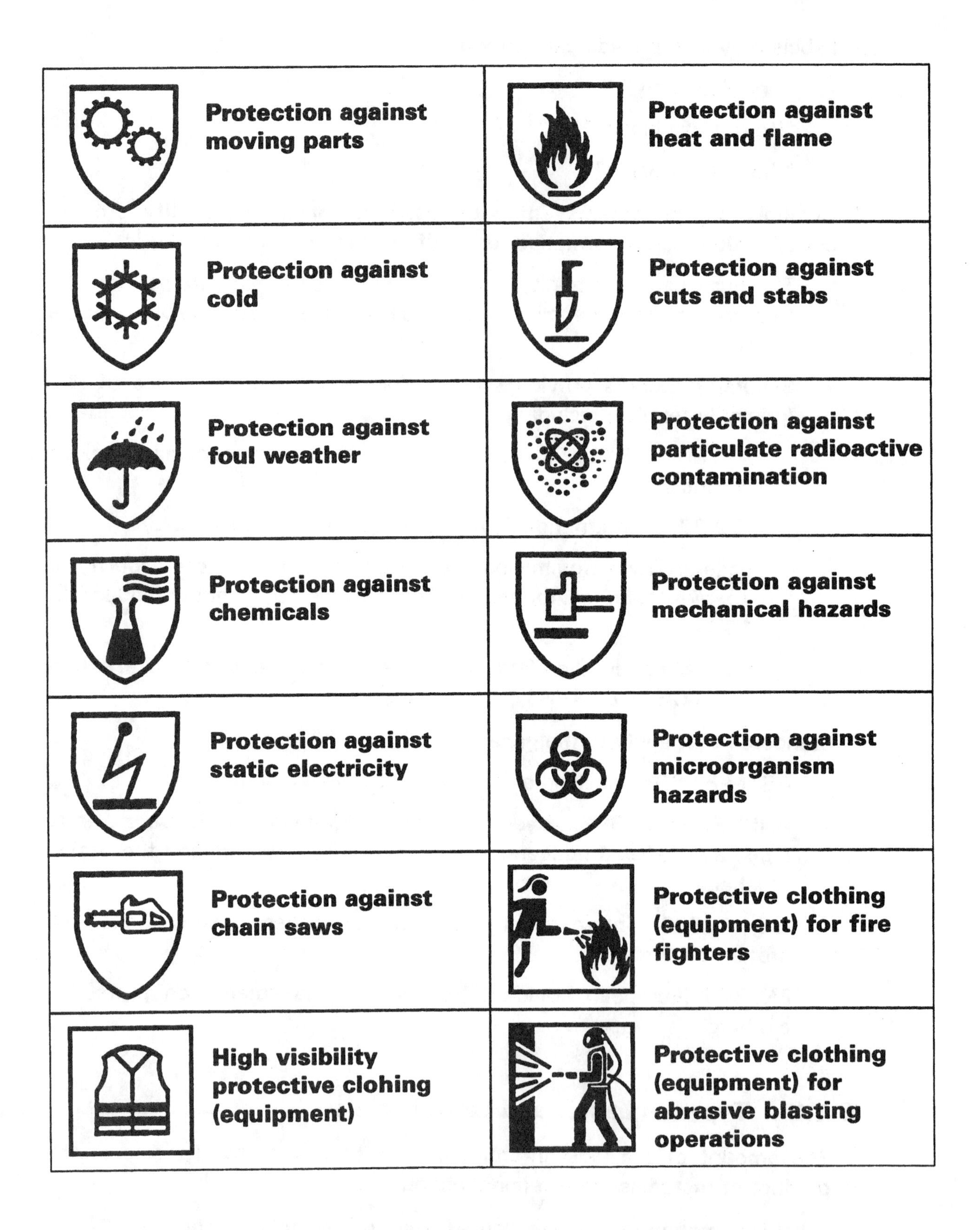

Figure 36-1. Pictograms Used for PPE in European Standards

- ☐ Labels may be:
  - Printed (stencilled, painted) or etched directly onto the product
  - Printed onto a textile material which is sewn to the product
  - Printing onto a piece or plastic, rubber, or metal which is laminated, glued, or otherwise bonded to the product
- ☐ Some labels may be affixed on:
  - All four sides, or
  - One side ("hang tag")
- ☐ The position of the label will often depend on the specific item of PPE but is usually located in a highly visible area or an area to protect the label from physical abuse in the product's use:
  - Garment labels typically appear in the inside collar for shirts, jackets, or coveralls, back upper inside waist band for trousers, or inside the item; larger labels will be on the interior of the clothing such that the wearer sees the label when donning the item.
  - Glove and footwear labels may be inside the glove or outside the glove
  - Headwear labels are typically on the inside of the item near the suspension
  - Labels for other types of PPE will usually appear on the exterior in a protected location.
- ☐ For small items such as ear protectors, some safety glasses, and rubber gloves, the product label may appear on the product's packaging or carton.

36.1.3 Label Legibility and Durability

- ☐ The chief design and performance properties for labels are:
  - Legibility
  - Durability
- ☐ Legibility refers to ability of the wearer to read the label information and is primarily a function of type size:
  - Typically type size greater than 1/16 of inch (1.6 mm) is needed to be visible to the ordinary person with corrected vision at a distance of 12 inches (30 cm).

- Readability tests may be performed where a determination of legibility is made using tests subjects under selected lighting conditions and at specific reading distances.

☐ Label information is subject to wear from cleaning, sterilization, decontamination, abrasion, and ordinary wear:

- Specific tests may be prescribed to determine legibility of label information following a conditioning technique (*see Section 33*)

36.2 Product Information

☐ Product information refers to additional information provided by the manufacturer that is intended to assist the wearer:

- In correctly using the product, or
- By providing details about the product

☐ Product information may encompass the following areas:

- Instructions for use, care, and maintenance
- Product specifications and test data
- Manufacturer warranty information

36.2.1 Instructions

☐ Instructions for the PPE should address all aspects of the product's life cycle, including:

- Inspection
- Donning
- Use limitations and safety considerations
- Adjustment
- Doffing
- Cleaning
- Sterilization or decontamination
- Testing
- Repairs
- Storage

- Retirement criteria
- Disposal

☐ Instructions should be written so that the anticipated user groups can easily understand the use of the product.

☐ Instructions may contain several warning or caution statements that point out areas of concern for using, caring for, or maintaining the product.

☐ Many instructions may be supplemented by diagrams, photographs, or videotapes to provide this information in an easier to understand format.

36.2.2 Product Specifications

☐ Product specifications provide a description of the product in terms of the PPE item's construction, design, and performance.

☐ Product specifications may range from indicating the model name and number of the PPE item to a full description of all parts and design aspects.

☐ Information usually provided as part of product specifications include:

- Model name and number
- Intended use
- Product configuration or design
- Composition or parts
- Features
- Options or accessories
- Test data indicating compliance to a specific standard

36.2.3 Manufacturer Warranties

☐ Manufacturer warrranties are agreements of the manufacturer with the purchaser relating to the product's:

- Workmanship
- Performance in use
- Longevity or service life

- ☐ Most manufacturers of PPE offer only limited warranties covering workmanship of the product since use conditions can vary significantly affect protective performance and because varying use can significantly affect service life.

Section 37

Overview of PPE Selection

The purpose of this section is to provide an overview of the PPE selection process, specifically focusing on using information from the risk assessment, choosing appropriate PPE designs, and setting minimum design and performance criteria. This section also discusses the types of product specifications and provides methods for evaluating candidate PPE in terms of manufacturer technical information, certification processes, and field trials.

37.1 The PPE Selection Process

☐ The general PPE selection process encompasses three principal steps:

- ***Step 1:*** Conduct a risk assessment to:
  - -- Identify the hazards present
  - -- Determine the affected body areas or body systems by the identified hazards
  - -- Assess the risks associated with each hazard
- ***Step 2:*** Use information from the risk assessment to:
  - -- Determine which hazards must be protected against
  - -- Decide which general types of PPE must be used
  - -- Determine which performance properties are relevant for selecting PPE
- ***Step 3:*** For the general types of PPE items determined to be necessary:
  - -- Choose a specific design type
  - -- Choose appropriate design features
  - -- Set minimum performance levels for relevant properties
- ***Step 4:*** Prepare the PPE specification:

-- Select an existing standard

-- Modify or supplement an existing standard

-- Prepare a minimum design and performance specification

- ***Step 5:*** Evaluate candidate PPE against the minimum acceptance criteria:

-- Require manufacturers to supply product information and technical data

-- Use PPE evaluated by independent test laboratories or certified to specific standards

-- Conduct field trials

- ***Step 6:*** Establish a PPE program within the organization:

-- Set responsibilities within the organization for overseeing the PPE program

-- Train workers in the selection, use, care, and maintenance of PPE

-- Provide a system for care and maintenance of PPE

- ***Step 7:*** Periodically conduct a new risk assessment and evaluate how selected PPE is working to provide adequate levels of protection

☐ This process is illustrated in Figure 37-1.

37.2 STEP 1 - Conduct the Risk Assessment

☐ Sections 4 through 14 provide guidance for conducting the risk assessment.

☐ The risk assessment will provide:

- The type of hazards presents
- The body areas or body system which will be affected
- The level of risk associated with each hazard

☐ PPE must be selected based on a risk assessment that is specific to the task, job, or application.

☐ PPE must be fit to the task, job, or application; the task, job, or application should not be fit to the available PPE unless the needed PPE is not offered by the PPE industry.

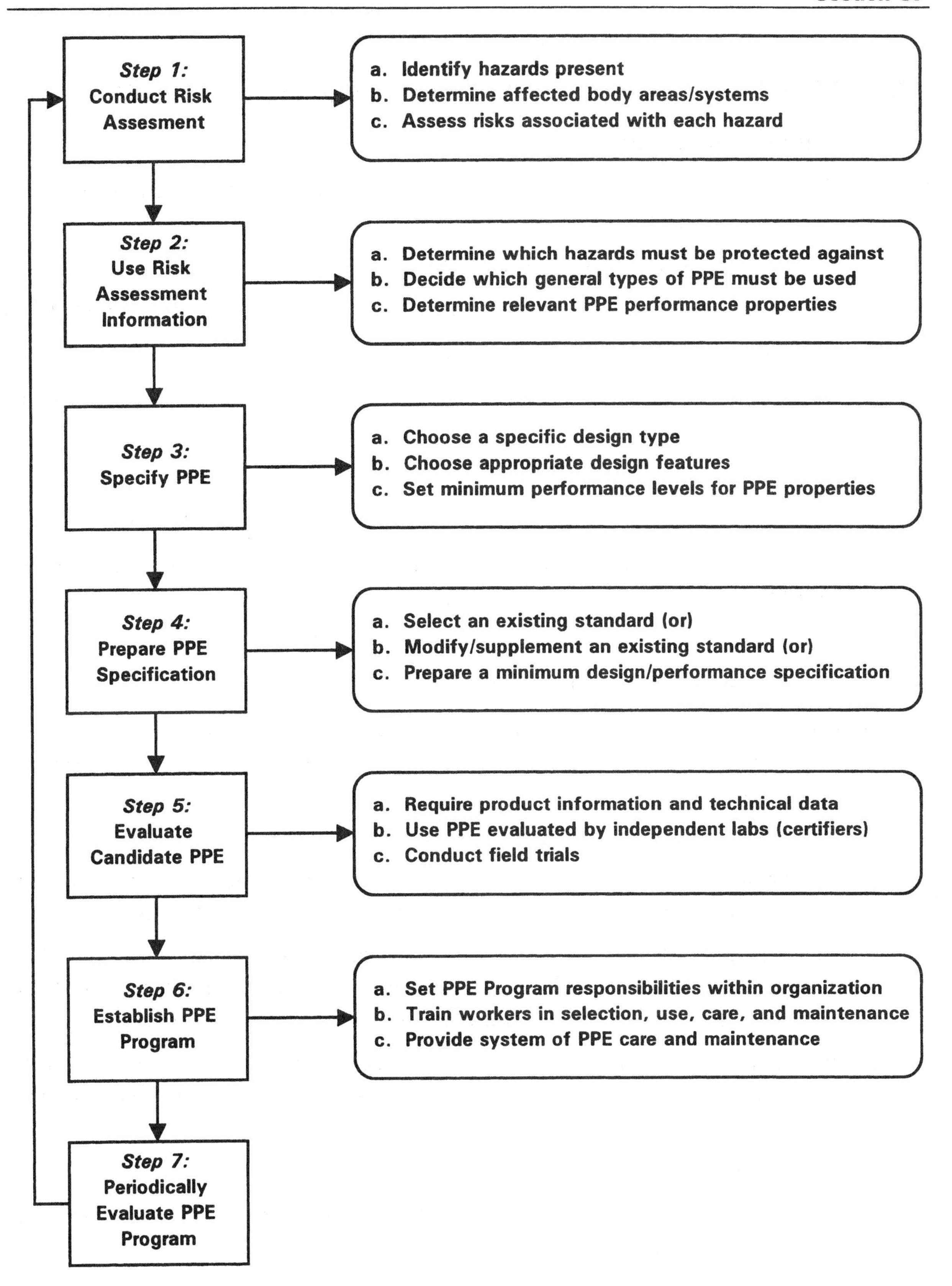

Figure 37-1. Detailed PPE Selection Process.

37.3 STEP 2 - Use Information from the Risk Assessment to Make Selection Decisions

☐ The PPE selection process uses input from the risk assessment by answering the following questions:

- What types of hazards are present?
- Which body areas and body systems will be affected by those hazards?
- What is the risk of exposure for each hazard and body area/system?

☐ The selection process uses the input from the answers to these questions in reverse order.

37.3.1 Selection Decisions based on the Risk of Exposure

☐ The risk of exposure for each hazard and body area/system is used to determine:

- Is protection needed?
- How much protection is needed?

☐ Based on its risk assessment of a task, job, or application, each organization should determine the level of risk or "action level" at which PPE is warranted.

☐ The "action" level can be determined by:

- Using the numerical ratings from the risk assessment and assigning a maximum acceptable risk level, above which PPE should be used; or
- Making individual subjective decisions as to which risks are acceptable with or without PPE (These decisions must be made by the authority that is responsible for PPE decisions within the organization).

☐ At risk levels higher than the action level, PPE must be selected which offers greater protection as demonstrated by PPE design and performance properties.

37.3.2 Selection Decisions based on the Affected Body Areas/ Body Systems

☐ By virtue of the affected body areas and body systems, the type of PPE can be chosen:

- Entire body: combination of clothing or suit and other equipment that encloses the entire body
- Trunk or torso: full body or partial body protective clothing
- Head: partial body protective clothing or headwear
- Eyes and face: partial body protective clothing or face and eyewear
- Arms: partial body protective clothing
- Hands: protective gloves
- Legs: partial body protective clothing
- Feet: protective footwear
- Respiratory system: respirators
- Hearing: hearing protectors

☐ In some case, certain types of PPE may protect multiple parts of the body, such as a full facepiece respirator, which provides respiratory system protection concurrent with eye and face protection.

37.3.3 Selection Decisions based on the Hazards Present

☐ The types of hazards present indicate which performance properties are relevant for qualifying PPE items or their materials and components as providing protection against those hazards (Table 24-1 in Section 24, *Overview of PPE Performance Properties*, lists relevant performance properties for different hazards).

☐ Performance properties may apply to:

- The overall PPE item
- Certain features or components of the PPE item
- Materials used in the construction of the PPE item

37.4 STEP 3 - Choose the PPE Design, Design Features and Minimum Performance

☐ Step 2 of the selection process yields:

- Which hazards should be protected against
- The general types of PPE which should be worn

- The general performance properties that the selected PPE should meet

☐ The selection of appropriate PPE encompasses specifying:

- The specific PPE design or type
- Design features associated with the specific PPE design
- Minimum performance for the specific PPE item

☐ The combination of specifying PPE designs, design features, and minimum performance results in the development of minimum design and performance criteria.

☐ In selecting PPE, particular designs can be specified or performance criteria can be used.

- Performance criteria are preferred over design criteria since different manufacturer designs may offer the same level of performance (performance criteria permit a wide range of designs to be considered).
- Design criteria are justified when:
  - -- Performance tests do not capture the specific properties desired for the PPE.
  - -- The organization has experience with specific designs that are compatible with other equipment or uses processes which warrant specifying a particular design or product.
- Usually it makes the best sense to specify some design criteria in combination with performance criteria:
  - -- The design criteria specify the type of design that the organization has experience with and wants to use.
  - -- The performance criteria specify the level of protection that the product must be capable of providing.

37.4.1 Choice of PPE Designs

☐ For each general type of PPE, several potential designs exist which provide a range of coverage and performance.

- These designs or styles are generally intended for specific types of hazards.
- The level of performance for different designs of the PPE item can vary significantly.

- ☐ Most PPE is marketed for a specific application or a limited number of applications and may not account for all hazards for a specific situation.
- ☐ Each industry has its owns practices related to the selection and use of PPE; these practices are usually not consistent among various industries and applications.
- ☐ Sections 38 through 45 offer guidance on selecting specific PPE designs.

37.4.2 Choice of PPE Design Features

- ☐ For each specific design of PPE, there are a variety of different features that will affect the protective performance offered by the item.
- ☐ The types of design features available for each specific PPE item vary considerably and cannot be considered fully in this book. Identification of primary features for each type of PPE is provided in Sections 38 through 45.

37.4.3 Setting Minimum Performance

- ☐ The specific minimum design and performance criteria must be selected based on the severity of the hazards and the level of risk established in the risk assessment.
- ☐ The challenge in specifying PPE is setting minimum levels of performance for acceptable protection.
  - In some cases, some performance area criteria will be in conflict (e.g., increased thermal insulation from heat is likely to decrease thermal comfort, which can result in heat stress).
  - Some performance properties do not have easily identifiable end points for judging acceptable levels of protection.

37.5 STEP 4 - Prepare the PPE Specification

- ☐ Specifications may be developed in three different ways:
  - Using an established standard or specification
  - Modifying or supplementing an existing standard or specification
  - Creating a new specification
- ☐ Established specifications come from several different sources:

- Consensus standards from ASTM, NFPA, CEN, ISO and other organizations
- Trade organizations or labor unions (as model specifications)
- PPE manufacturers
- Other end user organizations

37.5.1 Types of Consensus Standard Specifications

☐ Consensus standard specifications are of two types:

- Minimum criteria
- Classification

☐ Minimum criteria-based standards

- Set specific design features in terms of:
    - -- PPE configuration
    - -- Design tolerances (dimensions)
    - -- PPE sizing
- Specify certain test methods for each performance property and set a minimum or maximum requirement based on test results
- Qualification of PPE to minimum criteria-based standards is either as pass or fail
    - -- Failure to meet any one requirement may disqualify a product from meeting the standard
    - -- Products may exceed minimum requirements

☐ Classification standards

- Use PPE design features to type classify PPE types
- Specify certain test methods but classify product performance by different levels according to the test results
- May designate the minimum acceptable level of the PPE to be that with the lowest defined level of performance or the PPE item having specific design features

37.5.2 Other Types of Specifications

☐ Trade organization or labor union specifications

- Specifications may appear similar to consensus standards; these standards are sometimes called "model" specifications.
- These standards differ from consensus standards in that they do not usually have the broad review of industry and may be biased towards the specific concerns of the organization.
- These specifications can be modified to provide additional flexibility for meeting organizational or end user needs.

☐ Manufacturer specifications

- Many manufacturers prepare specifications for their own products which provide design and performance criteria that only their products can satisfy.
- These specifications may provide insight into specific products' performance; however, such standards are often biased.
- Manufacturer specifications can be modified to provide additional competition, if desired.

☐ End user specifications

- Some organizations with related applications may have developed specifications which can be examined or serve as a starting point for creating new standards.

37.5.3 Using or Modifying Existing PPE Standards

☐ In using existing PPE standards, the organization must check to determine if the selected standard applies to their application.

☐ Existing standards may be modified or supplemented to meet all the protection requirements of the user organization.

- Performance levels may be lowered or raised based on the risk assessment.

  NOTE: lowering the minimum performance defined in a consensus standard may expose the user organization to additional liability).

- Additional criteria for both design and performance may be added when the specification or standard does not cover all protection needs (see next section for establishing new criteria).

37.5.4 Setting Performance Criteria in Absence of Existing Standards

- ☐ When no guidance is available from PPE standards, or standards do not exist for the type of PPE to be specified, minimum criteria must be developed.
- ☐ It is best to avoid specifying a brand product or equivalent, unless a list of salient characteristics can not be established that allows the purchaser to determine if other products are truly equivalent.
- ☐ A recommended practice for setting criteria includes:
  - Selecting both acceptable and unacceptable products from experience or end user surveys
  - Making measurements of the relevant performance properties for both acceptable and unacceptable products
  - Determining the cut off value between acceptable and unacceptable products to establish the minimum acceptable performance
- ☐ In order for the recommended practice to work, the selected test methods for measuring relevant performance properties must discriminate product performance consistent with field observations.
- ☐ Another practice is to select an acceptable product and write the specification based on that product.
  - This practice is accomplished by measuring the performance properties of the selected product and setting minimum criteria based on that product, taking into account that some measurements may have variation in their results.
  - The disadvantage of this practice is that some perfectly acceptable products may be prevented from qualifying to the standard since unacceptable performance has not been established.

37.5.5 Organizing the Specification

- ☐ Product specifications should have the following sections:
  - Scope
  - Purpose
  - Definitions
  - Referenced standards

- Design criteria
- Performance criteria
- Test methods
- Documentation requirements
- Labelling and packaging requirements
- Criteria for acceptance

☐ The scope should specify the type of product covered by the specification and its intended use.

☐ The purpose should dictate how the specification will be used, i.e., to qualify products for use in a particular application for the organization.

☐ Definitions should be provided on any specific terms used throughout the specification whose meaning may not be understood, or terminology that might be specific to the type of PPE being specified.

☐ Referenced standards should list all documents cited within the specification, including:

- The designation of the standard (or document)
- Title
- Edition or date of publication
- Source, if not readily known.

☐ Design criteria should provide:

- The minimum configuration of PPE item
- Any design features which are considered mandatory, including:
  - -- Dimensional considerations
  - -- Interfaces with other clothing or equipment
  - -- Sizing

☐ Performance criteria should specify:

- Measurement of a specific property in accordance with a specific test property
- The minimum or maximum acceptable value of the results produced by the test method

☐ A section on test methods may be necessary to

- Provide addition details for performing the test not listed in the reference method, such as how to interpret results (e.g., using the absolute value or an average for a measurement); or
- Provide details on methods for which no standards exist

☐ Documentation requirements should specify that information provided by the manufacturer, which can include:

- Safety considerations
- Instructions for use
- Care and maintenance instructions
- Warranty information

☐ The labeling and packaging requirements should specify

- Information to be on the label
- The size of printing on the label
- The language of the label
- Where the label should be located
- How the item is to be packaged
- Labeling of the package

☐ The acceptance criteria should list those practices the organization will use to determine the acceptability of the product; examples include:

- Provision of technical data
- Demonstration of certification
- Testing by the receiving organization

37.6 STEP 5 - Evaluate Candidate PPE Products

☐ A critical step of the selection process is to ensure that candidate products meet the established specification.

☐ Conformance of products to the organization's specification may be accomplished by:

- Reviewing manufacturer product information and technical data
- Using a third party testing laboratory or certification organization to verify product design, performance, and documentation

- Conducting a field trial to determine that the candidate product will meet the organization's needs.

- ☐ Reviewing manufacturer product information and technical data or using a third party lab or certification organization will only determine if the submitted or candidate product meets the specification; these processes will determine if a product meeting the specifications will provide adequate performance for the organization.

- ☐ Field trials are the most effective means for determining whether a specified product meets the organization's needs; feedback from the field trial can also be used to modify a specification.

- ☐ Methods used to demonstrate acceptable products should be the same as those included as part of the acceptance criteria within the product specification.

37.6.1 Review of Manufacturer Product Information and Technical Data

- ☐ As a minimum, the end user organization should require that the PPE manufacturer supply all performance data to demonstrate compliance with the specification.
    - To enhance comparison of data from different manufacturers, the end user organization should establish a standard format for manufacturers to submit this information.
    - The end user organization may require that test data be supplied by independent testing laboratories; however, unless the product order is sufficiently large, this may be added as a line item cost by the manufacturer as part of the bid.

- ☐ Manufacturers should also be required to submit technical specifications for their products which are not of a proprietary nature.

- ☐ Manufacturers should include all care and use instructions and any marketing claims as part of their technical package for review.

37.6.2 Use of a Third Party Laboratory or Certification Organizations to Qualify Products

- ☐ Testing of PPE by third party or independent laboratories offers one means for obtaining an unbiased assessment of the product's conformance to the end user organization's specification.

- ☐ Third party testing laboratories can evaluate prospective products against a specification, including an evaluation of the product design, performance, and documentation.

- ☐ The testing laboratory should be required to prepare a report which indicates conformance of the product to each element of the specification in terms of any measured performance, and a pass or fail determination.

- ☐ Certification may be used by some organizations to ensure that PPE meets a specification.

- ☐ Certification is usually limited to recognized standards for which a certification program has been established or is required by the standard itself.

- ☐ Independent third party certification usually entails:

  - A certification organization which is exclusively in the business of testing and certification (certification organizations may also be accredited by ANSI or other accreditation organizations)

  - Using a mark to signify certification of the product that is placed on the product or product label

  - Qualification testing, and sometimes follow-on testing, of the product

  - Manufacturer on-site quality assurance audits

  - List of the product in a list of certified products

- ☐ Certification generally ensures a higher level of product conformance with a standard but is not a guarantee that the product will provide any better protection.

37.6.3 Conduct of Field Trials

- ☐ Field trials with products may also be conducted to determine if the specification results in acceptable PPE performance and that candidate PPE meets the minimum protection requirements of the user organization.

- ☐ Field trials or tests may be organized to evaluate:

  - Specific products as compared to products currently in use by the organization

  - A wide range of products for a specification application

- New technologies for PPE

☐ Field trials are well suited to evaluate PPE designs, features, and other attributes that may not easily be demonstrated by test data such as:

- Ease of use
- Restriction of movement
- Interface with other clothing or equipment

☐ Field trials must be properly devised to provide an unbiased and fair evaluation of products by using the following guidelines:

- A field test plan should be developed prior to instituting any field trials of PPE.
- The field trial should have definite objectives and state areas for comparing and rating performance.
- Whenever possible, the same group that evaluates one item of PPE should evaluate other competing items of PPE (otherwise, it is difficult to ensure that the PPE items are compared on the same basis).
- PPE that is of an experimental nature or involves unproven protective capabilities should not be used in actual operations where hazardous exposures occur.
- The field trial should have a definite starting point and ending point.
- Workers participating in field trials should complete aware of that they are using new equipment and are part of a field trial.
- Responsibility for evaluating different PPE should be assigned to a single individual or committee.

☐ The field test plan should entail the following elements:

- The objective(s) of the field trial
- Identification of the PPE items to be evaluated
- Identification of the individual or committee which is responsible for conducting the field trial
- Identification of those workers or groups who will be involved in the evaluation
- The range of activities to be performed as part of the field test, including any limitations as to how the PPE should not be used
- The starting and ending dates of the field trial

- The performance areas or factors that are to be considered in evaluating the PPE by the field test participants
- Methods for providing written feedback from participant evaluations
- Methods for comparing different PPE items, including items currently in use
- The disposition of the items evaluating following the field trial
- Restrictions on how information from the field study will be distributed outside the organization

38.7 STEP 6 - Establish a PPE Program

☐ Selected PPE must be incorporated into a new or existing PPE Program.

☐ Section 47 provides an overview for putting together a PPE program.

38.8 STEP 7 - Periodically Review PPE Selections

☐ In order to ensure that selected PPE continues to meet the organization's needs, part of the PPE program must include a periodic review of PPE decisions.

☐ This review should be conducted by the individual or committee responsible for the PPE decisions and should include:

- A review of injury data for workers
- Specific complaints about selected PPE
- Examination of new technology available in the marketplace

☐ No change in the incidence of injuries may indicate that PPE is not working or that appropriate work practices are not in place; a determination should be made as to whether the PPE is providing protection as intended.

☐ Specific complaints about PPE should be investigated to determine if the selected PPE poses a risk to the employee or fails to perform as expected; these include complaints for worker productivity and comfort.

☐ With continued development of PPE technology, new products are becoming available all the time; these new products should be examined to determine if protection or cost benefits can be achieved by their acceptance.

☐ If the responsible individual or committee determines that the selected PPE is adequate and performing properly, then this decision should be documented.

Section 38

Selecting Full-Body and Partial Body Garments

The purpose of this section is to provide a systematic approach for selecting full body and partial body garments in terms of the overall design, intended service life, design features, material or material systems, and consideration of hazards created by the garment.

38.1 Overview of Garment Selection Parameters

☐ The basic approach for selecting garments encompasses:

- Selecting the overall design
- Deciding on the intended service life
- Choosing design features related to the design
- Choosing the material or material system having performance features related to the hazards identified as priorities during the risk assessment or meeting the organization's basic requirements
- Considering potential hazards from the selected garment

☐ OSHA 29 CFR Subpart I offers no specific guidance for selection of protective garments and does not cite specific standards for compliance of protective garments.

☐ Background on full body garments is provided in Section 16 while partial body protective garments are covered in Section 17.

38.2 Select the Specific Garment Design

☐ Decide whether full body protection is needed or if partial body protection is needed:

- Full body protection is warranted when the natures of the hazards or use environment are severe or cannot be totally anticipated.
- Partial body protection is acceptable if the exposure of the wearer is limited to a specific area of the body (torso, arms, or legs) and the means of exposure can be anticipated.

☐ The selected garment should provide protection to those areas of the body that can be or are likely to be exposed.

☐ The garment design should provide uniform minimum protection for all areas covered by the body, or should be designed such that those areas of the garment offering protection are clear and distinct from those areas that do not offer protection or offer relatively limited protection.

38.2.1 Full Body Garments

☐ For full body exposures, select full body protective garments which cover the torso, arms, and legs, such as

- Full body suits
- Jacket and trouser combinations
- Jacket and overall combinations
- Coveralls

☐ Where the respiratory equipment is at risk of exposure, use encapsulating suits, which enclose both the wearer and their respiratory equipment; Examples of situations requiring encapsulating suits include:

- Potential exposure to hazardous chemical vapors or gases
- Potential exposure to extremely dangerous biological agents or toxins
- Potential exposure to very high radiant heat environments

☐ Where protection must be provided for the head, use:

- A separate hood, or
- Suits, jackets, or coveralls which have an attached hood

☐ Where face and eye protection is required, but not provided by other equipment, use a separate or integrated hood, which incorporates a visor or faceshield.

- The visor may need to provide similar performance as a separate face or eye protective device.

38.2.2 Partial Body Garments

☐ Where the head area is at risk of exposure, use:

- A hood which covered the wearer's head with or without an integrated faceshield or visor
- A bouffant or head cover for protecting the upper portion of the head (usually the hair area) only

☐ Where portions of the torso are at risk of exposure, use:

- An apron for frontal exposure risks
  - Aprons are typically used in laboratory and maintenance situations where the risk of exposure is very low
- A gown for frontal and side exposure risks and exposure risks to the arms and tops of the legs
  - Gowns in some industries may have zoned protection, i.e., the gown may not offer uniform protection over the area of the garment; instead the gown has reinforcement or extra layers in those areas for which exposure is most likely.
- A smock or lab/shop coat for exposure risks to the torso, arms, and tops of the legs
  - Smocks or lab/shop coats are typically used in laboratory and maintenance situations where the risk of exposure is very low but are intended to offer greater protection to the front of the body than aprons
- Vests for protection of the upper front (and sometimes back) torso
  - Vests are typically used as accessories for high visibility or for specific torso protection such as in the case of projectile (bullet) or cut (stab) protection.

☐ Where portions of the arms are at risk of exposure, use sleeve protectors

- Sleeve protectors are generally used for controlled operations where the hazards are well characterized and are often used in conjunction with gloves.

☐ Where portions of the legs are at risk of exposure, use:

- Chaps for full leg exposure risks (usually frontal, sometime frontal and back protection)
- Leggings for lower leg and ankle exposure risks
- Spats for ankle and upper foot exposure risks

38.3 Decide on Intended Service Life

☐ Before choosing design features and material systems, the intended service life of the PPE must be determined since this decision will affect both the garment design and materials.

☐ Intended service life should account for:

- Garment durability
- Ease of garment reservicing
- Life cycle cost

☐ Garment service lives are usually classified in three general categories

- Durable
- Limited use
- Single use

☐ If durable or limited use service lives are desired, it is important to determine how the product will be used and maintained and:

- Set criteria for garment and material performance based on simulated use and maintenance, or
- Set criteria at levels that allow for decreases of performance but maintain the performance above minimum acceptable levels.

☐ Disposable clothing may be warranted if:

- The life cycle cost indicates that garments cannot be properly maintained economically, or
- Garments cannot be properly decontaminated (there are no methods to ensure that the garment contamination has been removed), or
- The type of contamination does not warrant decontamination (e.g., highly hazardous chemicals, biological agents, or radioactive materials).

☐ As suggested in Section 35, durability properties will be based on how well the PPE retains its performance properties after use, care, and maintenance

(Section 35 also provides guidance for calculating product life cycle costs).

38.4 Choose Design Features

☐ Potential design features for consideration include:

- Seams
- Closures
- Interface areas
- Accessories

☐ A large variety of design features exist for specific types of garments, while other may be common to several types of garments.

☐ Decide on overall integrity which the garment should have, if any, in preventing intrusion of:

- Particulates
- Liquids
- Vapors

38.4.1 Common Design Features

☐ Choose type of seam construction based on needed integrity, strength, and durability:

- Sewn seams for general garment construction where garment integrity is not needed
- Bound seams for slightly increased level of integrity from particles and incidental splash of some liquids but not from penetration by liquids or vapors
- Heat sealed, taped, or other types of bonding seams which offer high levels of integrity from particulates, liquids and some vapors
- Seams which include a combination of taping or sealing on both sides for the highest level of integrity from penetration of particulates, liquids, or vapors.
- Double stitching, reinforced stitching, or increased bond strength for applications where high seam strength and durability are needed.

- The number of seams and their location are also important for fit, integrity, strength, and durability

☐ Choose type, size, and location of closure system based on needed integrity, strength, and durability:

- Snaps, zippers, and similar hardware (hooks and dees) provide positive closure of the garment but do not provide integrity from penetrating particulates, liquids, or vapor
- Buttons are not positive closures and may easily come undone
- Closure hardware materials and quality can affect durability
- Two track closures and some special rubber or plastic based zippers provide integrity from penetrating agents, but may be subject to stiffening during cold temperature applications
- Hook and loop tapes provides convenience but lack positive closure properties and may have limited durability
- Adhesive strips or tapes are generally provided for one-time use only
- Longer closures can facilities ease of donning and doffing but can make garment design difficult
- Closure location also affects donning and doffing ease but can also be related to garment integrity for protecting wearer from exposure:
  - -- Front closures make garments easier to don but may be more easily exposed to hazards.
  - -- Side or rear closures may be more difficult to don but are kept away from areas of the garment likely to be exposed.

☐ Choose methods of interfacing with other clothing and equipment; consider:

- Overlap of garments when in use
  - -- Overlap between the coat (or upper torso garment) and the trouser (lower torso garment) can be extremely important for protecting the wearer; the amount of overlap should account for all wearer body positions during use.
  - -- Overlap is also important in the hood, neck, face area; in applications where barrier protection is required, the hood should overlap the top of the garment such that hazardous substances will flow away from the neck and face opening.
- Interface areas with gloves

- -- The type of sleeve end and glove end together with the amount of overlap will affect how the wrist area remains covered and protected.
- Interface areas with footwear
  - -- The type of lower torso garment end and the footwear type and height together with the amount of overlap will affect how the ankle area remains covered and protected.
- Interface areas with respirators, head protection, eye and face protection, and partial body protective clothing
  - -- The interface in the head area may have to account for several pieces of clothing and equipment.

☐ Choose type and location of other garment components and accessories

- Pockets (**ΔWARNING: Unless protected, pockets can become a reservoir for liquids)**
- Hardware
- Visibility materials or markings
- Labels

38.4.2 Protective Hoods and Head Protective Garments

☐ Choose area of head coverage

☐ Choose general styles

- Pullover
- Front closure
- Integrated visor and faceshield

☐ Choose type of face opening

- Elastic
- Stretchable material
- Drawstrings

☐ Choose size of face opening

- Full face
- Eye slit

- Eye holes

☐ Choose length and amount of overlap (for upper torso garments)

☐ For head covers and bouffants, choose area of head coverage and type of opening

38.4.3 Protective Aprons, Gowns, Smocks, and Vests

☐ Choose area of front and back torso coverage (consider potential for side exposure in protective vests)

- Length
- Coverage in upper torso region
- Coverage in lower torso region

☐ Choose types of ties or other closures

☐ Choose type and location of pockets **(WARNING: Unless protected, pockets can become a reservoir for liquids)**

38.4.4 Sleeve Protectors

☐ Choose length

☐ Choose type of arm end opening

- Elastic
- Straps
- Hook and loop closure tape
- Other types of fasteners

☐ Choose type of hand end finish

- Open with elastic, straps, hook and loop closure tape, or other fasteners
- Bar-tack or thumb hole
- Bar-tack for all fingers

38.4.5 Protective Chaps, Leggings and Spats

☐ For chaps, choose:

- Area of leg coverage

- Type of waist attachment
  - -- Individual waist belt
  - -- Attachment to wearer's belt
- Type of leg attachment

☐ For leggings and spats, choose:

- Areas of leg, ankle, and foot coverage
- Type of closures

38.5 Choose Material or Material System

☐ In choosing materials, consider the different types of performance that are required:

- Use the hazard category or categories that apply as based on the hazard assessment.
- In some cases, all types of protection may not be available in a single product and choices must be made in establishing protection priorities.

☐ When stronger materials are desired, choose materials that show high tensile strength and bursting resistance.

☐ Take into account needed integrity of garment, when selecting material systems.

☐ Table 38-1 provides a matrix of materials versus hazards for aiding selection of material types.

38.5.1 Physical Protection

☐ For protection from flying debris, choose materials such as:

- Tightly woven fabrics
- Coated fabrics
- Other fabrics which does not permit particulates to enter under the conditions of use

☐ For protection from projectile and ballistic threats, materials must be capable of absorbing the energy of the projectile; therefore, multiple layers or energy absorbing materials must be used such as:

- Para-aramid fabrics

Table 38-1. Matrix of Garment Materials versus Performance Needs

HAZARD/ MATERIAL	Repellent fabrics	FR treated fabrics	Meta-aramids or other FR materials	Para-aramids or other energy abs. mat'ls	Rubber materials (unsupported)	Plastic materials (unsupported)	Microporous laminate materials	Plastic laminates (non-woven support)	Coated fabrics	Aluminized mat'ls	Leathers	Insulated materials (fabric/batts)	FR insulated materials (fabric/batts)	Lead-impregnated materials
Abrasion				X	X				X		X			
Ballistic				X										
Cut/Puncture				X	X				X		X			
Electric Shock					X				X					
Electric Arc		X	X											
Flame		X	X							X			X	
Gases					X	X		X	X					
Hazardous Liquids					X	X	X	X	X	X				
High Energy Radiation														X
High Heat		X	X							X			X	
Molten Mat'ls		X	X								X			
Particles	X				X	X	X	X	X	X				
Radiant Heat		X	X							X			X	
Temperature Extremes	heat						heat					cold	cold	
Wetness	X				X	X	X	X	X	X				

WARNING: Selected garment material must demonstrate needed performance in selected tests.

- Ultra high molecular weight, high density polyethylene fabrics
- Other high energy absorbing fabrics

☐ For protection from abrasive surfaces, additional layers of material can be used at areas of abrasion such as shoulders, elbows, and knees; in addition, high abrasion resistant materials can be used, such as:

- Leather
- Heavy coated fabrics
- Thick plastic or rubber materials
- Other abrasion-resistant materials

☐ For protection from sharp edges, select cut resistant fabrics such as:

- Metal reinforced fabrics
- Leather
- Para-aramid fabrics
- Ultra high molecular weight high density polyethylene fabrics
- Heavy coated fabrics
- Thick plastic or rubber materials
- Other cut-resistant materials

☐ For protection from pointed objects, select puncture-resistant fabrics such as:

- Leather
- Heavy coated fabrics
- Thick plastic or rubber materials
- Other puncture-resistant materials

☐ Existing standard specifications include:

- EN 381 (protective clothing for chainsaw protection)
- EN 412 (protective aprons, trousers, and vests for use with hand knives)
- EN 1082 (gloves and arm guards for protection against cuts and stabs by hand knives)
- ISO 11393 (protective clothing for chainsaw protection)
- ISO 13998 (aprons, trousers, and vests for protection against cuts and stabs by hand knives)

- ISO 13999 (gloves and arm guards for protection against cuts and stabs by hand knives)
- ISO 14876 (bullet and stab-protective vests)
- ISO 14877 (abrasive blasting protective clothing)

38.5.2 Environmental Protection

☐ For reducing the effects of high heat and humidity, choose fabrics that are breathable and allow for moisture vapor transport; avoid materials that have continuous films or thick coatings.

☐ For protection against ambient cold, choose materials which provide thermal insulation usually in the form of multiple layers, batting, or other material constructions which offer a high degree of loft; also new phase-change materials are available for cold protection.

☐ For protection from wetness, select fabrics which repel water or resist penetration by water, such as:

- Microporous film laminated to fabrics
- Coated fabrics
- Plastic film-based nonwoven fabrics
- Unsupported plastics or rubber material
- Other water repellent and water penetration resistant fabrics

☐ For protection from wind, select fabrics which limit air flow, such as:

- Multilayer fabrics
- Microporous film laminated to fabrics
- Coated fabrics
- Plastic film-based nonwoven fabrics
- Unsupported plastics or rubber materials
- Other wind limiting fabrics

☐ Existing standard specifications include:

- CAN/CGSB 65.16-M88 (marine abandonment immersion suits)
- CAN/CGSB 65.21-95 (marine anti-exposure work suit)
- EN 342 (protective clothing against cold)

- EN 343 (protective clothing against foul weather)
- EN 1913 (survival suits)
- ISO 8096 (rubber- or plastic-coated fabrics for water-resistant clothing)

38.5.3 Chemical Protection

☐ For protection against particulates, select garments with particulate-tight integrity materials that prevent particulate penetration such as:

- Fabrics with limited porosity (some forms of polyethylene and polypropylene)
- Microporous film laminated to fabrics
- Coated fabrics
- Plastic film-based nonwoven fabrics
- Unsupported plastics or rubber materials
- Other particulate penetration resistant materials

☐ For protecting the environment from wearer particulates, select garments with particle-tight integrity and materials that prevent particulate penetration such as those above; in addition, choose materials which are non-linting and which do not create static charge.

☐ For protection against liquids, select garments with liquid-tight integrity and materials that prevent particulate penetration such as:

- Microporous film laminated to fabrics
- Coated fabrics
- Plastic film-based nonwoven fabrics
- Unsupported plastics or rubber materials
- Other liquid penetration resistant materials

☐ For protection against vapors, select garments with gas-tight integrity and materials that prevent particulate penetration such as:

- Microporous film fabrics in combination with adsorbents
- Coated fabrics
- Plastic film-based nonwoven fabrics
- Unsupported plastics or rubber materials

- Other vapor penetration resistant or permeation resistant materials

☐ For protection against vapors which may create a flammable atmosphere, select materials which also provide heat and flame resistance or use overcovers which protective inner chemical protective layers; examples include:

- Aluminized flame resistant materials
- Intrinsically flame resistant materials (such as meta-aramids)

☐ For protection against possible explosions associated with chemicals, select supplemental materials that provide projectile/ballistic protection described in 38.5.1 above.

☐ Existing standard specifications include:

- ANSI/ISEA 103 (proposed standard for chemical protective clothing)
- EN 465 (spray-tight chemical protective clothing)
- EN 466 (liquid-tight chemical protective clothing)
- EN 466-2 (liquid-tight chemical protective clothing for emergency teams)
- EN 467 (partial body chemical protective clothing)
- EN 943 (gas-tight chemical protective clothing)
- EN 943-2 (gas-tight chemical protective clothing for emergency teams)
- EN 1511 (limited use liquid-tight chemical protective clothing)
- EN 1512 (limited use spray-tight chemical protective clothing)
- EN 1513 (limited use partial body chemical protective clothing)
- EN 13982 (particulate-tight chemical protective clothing)
- ISO 10078 (protective clothing materials intended for contact with liquid or solid chemicals)
- IES-RP-CC-003 (clean room garments)
- NFPA 1991 (vapor-protective suits for hazardous chemical emergencies)
- NFPA 1992 (liquid splash protective suits for hazardous chemical emergencies)
- NFPA 1993 (support function protective clothing for hazardous chemical operations)

38.5.4 Biological Protection

☐ For protection from blood-borne pathogens, select garments with liquid-tight integrity and materials that limit or prevent the passage of blood and other body fluids:

- Fabrics with repellency characteristics (these fabrics will not prevent blood or body fluid passage under many use conditions)
- Microporous film laminated to fabrics
- Coated fabrics
- Plastic film-based nonwoven fabrics
- Unsupported plastics or rubber materials
- Other fluid penetration resistant materials

☐ For protection from biological toxins and allergens, select materials that limit or prevent the biological particulates or fluids:

- Fabrics with repellency characteristics
- Microporous film laminated to fabrics
- Coated fabrics
- Plastic film-based nonwoven fabrics
- Unsupported plastics or rubber materials
- Other fluid or particulate penetration resistant materials

☐ For protection from insects, select clothing material treated with insect repellent (non-irritating to the wearer's skin) or separately apply repellents to clothing and wearer's skin.

☐ Existing standard specifications include:

- AAMI TIR 11-94 (guidance document only for surgical gowns and drapes)
- CAN/CGSB 38.16-92 (surgical gowns)
- NFPA 1999 (emergency medical garments)

38.5.5 Thermal Protection

☐ For protection from high heat, select multilayer materials systems which provide a large degree of thermal insulation and resist the degradation effects of high heat exposure; examples include:

- Aramid-based fabrics
- Intrinsically heat and flame resistant fabrics
- Flame-retardant treated fabrics
- Aluminized heat resistant fabrics (for radiant high heat)
- Non-woven batting composed of heat resistant materials
- New generation phase change materials
- Other heat resistant materials

☐ For protection from flame contact, select materials systems that resist ignition when contacted by flame or which readily extinguish once removed from the flame; examples include:

- Aramid-based fabrics
- Intrinsically heat and flame resistant fabrics
- Flame-retardant treated fabrics
- Aluminized flame resistant fabrics (for radiant high heat)
- Other flame resistant materials

☐ For protection from hot liquids and gases, select multilayer material systems which provide thermal insulation by virtue of several thermally insulation layers in combination with a barrier against liquid or vapor penetration, such as:

- High temperature resistant microporous films laminated to heat resistant fabrics
- High temperature resistant plastic or rubber materials laminated to heat resistant fabrics

☐ For protection from molten substances, select fabrics which are heat resistant and allow runoff of molten substances; examples include:

- Thermally resistant leather
- Intrinsically heat and flame resistant fabrics
- Flame-retardant treated fabrics

☐ Existing standard specifications include:

- ASTM F 1002 (molten metal and related hazard protective clothing)
- CAN/CGSB 155.1-M88 (fire fighter protective clothing)
- EN 469 (fire fighter protective clothing)

- EN 470 (protective clothing for welding and allied processes)
- EN 531 (industrial heat protective clothing)
- EN 1486 (specialized fire fighting protective clothing)
- ISO 11611 (protective clothing for welding and allied processes)
- ISO 11612 (industrial heat protective clothing except welding)
- ISO 11613 (fire fighter protective clothing)
- ISO 14460 (protective clothing for racing car drivers)
- ISO 15384 (wildland fire fighting protective clothing)
- ISO 15538 (specialized fire fighting protective clothing)
- NFPA 1951 (protective garments and hoods for technical rescue operations)
- NFPA 1971 (protective garments and hoods for structural fire fighting)
- NFPA 1975 (station/work uniforms for fire fighters)
- NFPA 1976 (protective garments for proximity fire fighting)
- NFPA 1977 (protective clothing for wildland fire fighting)
- SFI 3.2 (protective garments for racing car drivers)

38.5.6 Electrical Protection

☐ For protection from electrical shock, select electrically insulative materials meeting the requirements of OSHA 29 CFR 1910.137, such as:

- Thick heavy, non-conductive rubber materials
- Other electrically insulative materials

☐ For protection from electrical arc flashover, select multilayer material systems which provide a high degree of insulation to electrical arc flashover, such as:

- Aramid-based fabrics
- Intrinsically heat and flame resistant fabrics
- Flame-retardant treated fabrics
- Non-woven batting composed of heat resistant materials
- Other heat resistant materials

☐ For protection from static charge build-up, select materials that dissipate static charge such as:

- Fabrics which incorporate metallic elements
- Conductive fabrics
- Fabrics with topical antistatic treatments or treatments applied by the user which are compatible with the selected clothing and approved by the clothing manufacturer (**ΔWARNING:** ***Topically treated fabrics may quickly lose their antistatic properties in use***).

☐ Existing standard specifications include:

- ASTM D 1051 (rubber insulating sleeves)
- ASTM F 1506 (Wearing apparel for electrical workers momentarily exposed to electric arc discharge)

38.5.7 Radiation Protection

☐ For protection against radioactive particulates (alpha and beta particles), select garments which have particle-tight integrity and materials which prevent penetration of particulates, such as:

- Microporous film laminated to fabrics
- Coated fabrics
- Plastic film-based nonwoven fabrics
- Unsupported plastics or rubber materials
- Other particulate penetration resistant materials

☐ For protection against gamma and X-ray radiation, used shielding materials such as:

- Lead and other high density, energy absorbing materials

☐ For protection from non-ionizing radiation as ultraviolet light, select fabrics that have demonstrated UV blocking.

☐ For protection against other forms of non-ionizing radiation, select fabrics that demonstrate attenuation of the exposure radiation.

☐ Existing standard specification includes:

- ISO 8194 (protective clothing against radioactive contamination)

38.5.8 Enhanced Visibility

☐ For enhanced visibility of wearer, particularly for garments that may be used around vehicle traffic or in large outdoor areas, choose garments that incorporate high visibility materials on in recognizable patterns for identifying workers:

- Use fluorescent materials for improved daytime visibility
- Use retroreflective materials for improved nighttime visibility

☐ As a minimum, apply high visibility materials near cuffs of upper torso garments and in horizontal bands around chest or bottom hem; and bottom cuffs of lower torso garments.

☐ Worker daytime recognition requires a minimum of 0.5 square yards of material over exposed area of garment(s).

☐ Alternatively, use vests or similar separate garments that incorporate high visibility materials.

☐ Existing standard specifications include:

- ANSI/ISEA 107 (proposed standard on high visibility garments)
- EN 471 (high visibility protective clothing)
- NFPA 1971 (structural fire fighter protective clothing)

38.6 Consider Other Use Factors

☐ Are garments being using in areas where wearer daytime or nighttime visibility is important? If so, consider using high visibility colors or materials on garments.

☐ Is work being conducted next to water? If so, ensure that the worker also has personal floatation device.

☐ Is work being conducted on an elevated platform? If so, ensure that worker also has adequate fall protection.

38.7 Consider Potential Hazards from Selected Garments

☐ Do protective garments create particles in particle-sensitive environments?

☐ Do protective garments create static electricity in potentially flammable or static-sensitive environments?

- ☐ Do garment materials irritate or sensitize the wearer's skin?
- ☐ Are garments likely to retain contamination after cleaning?
- ☐ Do garments have designs with loose straps or material that can be caught in moving machinery?
- ☐ Do garments significantly limit wearer mobility or range of motion?
- ☐ Do garments have faceshields or visor which impair the wearer's vision?
- ☐ Do garments covering the head make communications difficult?
- ☐ Do garments create back support problems?
- ☐ Are garments available in sufficient sizes to provide adequate fit of organization personnel?
- ☐ Are garments difficult to use and reservice?
- ☐ Do garments cause heat stress for the wearer under use conditions?
- ☐ Do garments have sufficient durability for meeting the intended service life?

Section 39

Selecting Gloves and Other Handwear

The purpose of this section is to provide a systematic approach for selecting gloves and other handwear in terms of the overall design, intended service life, design features, material or material systems, and consideration of hazards created by the handwear.

39.1 Overview of Glove and Handwear Selection Parameters

☐ The basic approach for selecting gloves and handwear encompasses:

- Selecting the handwear design
- Choosing design features related to the design
- Choosing the material or material system having performance features related to the hazards identified as priorities during the risk assessment or meeting the organization's basic requirements
- Considering potential hazards from the selected gloves

☐ OSHA 29 CFR 1910.138 requires selection of handwear that provides appropriate protection against the following specific standards but does not specify compliance with a particular standard or set of standards:

- skin absorption of harmful substances
- severe cuts or lacerations
- severe abrasions
- punctures
- chemical burns
- thermal burns
- harmful temperature extremes

☐ Background information on gloves is provided in Section 18.

39.2 Select the Handwear Design

☐ When full hand and wrist and protection and full hand function are needed, select full five-fingered gloves.

- Use full five-fingered gloves for all applications requiring barrier protection against highly hazardous liquids or gases.

☐ When protective performance requirements dictate high levels of protection (particularly for thermal and ambient cold protection) and some hand function can be sacrificed, select two fingered gloves or mittens.

- Two-fingered gloves and mittens may be required when insulation levels cannot be achieved in a five-fingered glove style

☐ When hazards only affect the palm or portions of the hand, select partial gloves, fingerless gloves, or finger guards.

- Partial gloves, fingerless gloves, or finger guard are usually only acceptable for protection from certain physical hazards.
- Partial gloves or handwear are generally only viable for very specific applications where hazards are slight or moderate and last for very short durations and uncovered portions of the hand must be free for task dexterity or tactility.

☐ When short-term finger end protection is needed, select finger cots.

- Finger cots should only be used for low-level hazards in controlled environments where fingertip contact is the only type of exposure anticipated.

☐ When temporary hand protection is needed and hand function is not needed, hand pads may be selected.

- Hand pads are generally only used in hot work that is well characterized.

39.3 Choose Handwear Design Features and Materials

☐ Handwear design will usually be a function of the material and the intended application (or industry).

☐ Common types of gloves and handwear include:

- Disposable two dimensional plastic gloves

- Disposable lightweight rubber gloves and handwear
- Unsupported rubber gloves
- Fabric gloves and handwear
- Knit fabric glove and handwear
- Leather gloves and handwear
- Rubber or plastic coated fabric gloves

☐ Specialized types of gloves and handwear include:

- Insulated gloves and handwear
- Multilayer gloves and handwear

☐ For two-dimensional plastic gloves, choose:

- Type of material
  - Polyethylene
  - Teflon
  - Other plastic materials
- Few design options are available.
- This type glove is for disposable use only.

☐ For disposable lightweight rubber gloves and handwear, choose:

- Type of material
- Natural rubber
- Polyvinyl chloride
- Nitrile rubber
- Other rubber materials or combinations
- Type of cuff end
  - Straight (least durable)
  - Rolled (most durable)
  - Pinked or serrated
- Type of glove treatment
  - Non-powdered
  - Powdered (may improve donning and prevent gloves from sticking to hands)

- Type of grip finish
  - -- Smooth finish (no texture)
  - -- Embossed or raised texture (improved grip)
- This type glove is for disposable use only.

☐ For unsupported rubber gloves, choose:

- Type of material
  - -- Natural rubber
  - -- Neoprene
  - -- Nitrile rubber
  - -- Polyvinyl chloride
  - -- Polyvinyl alcohol
  - -- Butyl rubber
  - -- Viton or other fluoroelastomers
  - -- Polymer combinations and other rubber materials
- Type of cuff end
  - -- Straight (least durable)
  - -- Rolled (most durable)
  - -- Pinked or serrated
- Type of grip finish
  - -- Smooth finish (no texture)
  - -- Embossed or raised texture (improved grip)

☐ For fabric gloves and handwear, choose:

- Type of material
  - -- Cotton fabric
  - -- Canvas fabric
  - -- Para-aramid
  - -- Ultra high molecular weight, high density polyethylene
  - -- Liquid crystal polymer
  - -- Steel mesh
  - -- Other fabric materials

- Type of construction
  - -- Clute cut (no seams in the palm; seams between each finger extending down the back of the glove; more comfortable)
  - -- Gunn cut (all seams are along the periphery of the glove; more durable)
- Type of cuff end
  - -- Knit wrist (keeps glove in place and prevents debris from entering glove, enhances warmth, but makes removal difficult)
  - -- Safety cuffs (allows easy donning)
  - -- Slip on cuffs (easy donning, eliminates seam for increased wear)
  - -- Closure ends (uses straps at glove end to secure glove around wrist)

☐ For knit fabric gloves and handwear, choose:

- Type of material
  - -- Cotton
  - -- Para-aramid
  - -- Ultra high molecular weight, high density polyethylene
  - -- Other materials
- Type of cuff end
  - -- Knit wrist
  - -- Closure ends

☐ For leather gloves and handwear, choose:

- Type of construction
  - -- Clute cut (no seams in the palm; seams between each finger extending down the back of the glove; more comfortable)
  - -- Gunn cut (all seams are along the periphery of the glove; more durable)
- Type of cuff end
  - -- Safety cuffs (allows easy donning)
  - -- Closure ends (uses straps at glove end to secure glove around wrist)

☐ For rubber or plastic coated fabric gloves and handwear, choose:

- Type of fabric substrate
  - -- Light cotton fabric

-- Canvas fabric

-- Para-aramid

-- Ultra high molecular weight, high density polyethylene

-- Other fabric materials

- Type of coating materials

-- Natural rubber

-- Neoprene

-- Nitrile rubber

-- Polyvinyl chloride

-- Polymer combinations and other rubber materials

- Type of liner construction

-- Two-piece (limited fit)

-- Five piece (more formed fit)

- Type of cuff end

-- Knit wrist (keeps glove in place and prevents debris from entering glove, enhances warmth, but makes removal difficult)

-- Safety cuffs (allows easy donning)

-- Slip on cuffs (easy donning, eliminates seam for increased wear)

-- Gauntlets (provides additional protection to lower arm)

-- Closure ends (uses straps at glove end to secure glove around wrist)

☐ For specialty gloves, different material layers are used to provide a combination of properties or additional insulation; these gloves generally offer design features consistent with the type of outer materials used in glove construction.

☐ Table 39-1 provides a matrix of glove types and hazards for aiding in the selection of glove types for specific hazards.

39.3.1 Physical Protection

☐ For protection from abrasive surfaces, choose:

- Heavyweight unsupported rubber gloves
- Fabric gloves and handwear

Table 39-1. Matrix of Glove/Handwear Types versus Performance Needs

HAZARD/ STYLE	Disposable 2-dimensional plastic gloves	Disposable lightweight rubber gloves	Unsupported rubber gloves	Woven fabric gloves (e.g., cotton)	Knit fabric gloves (e.g., cotton)	Para-aramid/high energy abs. mat'l gloves	Leather gloves	Plastic/rubber coated fabric gloves	Multilayer gloves	Insulated gloves	FR insulated gloves	Lead-impregnated rubber gloves
Abrasion				X			X	X	X			
Bloodborne Pathogens	X	X	X					X				X
Cold										X		
Cut				X	X	X	X	X	X			
Electric Shock			X									
Flame						some					X	
Gases	some	some	X					X				X
Hazardous Liquids	X	X	X					X				X
High Energy Radiation												X
High Heat						some					X	
Molten Mat'ls							X				X	
Particles	X	X	X					X				X
Puncture			X	X			X	X				X
Vibration									X	X		
Wetness	X	X	X					X				X

WARNING: Selected glove/handwear must demonstrate needed performance in selected tests.

- Leather gloves and handwear
- Rubber or plastic coated fabric gloves

☐ For protection from sharp edges, select cut resistant fabrics such as:

- Heavyweight unsupported rubber gloves
- High performance fabric gloves and handwear
- High performance knit fabric glove and handwear
- Leather gloves and handwear
- Rubber or plastic coated fabric glove
- Other cut-resistant gloves

☐ For protection from pointed objects, select puncture-resistant fabrics such as:

- Heavyweight unsupported rubber gloves
- Leather gloves and handwear
- Rubber or plastic coated fabric glove
- Other puncture-resistant gloves

☐ For improved grip when handling wet or oily substances, select gloves which provide slip resistant finishes such as rough or textured surfaces or raised elastic beads or similar constructions.

☐ For protection from vibration, select gloves that include cushioning materials in multilayer construction.

☐ Existing standard specifications for physical protection include:

- ANSI/ISEA 105 (proposed standard for hand protection)
- EN 388 (protective gloves against mechanical risks)
- EN 420 (general requirements for protective gloves)
- EN 10819 (protective gloves against vibration)

39.3.2 Environmental Protection

☐ For reducing the effects of high heat and humidity, choose gloves that are breathable and allow for moisture vapor transport; avoid materials that have continuous films or thick coatings.

☐ For protection against ambient cold, choose specialty gloves that include multiple layers, batting, and other material constructions that offer a high degree of loft.

☐ For protection from wetness or wind, choose:

- Unsupported rubber gloves
- Rubber or plastic coated fabric glove
- Other styles of gloves which offer liquid-tight integrity and incorporate a barrier material which repel water or resist penetration by water, rubber or plastic films, coated layers, and microporous film laminated to fabrics

☐ Existing standard specifications include:

- EN 511 (protective gloves against cold)

39.3.3 Chemical Protection

☐ For protection against particulates, choose particulate-tight gloves that include:

- Disposable two dimensional plastic gloves
- Disposable lightweight rubber gloves and handwear
- Unsupported rubber gloves
- Fabric gloves and handwear
- Leather gloves and handwear
- Rubber or plastic coated fabric glove
- Other types of gloves which resist penetration by particulates

☐ For protection against liquids, choose liquid-tight gloves that include:

- Disposable two dimensional plastic gloves
- Disposable lightweight rubber gloves and handwear
- Unsupported rubber gloves
- Rubber or plastic coated fabric glove
- Other types of gloves which resist penetration or permeation by liquids

☐ For protection against vapors, choose gas-tight gloves that include:

- Disposable two dimensional plastic gloves
- Disposable lightweight rubber gloves and handwear

- Unsupported rubber gloves
- Rubber or plastic coated fabric glove
- Other types of gloves which resist penetration or permeation by vapors

☐ For protection against vapors which may create a flammable atmosphere, choose gloves which also provide heat and flame resistance or use overgloves such as aluminized or similar flame resistant materials.

☐ For protection against cryogenic or liquified gases, choose gloves which provide multilayer constructions with insulating materials but use shell and inner layer materials which are resistant to fracture or cracking at use temperatures.

☐ Existing standard specifications include:

- ANSI/ISEA 105 (proposed standard for hand protection)
- ASTM D 3772 (rubber finger cots)
- ASTM D 4679 (rubber household or beautician gloves)
- EN 374 (protective gloves against chemicals)
- IES-RP-CC-004 (clean room gloves and finger cots)
- NFPA 1991 (gloves for emergency vapor protection)
- NFPA 1992 (gloves for emergency liquid-splash protection)
- NFPA 1993 (protective gloves for hazardous chemical operations)

39.3.4 Biological Protection

☐ For protection from blood borne pathogens, choose:

- Disposable two dimensional plastic gloves
- Disposable lightweight rubber gloves and handwear
- Unsupported rubber gloves
- Rubber or plastic coated fabric glove
- Other gloves which resist penetration by blood-borne pathogens or gloves which incorporate a liquid barrier layer

☐ For protection from biological toxins and allergens, select materials that limit or prevent the biological particulates or fluids:

- Disposable two dimensional plastic gloves

- Disposable lightweight rubber gloves and handwear
- Unsupported rubber gloves
- Fabric gloves and handwear
- Knit fabric glove and handwear
- Leather gloves and handwear
- Rubber or plastic coated fabric gloves
- Other gloves which resist penetration by biological particulates or fluids

☐ For protection from animal bites and microorganisms, select heavy work gloves in combination with rubber gloves or heavy coated work gloves.

☐ Existing standard specifications include:

- ANSI/ISEA 105 (proposed standard for hand protection
- ASTM D 3577 (rubber surgical gloves)
- ASTM D 3578 (rubber examination gloves)
- ASTM D 3772 (rubber finger cots)
- ASTM D 4679 (rubber household or beautician gloves)
- ASTM D 5250 (PVC gloves for medical application)
- CAN/CGSB 20.25-M91 (sterile surgical gloves)
- CAN/CGSB 20.27-M91 (sterile/non-sterile medical examination gloves)
- EN 374 (protective gloves against microorganisms)
- ISO 10282 (single-use sterile surgical rubber gloves)
- ISO 11193 (single-use rubber examination gloves)
- NFPA 1999 (emergency medical gloves)

39.3.5 Thermal Protection

☐ For protection from high heat, select multilayer gloves which provide a large degree of thermal insulation and resist the degradation effects of high heat exposure; examples include:

- Fabric gloves and handwear
- Knit fabric glove and handwear
- Leather gloves and handwear

- ☐ For protection from flame contact, select glove systems that resist ignition when contacted by flame or which readily extinguish once removed from the flame; examples include:
  - Fabric gloves and handwear
  - Knit fabric glove and handwear
  - Leather gloves and handwear
- ☐ For protection from hot liquids and gases, select multilayer material systems that provide thermal insulation by virtue of several thermally insulative layers in combination with a barrier against liquid or vapor penetration.
- ☐ For protection from molten substances, select fabrics which are heat resistant and allow runoff of molten substances; examples include:
  - Fabric gloves and handwear
  - Leather gloves and handwear
- ☐ Existing standard specifications include:
  - EN 407 (protective gloves against thermal risks)
  - EN 659 (fire fighter protective gloves)
  - EN 12477 (protective gloves for welders)
  - ISO 15383 (fire fighting protective gloves)
  - NFPA 1951 (protective gloves for technical rescue operations)
  - NFPA 1971 (protective gloves for structural fire fighting)
  - NFPA 1976 (protective gloves for proximity fire fighting)
  - NFPA 1977 (protective gloves for wildland fire fighting)

39.3.6 Electrical Protection

- ☐ For protection from electrical shock, select unsupported rubber gloves meeting the requirements of ASTM D 120 and those specified in OSHA 29 CFR 1910.137.
- ☐ For protection from static charge build-up, select specialized gloves that dissipate static charge such as:
  - Unsupported gloves
  - Gloves which incorporate metallic elements

- Gloves constructed of conductive fabrics
- Gloves with topical antistatic treatments (**WARNING: Topically treated gloves may quickly lose their antistatic properties in use**).

☐ Existing standard specifications include:

- ASTM D 120 (rubber insulating gloves)
- ASTM F 696 (leather protectors for rubber insulating gloves and mittens)

39.3.7 Radiation Protection

☐ For protection against radioactive particulates (alpha and beta particles), select gloves which have particle-tight integrity and materials which prevent penetration of particulates, such as:

- Unsupported rubber gloves
- Rubber or plastic coated fabric gloves

☐ For protection against gamma and X-ray radiation, used shielding materials in conjunction with gloves made of material such as:

- Lead and other high density, energy absorbing materials impregnated into rubber.

☐ For protection from non-ionizing radiation as ultraviolet light, select gloves that have demonstrated UV blocking.

☐ For protection against other forms of non-ionizing radiation, select gloves that demonstrate attenuation of the exposure radiation.

39.3.8 Enhanced Visibility

☐ For enhanced visibility of wearer, particularly for gloves that may aid in hand signalling, choose gloves that incorporate high visibility materials on outer surface of gloves:

- Use fluorescent materials for improved daytime visibility
- Use retroreflective materials for improved nighttime visibility

39.4 Consider Potential Hazards from Selected Handwear

☐ Do protective gloves create particles in particle-sensitive environments?

- ☐ Do protective gloves create static electricity in potentially flammable or static-sensitive environments?
- ☐ Do glove materials irritate or sensitize the wearer's skin? (Many forms of natural rubber are known to cause skin allergic reactions)
- ☐ Are gloves likely to retain contamination after cleaning?
- ☐ Do gloves have a design with loose straps or material that can be caught in moving machinery?
- ☐ Do gloves significantly limit wearer hand function in terms of dexterity, tactility and grip?
- ☐ Are gloves available in sufficient sizes to provide adequate fit of organization personnel?
- ☐ Are gloves difficult to use and reservice?
- ☐ Are gloves uncomfortable for the wearer under use conditions?
- ☐ Do gloves have sufficient durability for meeting the intended service life?

Section 40

Selecting Footwear

The purpose of this section is to provide a systematic approach for selecting footwear in terms of the footwear type, design features, material or material systems, and consideration of hazards created by the footwear.

40.1 Overview of Footwear Selection Parameters

☐ The basic approach for selecting footwear encompasses:

- Selecting the footwear design
- Choosing design features and materials or material systems having performance related to the hazards identified as priorities during the risk assessment or meeting the organization's basic requirements
- Considering potential hazards from the selected footwear

☐ Most footwear with the exception of disposable shoe and boot covers is intended for reuse.

☐ Selection of many footwear types is governed by ANSI Z41-1991, American National Standard for Personal Protection on Protective Footwear.

☐ OSHA 29 CFR 1910.136 requires selection of footwear that complies with the relevant edition of ANSI Z41.

☐ Background information on footwear is provided in Section 19.

40.2 Select the Footwear Design

☐ ANSI Z41-1991 defines five types of protective footwear:

- Metatarsal footwear
- Conductive footwear

- Electrical hazard footwear
- Sole puncture resistant footwear
- Static dissipative footwear

☐ General footwear designs include:

- Shoes
- Boots
- Overshoes or overboots
- Shoe or boot covers
- Toe caps

☐ Select protective shoes when the primary concern is physical hazards to the foot from impact

☐ Select protective boots for most applications where several different hazards exist for the wearer's feet.

☐ Select overshoes or overboots to wear over shoes or boots for additional barrier protection; overshoes or overboots may also be used to supplement physical protection for the wearer's feet.

☐ Select shoe or boot covers only when a disposable shield for preventing contamination of the wearer's primary footwear is needed.

☐ Select toecaps only as a temporary measure for providing impact resistance to normal wearer footwear.

40.3 Select Footwear Design Features and Materials

☐ Footwear design will usually be a function of the material and the intended application (or industry).

☐ Footwear is usually constructed of:

- Leather
- Rubber
    - Neoprene
    - Polyvinyl chloride (PVC)
    - Polyurethane
    - Butyl rubber

  - -- Rubber blends
- Composite materials

☐ Design features are very similar between footwear materials.

☐ Choose footwear height:

- Shoe height
- Ankle height
- Calf height

☐ Choose footwear sole

- Plain (least slip resistance)
- Chevron (for improved slip resistance)
- Safety Loc (for slip resistance in rough environments)
- Cleated (for slip resistance in rough environments)

☐ For leather footwear and some rubber footwear, choose closure system:

- None
- Laces and eyelets
- Laces and stud hooks (better footwear integrity)
- Zippers (positive closure, less integrity)
- Straps and buckles (positive closure)
- Buckles (positive closure)

☐ Choose lining system depending on the application:

- Insulation from cold
- Insulation from high heat
- Barrier layers for preventing penetrations or other agents
- Additional layers for absorption of sweat and comfort

40.3.1 Physical Protection

☐ For impact resistance protection to the toe, choose footwear that conforms to the general requirements for footwear in ANSI Z41-1991.

- Alternatively, select toe caps for visitors or jobs where temporary toe impact protection is needed (selected toecaps should be properly size for wearer footwear to prevent tripping hazards).

☐ For impact resistance protection to the top of the foot, choose footwear that conforms to the requirements for metatarsal footwear in ANSI Z41-1991.

☐ For protection from puncture through the sole, choose footwear that meets the requirements for sole puncture resistant footwear in ANSI Z41-1991.

☐ For protection from slipping hazards, choose footwear that has cleated or other type of sole that will provide traction for wet or smooth surfaces.

- Alternatively, select outer footwear which provides sole treatments design to increase footwear traction (ensure that outer footwear properly fits primary footwear).

☐ Existing standard specifications include:

- ANSI Z41-1991 (personal protective footwear)
- EN 344 (safety, protective, and occupational footwear)
- EN 345 (safety footwear for professional use)
- EN 346 (protective footwear for professional use)
- EN 347 (occupational footwear for professional use)
- ISO 2023 (lined industrial rubber footwear)

40.3.2 Environmental Protection

☐ For protection from high humidity or high ambient heat, select footwear that is breathable such as leather footwear or footwear that include breathable liners such as microporous membranes.

☐ For protection from ambient cold, select footwear that includes multiple layers of insulation and includes barriers to keep the feet dry.

☐ For protection from wetness, select footwear which provides a barrier from water either as an:

- outer layer (rubber footwear), or
- with an internal water barrier (multilayer footwear)

☐ Existing standard specifications include:

- ISO 2252 (lined industrial rubber footwear for use at low temperatures)
- ISO 4643 (industrial polyvinyl chloride boots)
- ISO 5423 (lined/unlined industrial polyurethane boots)

40.3.3 Chemical Protection

☐ For protection from particulates, choose leather or rubber footwear which

- Is constructed of materials which resist particle penetration
- Has overall particle-tight integrity
- Alternatively, use footwear covers to protect existing footwear from contamination

☐ For protection from liquids, choose:

- Footwear which uses materials which resist liquid penetration and haveliquid-tight integrity
- Leather footwear that incorporates a barrier material
- Rubber footwear or footwear with a rubber outer layer
- Alternatively, use footwear covers to protect existing footwear from contamination

☐ For protection from vapors, choose rubber footwear that is integral to clothing which:

- Resists vapor penetration or permeation
- Has gas-tight integrity

☐ Existing standard specifications include:

- EN 346 (protective footwear for professional use)
- ISO 2025 (lined industrial rubber boots with general purpose oil resistance)
- ISO 4643 (industrial polyvinyl chloride boots)
- ISO 5423 (lined/unlined industrial polyurethane boots)
- ISO 6110 (lined/unlined industrial polyvinyl chloride boots with chemical resistance)
- ISO 6111 (lined/unlined industrial rubber boots with chemical resistance)

- ISO 6112 (lined/unlined industrial polyvinyl chloride boots with oil/fat resistance)
- NFPA 1991 (footwear for emergency vapor protection)
- NFPA 1992 (footwear for emergency liquid-splash protection)
- NFPA 1993 (protective footwear for hazardous chemical operations)

40.3.4 Biological Protection

☐ For protection from bloodborne pathogens, choose:

- Footwear which uses materials which resist liquid penetration and have liquid-tight integrity
- Leather footwear that incorporates a barrier material
- Rubber footwear or footwear with a rubber outer layer
- Alternatively, use footwear covers to protect existing footwear from contamination

☐ For protection from biological toxins and allergens, select materials that limit or prevent penetration by the biological particulates or fluids:

- Footwear which uses materials which resist liquid penetration and have liquid-tight integrity
- Leather footwear that incorporates a barrier material
- Rubber footwear or footwear with a rubber outer layer
- Alternatively, use footwear covers to protect existing footwear from contamination

☐ For protection from animal bites and microorganisms, choose:

- Footwear which uses materials which resist liquid penetration and have liquid-tight integrity
- Leather footwear that incorporates a barrier material
- Rubber footwear or footwear with a rubber outer layer

☐ Existing standard specifications include:

- EN 346 (protective footwear for professional use)
- ISO 4643 (industrial polyvinyl chloride boots)
- ISO 5423 (lined/unlined industrial polyurethane boots)

40.3.5 Thermal Protection

☐ For protection from high heat and flame, choose footwear which:

- Uses high heat and flame resistant outer materials (leather or certain rubbers)
- Has multiple heat resistant layers for insulation

☐ For protection from hot liquids, choose footwear which:

- Uses high heat and flame resistant outer materials (leather or certain rubbers)
- Has multiple heat resistant layers for insulation
- Incorporates a barrier layer which prevents the penetration of liquids
- Has liquid-tight integrity

☐ For protection from molten substances, choose footwear which:

- Uses high heat and flame resistant outer materials (leather or certain rubbers) that also permit runoff of the molten substance
- Has multiple heat resistant layers for insulation

☐ Existing standard specifications include:

- EN 345 (safety footwear for professional use)
- NFPA 1951 (protective footwear for technical rescue operations)
- NFPA 1971 (protective footwear for structural fire fighting)
- NFPA 1976 (protective footwear for proximity fire fighting)
- NFPA 1977 (protective footwear for wildland fire fighting)

40.3.6 Electrical Protection

☐ For protection from high voltage contact on surfaces that are substantially insulated, choose footwear that meets the requirements for electrical hazard footwear in ANSI Z41-1991.

☐ For protection from high voltage contact on conductive surfaces and for environments with the potential for static charge buildup, choose footwear which meets the requirements for conductive footwear in ANSI Z41-1991.

☐ For protection from hazards associated with static charge, choose footwear that meets the requirements for electrical hazard footwear in ANSI Z41-1991.

- ☐ Existing standard specifications include:
  - ANSI Z41-1991 (personal protective footwear)
  - ASTM F 1117 (dielectric overshoe footwear)
  - EN 345 (safety footwear for professional use)
  - ISO 2024 (conductive electrical lined footwear)
  - ISO 2025 (lined industrial rubber boots with general purpose oil resistance)
  - ISO 2251 (lined antistatic rubber footwear)
  - ISO 7232 (antistatic sandals, sabots, and clogs)

40.3.7 Radiation Protection

- ☐ For protection against radioactive particulates (alpha and beta particles), select rubber footwear which has particle-tight integrity and materials that prevents penetration of particulates.
- ☐ For protection against gamma and X-ray radiation, used shielding materials in conjunction with footwear such:
  - Lead and other high density, energy absorbing materials.
- ☐ For protection from non-ionizing radiation such as ultraviolet (UV) light, select footwear that have demonstrated UV blocking.
- ☐ For protection against other forms of non-ionizing radiation, select footwear that demonstrates attenuation of the exposure radiation.

40.4 Consider Potential Hazards from Selected Footwear

- ☐ Does the protective footwear create particles in particle-sensitive environments?
- ☐ Does the footwear materials irritate or sensitize the wearer's skin?
- ☐ Is the footwear likely to retain contamination after cleaning?
- ☐ Does the footwear have a design with loose straps, laces, or material that can be caught in moving machinery?
- ☐ Is the footwear available in a sufficient number of sizes or can the footwear be adjusted to fit organization personnel?

- ☐ Does the footwear provide adequate support to the ankle area?
- ☐ Is the footwear difficult to use and reservice?
- ☐ Is the footwear uncomfortable for the wearer under use conditions?
- ☐ Does the footwear have sufficient durability for meeting the intended service life?

Section 41

Selecting Headwear

The purpose of this section is to provide a systematic approach for selecting headwear in terms of the design type, design features, material or material systems, and consideration of hazards created by the headwear.

41.1 Overview of Headwear Selection Parameters

☐ The basic approach for selecting headwear encompasses:

- Selecting the type of headwear design
- Choosing design features and materials having performance related to the hazards identified as priorities during the risk assessment or meeting the organization's basic requirements
- Considering potential hazards from the selected headwear

☐ Most headwear, with the exception of disposable head covers, is intended to be reusable.

☐ Selection of many headwear types is governed by ANSI Z89.1-1997, *American National Standard for Industrial Head Protection*.

☐ OSHA 29 CFR 1910.135 requires selection of headwear that complies with the earlier editions of ANSI Z89.

☐ Background information on headwear is provided in Section 20.

41.2 Select the Headwear Type

☐ Headwear types include:

- Bump caps
- Helmets for impact and electrical protection

- Helmets for specialized protection
- Head covers
- Hoods

☐ ANSI Z89.1-1997 defines two impact types of protective headwear:

- Type I (protects from top impact)
- Type II (protects from off center impact)

☐ ANSI Z89.1-1997 defines three electrical classes of protective headwear:

- Class G - General (protects from low voltage exposures)
- Class E - Electrical (protects from high voltage exposures)
- Class C - Conductive (provides no electrical protection)

☐ Headwear type decisions will be based on the type of protection required.

41.2.1 Physical Protection

☐ For protecting the head from bumps or rough surfaces, but not impact choose a bump cap.

☐ For protecting the head from top impact, choose a helmet that meets the Type I requirements in ANSI Z89.1-1997.

☐ For protecting the head from off center, top impacts, choose a helmet which meets the Type II requirements in ANSI Z89.1-1997.

☐ For protecting the head from impact on the top and sides, choose a specialized helmet that exceeds impact requirements in ANSI Z89.1-1997.

41.2.2 Electrical Protection

☐ For protecting the head from low voltage conducts, choose a helmet which meets the Class G requirements in ANSI Z89.1-1997.

☐ For protecting the head from high voltage conducts, choose a helmet which meets the Class E requirements in ANSI Z89.1-1997.

41.2.3 Environmental, Chemical, Biological, Thermal, and Radiation Protection

☐ For protecting the head from environmental, chemical, biological, thermal, and radiation hazards, select hoods which conform to the same guidelines as those provided for garments (see Section 38).

☐ For protecting the head from thermal hazards, choose helmets are constructed of high heat and flame resistance materials.

☐ Choose headcovers to prevent falling or loose hair in clean or sensitive processes.

41.2.4 Enhanced Visibility

☐ For enhanced visibility of wearer, particularly in areas of high vehicular traffic, choose headwear that incorporates high visibility materials on outer visible parts of the headwear:

- Use fluorescent materials for improved daytime visibility
- Use retroreflective materials for improved nighttime visibility

41.2.5 Available Specifications

☐ Existing standard specifications include:

- ANSI Z89.1-1997 (industrial head protection)
- EN 443 (fire fighter protective helmets)
- ISO/R 1511 (protective helmets for road users)
- ISO 3873 (industrial safety helmets)
- NFPA 1951 (protective helmets for technical rescue operations)
- NFPA 1971 (protective helmets for structural fire fighting)
- NFPA 1976 (protective helmets for proximity fire fighting)
- NFPA 1977 (protective helmets for wildland fire fighting)

41.3 Select Headwear Design Features and Materials

☐ Choose type of helmet design

- With brim (Type I)
- Without brim (Type II)

☐ Choose shell material

- Plastic

- Fiberglass
- Aluminum (conductive)

☐ Choose type of suspension

- Self-adjusting suspension
- Ratchet type adjustment
- 4, 6, or 8 point suspensions
- Brow band

☐ Choose mountable accessories

- Faceshields
  -- Thickness
  -- Length
  -- Width
  -- Materials (polyester, polycarbonate, mesh)
  -- Brackets
- Sun shades
- Goggle clips
- Welding helmets
- Ear muffs
- Shroud
- Lamp brackets
- Visibility stripping or lettering
- Chin straps
- Winter liners

41.4 Consider Potential Hazards from Selected Headwear

☐ Does protective headwear create particles in particle-sensitive environments?

☐ Do headwear materials (especially those in the suspension, chinstrap, or brow pad) irritate or sensitize the wearer's skin?

☐ Is the headwear likely to retain contamination even after cleaning?

☐ Does the headwear have a design with loose straps or material that can be caught in moving machinery?

- ☐ Does the headwear interfere with the wearer's vision?
- ☐ Is the headwear available in a sufficient number of sizes or can the headwear be adjusted to fit organization personnel?
- ☐ Is the headwear difficult to use and reservice?
- ☐ Is the headwear uncomfortable for the wearer under use conditions?
- ☐ Does the headwear have sufficient durability for meeting the intended service life?

Section 42

Selecting Face and Eyewear

The purpose of this section is to provide a systematic approach for selecting face and eyewear in terms of the design type, design features, and consideration of hazards created by the face and eyewear.

42.1 Overview of Face and Eyewear Selection Parameters

- ☐ The basic approach for selecting face and eyewear encompasses: assessment will provide:
  - Selecting the type of face and eyewear design
  - Choosing design features and materials having performance related to the hazards identified as priorities during the risk assessment or meeting the organization's basic requirements
  - Considering potential hazards from the selected face and eyewear
- ☐ Most face and eyewear, with the exception of certain disposable faceshields, are intended to be reusable.
- ☐ Selection of most face and eyewear types is governed by ANSI Z87.1-1989, American National Standard Practice for Occupational and Educational Eye and Face Protection.
- ☐ OSHA 29 CFR 1910.133 requires selection of face and eyewear that complies with the relevant edition of ANSI Z87.1.
- ☐ Background information on face and eyewear is provided in Section 21.

42.2 Select the Face and Eyewear Type

- ☐ Face and eyewear types, defined by ANSI Z87.1-1989, include:

- Spectacles
- Faceshields
- Goggles
- Welding helmets
- Hand shields

☐ ANSI Z87.1-1989 differentiates between primary protectors and secondary protectors:

- Primary protectors can be worn alone or in conjunction with a secondary protector for protection against a specific hazard.
- Secondary protectors can only be worn in conjunction with a primary protector.

☐ ANSI Z87.1-1989 provides a very specific selection table that guides the selection of most types of face and eye protection. This selection chart is repeated as Table 42-1.

☐ The selection chart in Table 42-1 defines seventeen types of face and eye protective devices:

- Spectacle, no sideshield
- Spectacle, half sideshield
- Spectacle, full sideshield
- Spectacle, detachable sideshield
- Spectacle, non-removable lens
- Spectacle, lift front
- Cover goggles, no ventilation
- Cover goggles, indirect ventilation
- Cover goggles, direct ventilation
- Cup goggles, direct ventilation
- Cup goggles, indirect ventilation
- Spectacle, headband temple
- Cover welding goggle, indirect ventilation
- Faceshield
- Welding helmet, hand-held

Table 42-1. Selection Logic for Face and Eyewear

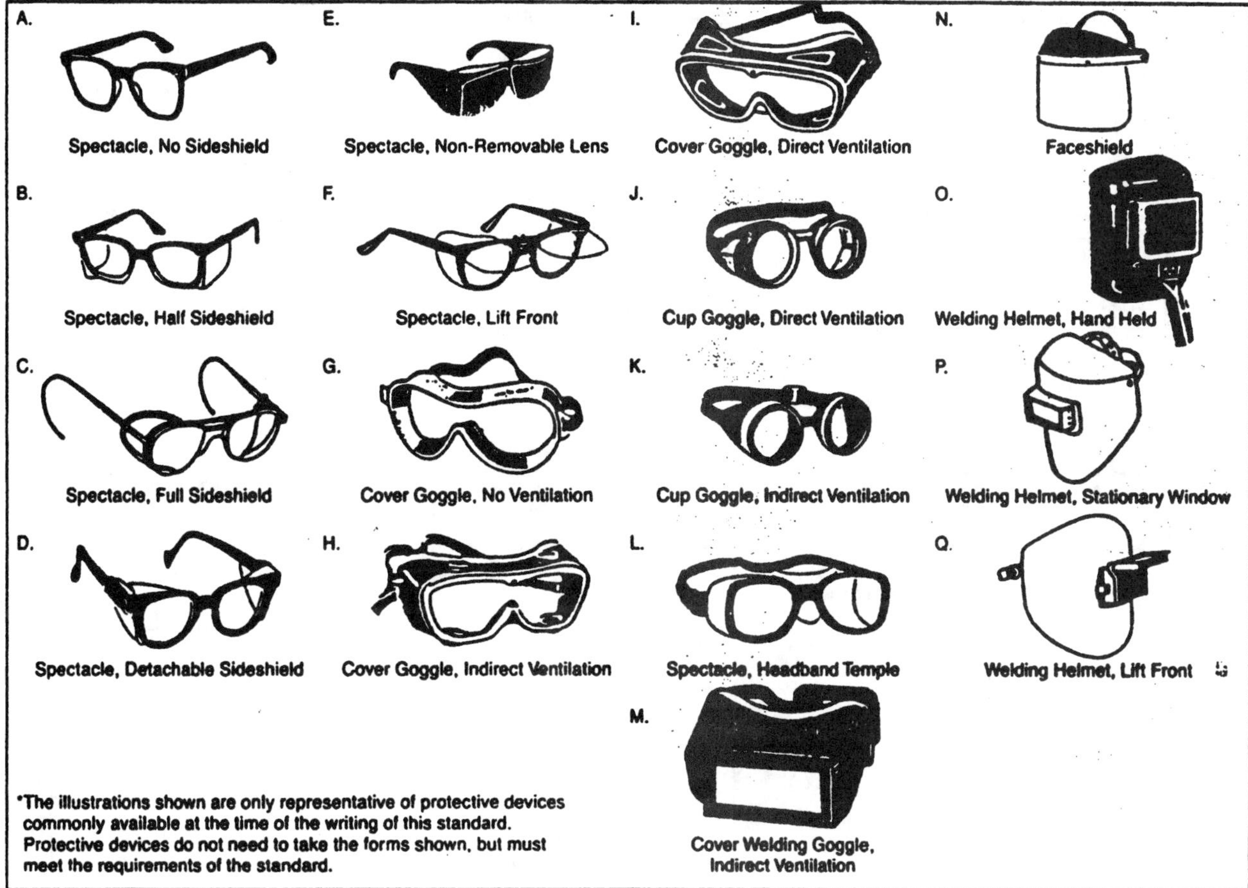

NOTES:

(1) Care shall be taken to recognize the possibility of multiple and simultaneous exposure to a variety of hazards. Adequate protection against the highest level of each of the hazards must be provided.

(2) Operations involving heat may also involve optical radiation. Protection from both hazards shall be provided.

(3) Faceshields shall only be worn over primary eye protection.

(4) Filter lenses shall meet the requirements for shade designations in Table 1.

(5) Persons whose vision requires the use of prescription (Rx) lenses shall wear either protective devices fitted with prescription (Rx) lenses or protective devices designed to be worn over regular prescription (Rx) eyewear.

(6) Wearers of contact lenses shall also be required to wear appropriate covering eye and face protection devices in a hazardous environment. It should be recognized that dusty and/or chemical environments may represent an additional hazard to contact lens wearers.

(7) Caution should be exercised in the use of metal frame protective devices in electrical hazard areas.

(8) Refer to Section 6.5, Special Purpose Lenses.

(9) Welding helmets or handshields shall be used only over primary eye protection.

(10) Non-sideshield spectacles are available for frontal protection only.

Table 42-1. (continued)

		ASSESSMENT SEE NOTE (1)	PROTECTOR TYPE	PROTECTORS	LIMITATIONS	NOT RECOMMENDED
IMPACT	Chipping, grinding, machining, masonry work, riveting, and sanding.	Flying fragments, objects, large chips, particles, sand, dirt, etc.	B,C,D, E,F,G, H,I,J, K,L,N	Spectacles, goggles faceshields SEE NOTES (1) (3) (5) (6) (10) For severe exposure add N	Protective devices do not provide unlimited protection. SEE NOTE (7)	Protectors that do not provide protection from side exposure. SEE NOTE (10) Filter or tinted lenses that restrict light transmittance, unless it is determined that a glare hazard exists. Refer to OPTICAL RADIATION.
HEAT	Furnace operations, pouring, casting, hot dipping, gas cutting, and welding.	Hot sparks	B,C,D, E,F,G, H,I,J, K,L,*N	Faceshields, goggles, spectacles *For severe exposure add N SEE NOTE (2) (3)	Spectacles, cup and cover type goggles do not provide unlimited facial protection. SEE NOTE (2)	Protectors that do not provide protection from side exposure.
		Splash from molten metals	*N	*Faceshields worn over goggles H,K SEE NOTE (2) (3)		
		High temperature exposure	N	Screen faceshields, Reflective faceshields. SEE NOTE (2) (3)	SEE NOTE (3)	
CHEMICAL	Acid and chemicals handling, degreasing, plating	Splash	G,H,K *N	Goggles, eyecup and cover types. *For severe exposure, add N	Ventilation should be adequate but well protected from splash entry	Spectacles, welding helmets, handshields
		Irritating mists	G	Special purpose goggles	SEE NOTE (3)	
DUST	Woodworking, buffing, general dusty conditions.	Nuisance dust	G,H,K	Goggles, eyecup and cover types	Atmospheric conditions and the restricted ventilation of the protector can cause lenses to fog. Frequent cleaning may be required.	
OPTICAL RADIATION	WELDING: Electric Arc		O,P,Q	TYPICAL FILTER LENS SHADE / PROTECTORS SEE NOTE (9) 10-14 Welding Helmets or Welding Shields	Protection from optical radiation is directly related to filter lens density. SEE NOTE (4). Select the darkest shade that allows adequate task performance.	Protectors that do not provide protection from optical radiation. SEE NOTE (4)
	WELDING: Gas		J,K,L, M,N,O, P,Q	SEE NOTE (9) 4-8 Welding Goggles or Welding Faceshield		
	CUTTING			3-6		
	TORCH BRAZING			3-4	SEE NOTE (3)	
	TORCH SOLDERING		B,C,D, E,F,N	1.5-3 Spectacles or Welding Faceshield		
	GLARE		A,B	Spectacle SEE NOTE (9) (10)	Shaded or Special Purpose lenses, as suitable. SEE NOTE (8)	

16

- Welding helmet, stationary window
- Welding helmet, lift-front

42.2.1 Physical Protection

☐ For protecting the eyes and/or face from impact, choose one of the following face and eye protective devices:

- Spectacle, half sideshield
- Spectacle, full sideshield
- Spectacle, detachable sideshield
- Spectacle, non-removable lens
- Spectacle, lift-front
- Cover goggles, no ventilation
- Cover goggles, indirect ventilation
- Cover goggles, direct ventilation
- Cup goggles, direct ventilation
- Cup goggles, indirect ventilation
- Spectacle, headband temple
- Faceshield (must be worn over other types of face and eyewear)

☐ In selecting between multiple types of face and eyewear, examine protection differences outlined in Section 42.2.7.

☐ Persons requiring prescription lenses should wear protective devices fitting with corrective lenses or protective devices which are designed to be worn over prescription eyewear

42.2.2 Environmental Protection

☐ For protecting the eyes from glare, choose one of the following eye protective devices:

- Spectacle, no sideshield
- Spectacle, half sideshield

☐ Special lenses can be used such as photochromic lenses which darken when exposed to sunlight.

42.2.3 Chemical Protection

- ☐ For protecting the eyes from dust or nuisance particulates, choose one of the following eye protective devices:
  - Cover goggles, no ventilation
  - Cover goggles, indirect ventilation
  - Cup goggles, indirect ventilation
- ☐ For protecting the eyes from irritating chemical mists, choose the following eye protective devices:
  - Cover goggles, no ventilation
- ☐ For protecting the eyes from chemical splashes, choose one of the following eye protective devices:
  - Cover goggles, no ventilation
  - Cover goggles, indirect ventilation
  - Cup goggles, indirect ventilation
  - Also use faceshield for severe exposures

42.2.4 Biological Protection

- ☐ For protecting the eyes and/or face from impact, choose one of the following face and eye protective devices:
  - Cover goggles, no ventilation
  - Cover goggles, indirect ventilation
  - Cup goggles, indirect ventilation
  - Also use faceshield for severe exposures

42.2.5 Thermal Protection

- ☐ For protecting the eyes and/or face from hot sparks, choose one of the following face and eye protective devices:
  - Spectacle, half sideshield
  - Spectacle, full sideshield
  - Spectacle, detachable sideshield

- Spectacle, non-removable lens
- Spectacle, lift-front
- Cover goggles, no ventilation
- Cover goggles, indirect ventilation
- Cover goggles, direct ventilation
- Cup goggles, direct ventilation
- Cup goggles, indirect ventilation
- Spectacle, headband temple
- Faceshield (must be worn over other types of face and eyewear)

☐ For protecting the eyes and/or face from splashes of molten metal, choose one of the following face and eye protective devices:

- Cover goggles, indirect ventilation with faceshield
- Cup goggles, indirect ventilation with faceshield

☐ For protecting the eyes and/or face from high temperature exposure, choose the following face and eye protective devices:

- Faceshields using screen or reflective materials

42.2.6 Radiation Protection

☐ For protecting the eye from different types of optical radiation, lenses or filters of specific shades will be required. Refer to Table 42-2 for information on transmittance requirements for clear and general purpose lenses and filters.

☐ For protecting the eyes from electrical arc welding radiation, choose one of the following face and eye protective devices:

- Welding helmet, hand-held
- Welding helmet, stationary window
- Welding helmet, lift-front
- Use filter lens shades 10 to 14

☐ For protecting the eyes from gas welding radiation, choose one of the following face and eye protective devices:

Table 42-2. Transmittance Requirements for Clear Lenses and General-Purpose Filters

Shade Number	Luminous Transmittance			Maximum Effective Far-Ultraviolet Average Transmittance %	Maximum Infrared Average Transmittance %
	Maximum %	Nominal %	Minmal %		
CLEAR	100	-	85	-	-
1.5	67	61.5	55	0.1	25
1.7	55	50.1	43	0.1	20
2.0	43	37.3	29	0.1	15
2.5	29	22.8	18.0	0.1	12
3.0	18.0	13.9	8.50	0.07	9.0
4	8.50	5.18	3.16	0.04	5.0
5	3.16	1.93	1.18	0.02	2.5
6	1.18	0.72	0.44	0.01	1.5
7	0.44	0.27	0.164	0.007	1.3
8	0.164	0.100	0.061	0.004	1.0
9	0.061	0.037	0.023	0.002	0.8
10	0.023	0.0139	0.0085	0.001	0.6
11	0.0085	0.0052	0.0032	0.0007	0.5
12	0.0032	0.0019	0.0012	0.0004	0.5
13	0.0012	0.00072	0.00044	0.0002	0.4
14	0.00044	0.00027	0.00016	0.0001	0.3

NOTES: (1) The near-ultraviolet average transmittance shall be less than one-tenth of the lumnous transmittance.

(2) The blue-light transmittance shall be less thant he luminous transmitance.

- Cup goggles, direct ventilation used with faceshield, hand-held welding helmet, stationary window welding helmet, lift-front welding helmet, or hand shield
- Cup goggles, indirect ventilation used with faceshield, hand-held welding helmet, stationary window welding helmet, lift-front welding helmet, or hand shield
- Spectacle, headband temple used with faceshield, hand-held welding helmet, stationary window welding helmet, lift-front welding helmet, or hand shield
- Cover welding goggle, indirect ventilation
- Use filter lens shades 4 to 8

☐ For protecting the eyes from cutting operation radiation, choose one of the following face and eye protective devices:

- Cup goggles, direct ventilation used with faceshield, hand-held welding helmet, stationary window welding helmet, lift-front welding helmet, or hand shield
- Cup goggles, indirect ventilation used with faceshield, hand-held welding helmet, stationary window welding helmet, lift-front welding helmet, or hand shield
- Spectacle, headband temple used with faceshield, hand-held welding helmet, stationary window welding helmet, lift-front welding helmet, or hand shield
- Cover welding goggle, indirect ventilation
- Use filter lens shades 3 to 6

☐ For protecting the eyes from torch brazing operation radiation, choose one of the following face and eye protective devices:

- Cup goggles, direct ventilation used with faceshield, hand-held welding helmet, stationary window welding helmet, lift-front welding helmet, or hand shield
- Cup goggles, indirect ventilation used with faceshield, hand-held welding helmet, stationary window welding helmet, lift-front welding helmet, or hand shield
- Spectacle, headband temple used with faceshield, hand-held welding helmet, stationary window welding helmet, lift-front welding helmet, or hand shield

- Cover welding goggle, indirect ventilation
- Use filter lens shades 3 to 4

☐ For protecting the eyes from torch soldering operation radiation, choose one of the following face and eye protective devices:

- Spectacle, half sideshield
- Spectacle, full sideshield
- Spectacle, detachable sideshield
- Spectacle, non-removable lens
- Spectacle, lift-front
- Faceshield (must be worn over other types of face and eyewear)
- Use filter lens shades 3 to 4

☐ For protecting the eyes from laser radiation, choose spectacles, with sideshields, or goggles that use lenses which demonstrate absorption of laser radiation in the specific laser wavelengths encountered.

☐ For protecting the eyes from X-ray radiation, choose spectacles, with sideshields, or goggles that use lead glass lenses that demonstrate absorption of X-rays under use conditions.

42.2.7 Selecting between Multiple Types of Face and Eyewear

☐ In choosing between different acceptable types of spectacles:

- Choose full sideshield spectacles over half sideshield spectacles for greater eye protection
- Choose detachable sideshields for convenience of using ordinary spectacles that use prescription lenses as routine eyewear in non-hazardous applications.
- Choose non-removable lens type spectacles when no modification of spectacles is desired and spectacles are used for specific applications.
- Choose lift-front spectacles for allowing use of special purpose filters when needed.

☐ In choosing between different acceptable types of goggles:

- Choose direct ventilation for greater eye comfort and minimization of fogging (direct ventilation goggles offer less protection than indirect or no ventilation goggles).
- Choose indirect ventilation for less eye comfort, minimization of fogging, but better protection than direct ventilation goggles (unventilated goggles offer more protection than indirect ventilation goggles).
- Choose cover goggles over eyecup goggles to better accommodate different facial sizes and provide for easier use of corrective lenses.

☐ In choosing between different acceptable types of spectacles and goggles, choose goggles when greater eye protection and coverage is required.

☐ In choosing between different types of welding helmets:

- Choose lift-front welding helmets for greatest flexibility for use in applications where repeated welding operations will be used and the wearer must use hands and switch between hazardous and non-hazardous conditions to perform work.
- Choose stationary window welding helmets when the worker performs long term welding and can safety remove welding helmet to perform non-hazardous operations.
- Choose hand-held welding helmets only when temporary, short-term welding operations are performed.

42.2.8 Available Specifications

☐ Existing standard specifications include:

- ANSI Z87.1-1989 (occupational and education eye/face protection)
- ANSI Z80.1-1987 (prescription ophthalmic lenses)
- ISO 4849 (eye-protector specifications)
- ISO 4850 (eye-protectors for welding and related techniques)
- ISO 4851 (ultraviolet filters)
- ISO 4852 (infrared filters)
- ISO 6161 (eye-protectors and filters against laser radiation)

42.3 Select Face and Eyewear Design Features and Materials

☐ Choose design features and materials based on face and eyewear type.

42.3.1 Spectacles

☐ Frame materials

- Plastic (lightweight)
- Metal (may pose electrical hazard)

☐ Choose temple type:

- Spatula (angle around ears)
  -- Fixed (cannot be adjusted for fit)
  -- Adjustable
- Cable (curve around ears)
  -- Fixed (cannot be adjusted for fit)
  -- Adjustable
- Headband (uses a strap to secure eyewear to wearer's head)

☐ Choose front type:

- Removable lenses (allow replacement with different shaded lens or prescription lenses)
- Non-removable lenses (must be used for specific application)

☐ Choose lens type:

- Plano
  -- Clear
  -- Filter (see Table 42-1)
  -- Tinted
- Prescription
  -- Clear
  -- Filter (see Table 42-1)
  -- Tinted

☐ Choose lens material and treatments

- Glass
- Plastic (lightweight)
- Scratch-resistant treatments

- Heat-resistant treatments

☐ Choose bridge type:

- Fixed bridge type (requires fitting frame size to specific wearer nose)
- Adjustable nose pad (allows adjustment to custom fit wearer nose)
- Adjustable bridge (allows adjustment to custom fit wearer nose)

☐ Choose sideshields:

- Flatfold or semi-sideshield
- Full sideshields (more protection)
- Detachable (allows removal for non-hazardous applications; may be lost)

42.3.2 Goggles

☐ Choose protective area:

- Cover (greater protection to eyes; easier to fit)
- Cup

☐ Choose type of frame:

- Rigid (cushion fitting)
- Flexible

☐ Choose type of ventilation:

- None
- Indirect
- Direct

☐ Choose type of lenses:

- Clear
- Filter (see Table 42-1)
- Tinted

☐ Choose antifogging design:

- None
- Sealed double lenses
- Chemical bonded lenses

42.3.3 Faceshields

- ☐ Choose faceshield material:
  - Heavy
  - Plastic
  - Plastic (heat-resistant)
  - Plastic (chemical-resistant)
  - Fiberglass
  - Wire mesh or screen
- ☐ Choose faceshield material thickness:
  - Glass must be minimum of 3 mm thick
  - Plastic must be minimum of 1 mm thick
- ☐ Choose lens treatment:
  - Clear
  - Filter (see Table 42-1)
  - Tinted
  - Antifog
- ☐ Choose size of faceshield (length and area of coverage)
- ☐ Choose type of headgear
  - Band attachment around sides and back of head
    - -- Adjustable
    - -- Elastic
  - Head suspension
  - Multipoint suspension
  - Ratchet-like adjustment
  - Helmet attachment
  - Hood attachment
- ☐ Choose accessories
  - Chin strap
  - Ear muffs

42.3.4 Welding Helmets and Handshields

☐ Choose general type

- Stationary lens
- Lift-front helmet
- Handshield

☐ Choose helmet material

- Aluminum
- Plastic
- Fiberglass
- Vulcanized fiber

☐ Choose type of lens or window

- Single fixed
- Second interior lens
- Magnifying
- Auto-darkening

☐ Choose size of window

☐ Choose accessories

- Hood attachment
- Bib or apron attachment
- High temperature helmet covers

42.4 Consider Potential Hazards from Selected Face and Eyewear

☐ Do face and eyewear materials (especially those in the nose pads, headbands or chin straps) irritate or sensitize the wearer's skin?

☐ Is face and eyewear likely to retain contamination even after cleaning?

☐ Does the face and eyewear have a design with loose bands, straps, or material which can be caught in moving machinery?

☐ Does face and eyewear provide clear and unobstructed vision for performing required tasks?

☐ Is the face and eyewear available in a sufficient number of sizes or the face and eyewear can be adjusted to fit organization personnel?

- ☐ Is face and eyewear difficult to use and reservice?
- ☐ Is face and eyewear uncomfortable for the wearer under use conditions?
- ☐ Does face and eyewear have sufficient durability for meeting the intended service life?

Section 43

Selecting Respirators

The purpose of this section is to provide a systematic approach for selecting respirators in terms of the type information required, respirator types, design features, and consideration of hazards created by the respirators.

43.1 Overview of Respirator Selection Parameters

☐ The basic approach for selecting respirators encompasses:

- Gathering the information necessary for proper selection of a respirator for the given situation
- Using a decision logic for selecting the respirator type
- Choosing design features having performance related to the hazards identified as priorities during the risk assessment or meeting the organization's basic requirements
- Considering potential hazards from the selected respirators

☐ In the United States, respirators must be certified by the National Institute for Occupational Safety and Health (NIOSH) to the respective requirements in 42 CFR Part 84.

☐ OSHA 29 CFR 1910.134 (January 8, 1998) specifies selection of general respirator types. These regulations update previous selection practices specified by the regulation from ANSI Z88.2-1992, *American National Standard for Respiratory Protection*.

☐ OSHA 29 CFR Part 1910 provides for specific selection of respirators for protection against the following substances:

- Asbestos (OSHA 29 CFR 1910.1001)
- 4-Nitrobiphenyl (OSHA 29 CFR 1910.1003)

- alpha-Napthylamine (OSHA 29 CFR 1910.1004)
- Methyl chloromethyl ether (OSHA 29 CFR 1910.1005)
- 3,3'-Dichlorobenzidine (OSHA 29 CFR 1910.1006)
- bis-Chloromethyl ether (OSHA 29 CFR 1910.1007)
- beta-Napthylamine (OSHA 29 CFR 1910.1008)
- Benzidine (OSHA 29 CFR 1910.1010)
- Aminodiphenyl (OSHA 29 CFR 1910.1011)
- Ethyleneimine (OSHA 29 CFR 1910.1012)
- beta-Propiolactone (OSHA 29 CFR 1910.1013)
- 2-Acetylaminofluorene (OSHA 29 CFR 1910.1014)
- 4-Dimethylaminoazobenzene (OSHA 29 CFR 1910.1015)
- N-Nitrosodimethylamine (OSHA 29 CFR 1910.1016)
- Vinyl chloride (OSHA 29 CFR 1910.1017)
- Inorganic arsenic (OSHA 29 CFR 1910.1018)
- Lead (OSHA 29 CFR 1910.1025)
- Cadmium (OSHA 29 CFR 1910.1027)
- Benzene (OSHA 29 CFR 1910.1028)
- Coke oven emissions (OSHA 29 CFR 1910.1029)
- Cotton dust (OSHA 29 CFR 1910.1043)
- 1,2-dibromo-3-chloropropane (OSHA 29 CFR 1910.1044)
- Acrylonitrile (OSHA 29 CFR 1910.1045)
- Ethylene oxide (OSHA 29 CFR 1910.1047)
- Formaldehyde (OSHA 29 CFR 1910.1048)
- Methylenedianiline (OSHA 29 CFR 1910.1050)

☐ The National Institute for Occupational Safety and Health (NIOSH) also has "recommended practices" documents for several chemicals, which provide specific respirator selection guidelines.

☐ NIOSH also has developed specific guidelines for respirator selection contained in *Guide to Industrial Respiratory Protection* (DHHS/NIOSH Publication No. 87-116, 1987).

☐ Background information on respirators is provided in Section 22.

43.2 Conduct General Respiratory Risk Assessment to Select General Types of Respirator

☐ Figure 13 provides a decision logic for selecting respirators which is based on the requirements in OSHA 29 CFR 1910.134 (January 8, 1998).

☐ Information needed to conduct the specific respiratory hazard assessment includes:

- Identification of atmospheric contaminant(s)
- Determination of specific regulations or guidelines that may be available for identified contaminant(s)
- Measurement of concentration for specific contaminant(s)
- Determination of immediately dangerous to life and health (IDLH) concentrations for contaminant(s)
    - IDLH means an atmosphere that poses an immediate threat to life, would cause irreversible adverse health effects, or would impair an individual's ability to escape from a dangerous atmosphere.
- Measurement of oxygen concentration in atmosphere
    - Oxygen deficiency exists when an atmosphere has an oxygen content below 19.5% by volume.
- Determination if respirator use is for work or escape
- Determination of chemical and physical state of contaminant(s)
    - Gas or vapor
    - Particulates (aerosols, mists, fumes)

☐ The general types of respirators specified from this respiratory hazard assessment include:

- Full facepiece pressure demand self-contained breathing apparatus
- Combination full facepiece pressure-demand supplied-air respirator with auxiliary self-contained air supply
- Respirator NIOSH-certified for escape from the atmosphere in which it will be used
- Atmosphere-supplying respirator

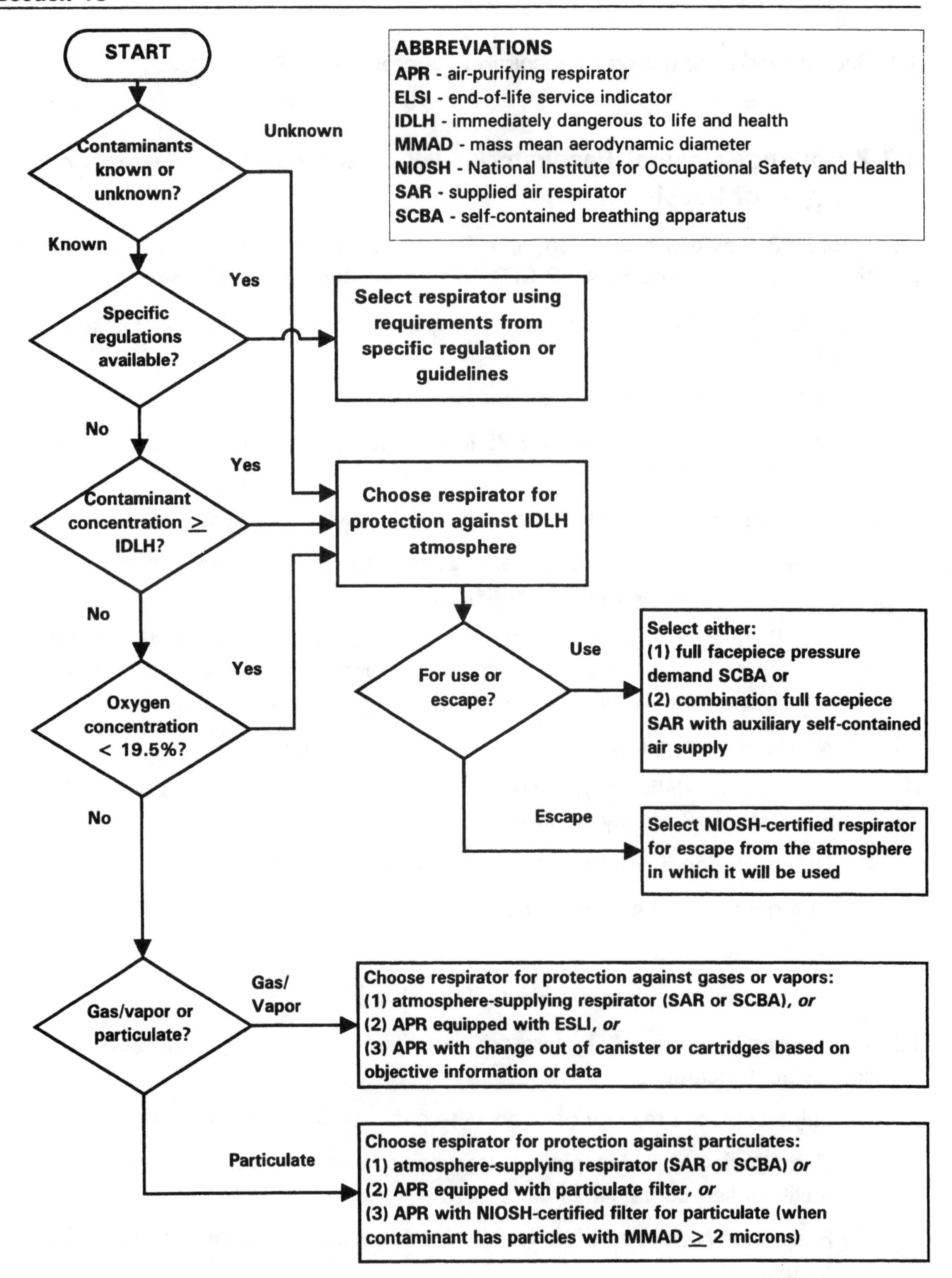

Figure 43-1. Respirator Decision Logic

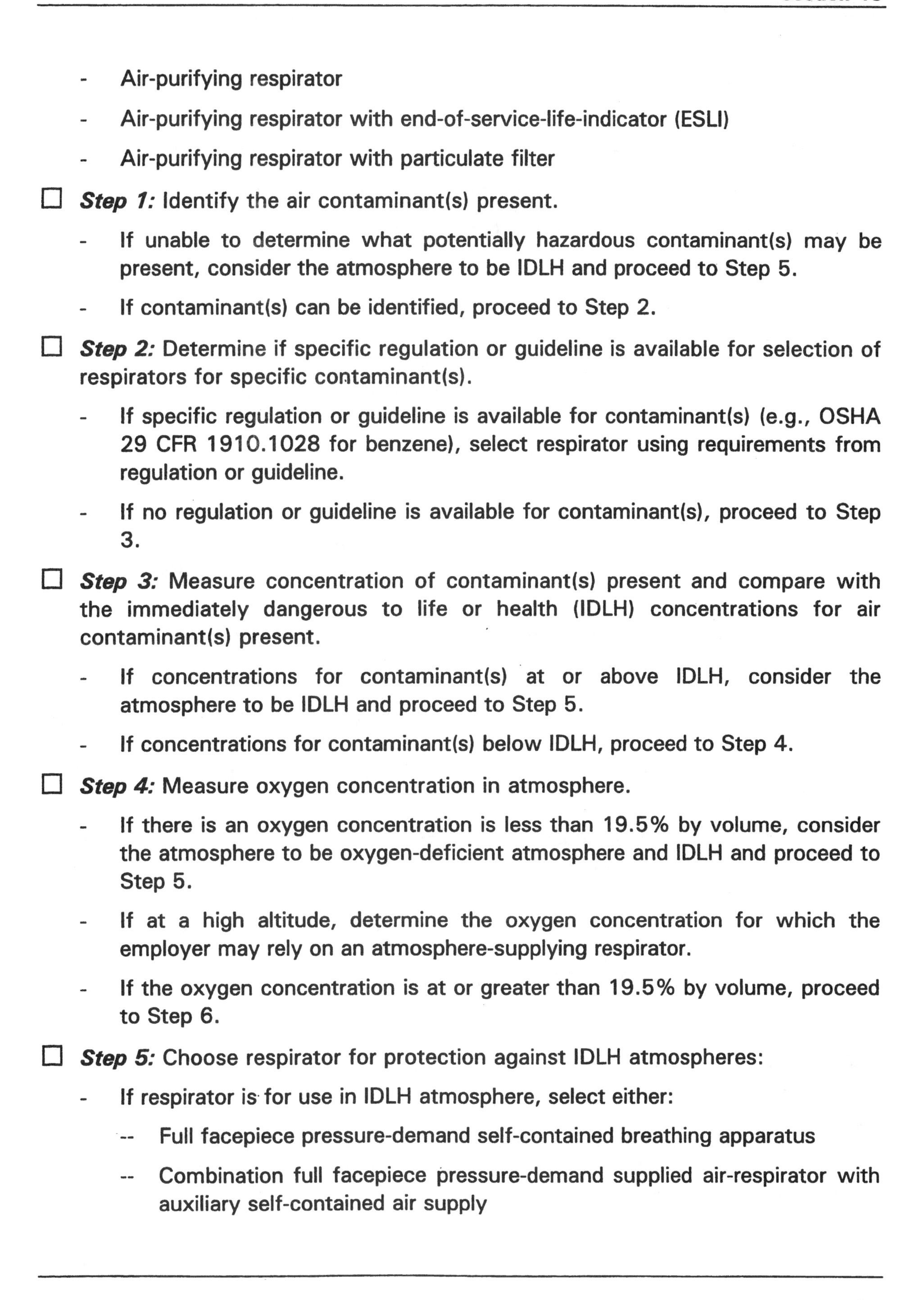

- Air-purifying respirator
- Air-purifying respirator with end-of-service-life-indicator (ESLI)
- Air-purifying respirator with particulate filter

☐ ***Step 1:*** Identify the air contaminant(s) present.

- If unable to determine what potentially hazardous contaminant(s) may be present, consider the atmosphere to be IDLH and proceed to Step 5.
- If contaminant(s) can be identified, proceed to Step 2.

☐ ***Step 2:*** Determine if specific regulation or guideline is available for selection of respirators for specific contaminant(s).

- If specific regulation or guideline is available for contaminant(s) (e.g., OSHA 29 CFR 1910.1028 for benzene), select respirator using requirements from regulation or guideline.
- If no regulation or guideline is available for contaminant(s), proceed to Step 3.

☐ ***Step 3:*** Measure concentration of contaminant(s) present and compare with the immediately dangerous to life or health (IDLH) concentrations for air contaminant(s) present.

- If concentrations for contaminant(s) at or above IDLH, consider the atmosphere to be IDLH and proceed to Step 5.
- If concentrations for contaminant(s) below IDLH, proceed to Step 4.

☐ ***Step 4:*** Measure oxygen concentration in atmosphere.

- If there is an oxygen concentration is less than 19.5% by volume, consider the atmosphere to be oxygen-deficient atmosphere and IDLH and proceed to Step 5.
- If at a high altitude, determine the oxygen concentration for which the employer may rely on an atmosphere-supplying respirator.
- If the oxygen concentration is at or greater than 19.5% by volume, proceed to Step 6.

☐ ***Step 5:*** Choose respirator for protection against IDLH atmospheres:

- If respirator is for use in IDLH atmosphere, select either:
  - -- Full facepiece pressure-demand self-contained breathing apparatus
  - -- Combination full facepiece pressure-demand supplied air-respirator with auxiliary self-contained air supply

- If respirator is for escape from IDLH atmosphere, select a respirator that is NIOSH-certified for escape from the atmosphere in which it will be used.

☐ ***Step 6:*** Determine the chemical state and physical form of the contaminant

- If contaminant(s) are gases or vapors, proceed to Step 7.
- If contaminant(s) are particulates, proceed to Step 8.

☐ ***Step 7:*** Choose a respirator for protection against gases and vapors:

- Select an atmosphere-supplying respirator; or
- Select an air-purifying respirator that:
  - -- Is equipped with an end-of-service-life indicator (ESLI) certified by NIOSH for the contaminant(s); or
  - -- If there is no ELSI appropriate for the conditions in the employer's workplace, use an air-purifying respirator for which the employer implements a change schedule for canisters and cartridges based on objective information or data that ensures that canisters and cartridges are changed before the end of their service life [must be part of the employer respirator program].

☐ ***Step 8:*** Choose a respirator for protection against particulates:

- Select an atmosphere-supplying respirator; or
- Select an air-purifying respirator equipped with a filter certified by NIOSH under 30 CFR as a high efficiency particulate air (HEPA) filter or an air-purifying respirator equipped with a filter certified for particulates by NIOSH under 42 CFR Part 84; or
- For contaminants consisting primarily of particles with mass median aerodynamic diameters (MMAD) of at least 2 micrometers, an air-purifying respirator equipment with any filter certified for particulates by NIOSH. (See *NIOSH Guide to the Selction and Use of Particulate Respirators*, DHHS/NIOSH No. 96-101; also ISEA *Use and Selection Guide for Non-Powered Air-Purifying PArticulate Respirators*).

43.3 Perform Detailed Respiratory Risk Assessment to Select Specific Respirator Type

☐ The general respirator risk assessment provides specific types of respirators for IDLH environments and only general respirator types for selection in non-IDLH environments.

- ☐ A more detailed analysis is required to allow decisions between different specific respirator types and features.
- ☐ This analysis consists of:
  - Determining specific exposure limits and characteristics of the contaminants
  - Evaluating workplace factors which affect respirator selection
  - Reviewing respirator features related to protection

43.3.1 Exposure Limits and Characteristics of Contaminants

- ☐ Determine the permissible exposure limit (PEL), threshold limit value (TLV), or other recommended exposure limit for each contaminant.
- ☐ Determine health effects or symptoms that may result from overexposure to each contaminant:
  - Skin absorption
  - Skin sensitization
  - Irritation or corrosiveness to eyes or skin
- ☐ Determine the warning properties associated from each contaminant:
  - Odor threshold
  - Irritation concentration
  - Taste

43.3.2 Workplace Factors Affecting Respirator Selection

- ☐ Review workplace factors for the specific areas of respirator use:
  - Ambient temperature and humidity
  - Location of the work area
  - Entry and exit procedures for location
  - Tight spaces or obstacles
  - Noise levels
  - Additional personal protective equipment required
  - Other conditions which may:
    - -- Reduce service life of cartridges or filters
    - -- Restrict mobility

-- Interfere with communication

-- Present tripping or other safety hazards

☐ Consider factors for use of respirator:

- Length of time required for respirator use
- Frequency of anticipated use
- Number of workers involved
- Variety of facial sizes and shapes for workers wearing respirators
- Workers who have beards or who wear contact lenses
- Means for controlling respirator distribution, cleaning, maintenance, use or other factors which are part of a respirator program

43.3.3 Respirator Features Related to Protection

☐ Review respirator feature related to protection and worker acceptance:

- Rated service life
- Face coverage
- Field of vision
- Weight
- Breathing resistance

☐ Determine the protection factor for specific type of respirator:

- Assigned protection factors for various types of respirators are listed in:

  -- ANSI Z88.2-1992, *American National Standard for Respiratory Protection*.

  -- National Institute for Occupational Safety and Health, *Guide to Industrial Respiratory Protection* (DHHS/NIOSH Publication No. 87-116, 1987)

  -- Respiratory Protection Handbook (W. H. Revoir and C. Bien, CRC Lewis Publishers, New York, 1997).

- Protection factors may be determined for a specific worker population for a specific type of respirator.

43.3.4 Specific Types of Respirators

☐ Specific types of respirators include:

- Air-purifying respirators (APR)
  - Disposable respirators

  Warning: Many disposable respirators do not provide an adequate seal on the user's face to prevent inward penetration of atmospheric contaminants and may not easily be evaluated by fit testing.

  - Particulate filter respirators
  - Cartridge or canister respirators (gas mask)
  - Cartridge or canister respirators (gas mask) with particulate filter
- Powered air-purifying respirators (PAPR)
- Supplied-air respirators (SAR)
  - Demand supplied-air respirators
  - Continuous flow supplied-air respirators
  - Pressure-demand supplied-air respirators
- Combination supplied-air/air-purifying respirators
  - Continuous SAR/APR
  - Pressure-demand SAR/APR
- Self-contained breathing apparatus (SCBA)
  - Demand self-contained breathing apparatus
  - Continuous flow self-contained breathing apparatus
  - Pressure-demand self-contained breathing apparatus
- Combination supplied air respirators with auxiliary self-contained air supply (SCBA/SAR)
  - Demand SCBA/SAR
  - Continuous SCBA/SAR
  - Pressure-demand SCBA/SAR

☐ Air-purifying respirators and supplied-air respirators offer different types of respiratory inlet covers.

- Air-purifying respirators may use:
  - Quarter-masks

- -- Half-masks
- -- Full facepieces

- Supplied air respirators may use:
  - -- Quarter-masks
  - -- Half-masks
  - -- Full facepieces
  - -- Helmets
  - -- Hoods
  - -- Blouses
  - -- Suits (**NOTE: Air-fed suits are not covered in a certification category by NIOSH)**
  - -- Loose-fitting facepieces
- The majority of self-contained breathing apparatus (SCBA), except some types of continuous flow SCBA used for escape purposes, are full facepiece respirators.

☐ Respirators may be either negative pressure or positive pressure respirators

- All non-powered air-purifying respirators are negative pressure respirators
- Other negative pressure respirators include:
  - -- Demand supplied-air respirators
  - -- Demand self-contained breathing apparatus
  - -- Combination continuous or pressure-demand supplied-air/air-purifying respirator (SAR/APR)
- Positive pressure respirators include:
  - -- Powered air-purifying respirators (PAPR)
  - -- Continuous flow supplied-air respirators
  - -- Pressure-demand supplied-air respirators
  - -- Continuous flow self-contained breathing apparatus
  - -- Pressure-demand self-contained breathing apparatus
  - -- Combination pressure-demand supplied air respirator with auxiliary self-contained air supply (SCBA/SAR)

43.3.5 Specific Respirator Selection Decisions

☐ Determine the Maximum Use Concentration (MUC) as the product of the protection factor (PF) times the permissible exposure limit (PEL):

MUC = PF (or APF) x PEL

- Select a respirator that has a MUC above the workplace concentration of the contaminant.
- For greater protection with air-purifying respirators, select:
  -- Full facepiece air-purifying respirators over half-mask or quarter-mask air-purifying respirators
  -- Powered air-purifying respirators over non-powered air-purifying respirators
  -- Positive pressure respirators over negative pressure respirators
- Assigned protection factors in ANSI Z88.2-1992 show powered air-purifying, continuous flow supplied-air, and pressure-demand supplied-air respirators to be equivalent for the same type of facepiece (e.g, half mask, full facepiece, helmet/hood).
- If a standard or manufacturer specifies a lower concentration than the calculated MUC, that concentration overrides the MUC.

☐ When using air-purifying respirators for protection against gaseous or vapor contaminants, use cartridge or canister respirators (gas masks) outfitted with the appropriate cartridge or canister:

- Acid gas
- Acid gas/ammonia gas
- Acid gas/formaldehyde
- Acid gas/organic vapor
- Acid gas/organic vapor/ammonia gas
- Acid gas/organic vapor/chlorine dioxide
- Acid gas/organic vapor/chlorine dioxide/formaldehyde
- Ammonia/methylamine
- Carbon monoxide
- Chlorine gas
- Hydrocyanic acid gas
- Hydrocyanic aid gas/chloropicrin vapor

- Organic vapor
- Radioactive materials, excepting tritium and noble gases

☐ When using air-purifying respirators for protection against particulates, select

- Particulate filter respirators with N-series filters for workplaces which are free of oil-aerosols
- Particulate filter respirators with R- and P-series filters for workplaces include any type of particulate, including oil-based liquid particles

  NOTE: The National Institute for Occupational Safety and Health offers two publications on respirator decision logic: *Guide to Industrial Respiratory Protection*, (DHHS/NIOSH Publication No. 87-116, 1987) and *Guide to the Selection and Use of Particulate Respirators* (DHHS/NIOSH Publication No. 96-101, 1996).

☐ When using air-purifying respirators for protection against both gaseous or vapor contaminants and particulates, use cartridge or canister respirators (gas masks) outfitted with cartridges or canisters that include particulate pre-filters.

☐ For contaminants which absorb into or sensitize the skin, wear appropriate protective garments (see Section 38).

☐ For contaminants that cause eye irritation at concentrations present in the workplace, select full facepiece respirators.

☐ For contaminants that do not have warning properties at concentration below recommended exposure levels, select:

- Continuous flow supplied-air respirators
- Pressure-demand supplied-air respirators
- Continuous flow self-contained breathing apparatus
- Pressure-demand self-contained breathing apparatus
- Combination continuous flow or pressure-demand supplied air respirator with auxiliary self-contained air supply
- Air-purifying, powered air-purifying respirators, or combination air-purifying/air-supplying respirators only when sufficient information and data have demonstrated that the cartridge or canister changing schedules ensure that contaminant concentrations remain below recommended exposure levels

☐ For contaminants encountered in low temperature environments, select:

- Respirators which use:
  - -- Antifogging agent on the inner surface of facepiece lens (for temperature down to 32°F (0°C)
  - -- An inlet air deflector or nose cup for temperatures down to -34°F (-30°C)
- A source of respirable air with a dew point is at least 18°F (10°C) lower than the coldest temperature expected [for SCBA, use respirable air with a dewpoint of -67°F (-55°C), and when used below -25°F (-32°C), use respirable air with a dewpoint of -103°F (-75°C)]
- SCBA with a minimum use temperature below the intended use temperature

☐ For contaminants encountered in high temperature environments, select respirators which are lightweight and have low breathing resistance to minimize the physical stress on the wearer:

- Supplied-air respirators are generally lightweight.
  - -- Continuous flow and pressure demand SAR offer minimum breathing resistance.
- Powered air-purifying respirators are usually lightweight and offer minimum resistance to breathing.

☐ If air-purifying respirators are used for contaminants encountered in high temperature and high humidity environments, ensure the effectiveness of cartridges, canisters, or filters at use conditions.

☐ For contaminants encountered where a high degree of mobility is required, select:

- Air-purifying respirators or self-contained breathing apparatus over supplied-air respirators
- Combination pressure-demand supplied air respirator with auxiliary self-contained air supply (SCBA/SAR) which offer ability to work off either airline or self-contained air supply

☐ For contaminants encountered in applications where long term wearing of a respirator is required, select:

- Air-purifying respirators or supplied-air respirators over self-contained breathing apparatus

- Self-contained breathing apparatus only when the rated service life is longer than the maximum anticipated use time for the respirator

☐ For choosing between air-purifying respirators and supplied-air respirators or self-contained breathing apparatus for environments where both types of respirators have been determined suitable, select air-purifying respirators over supplied-air respirators, when a suitable source of respirable air cannot be supplied.

☐ For noisy environments, use hearing protection (see Section 43) or special communication systems (see Section 44) in conjunction with selected respirators:

- Some respirator manufacturers offer communications options for full facepiece and other types of respirators.
- Auxiliary equipment may be worn with loose-fitting respiratory coverings (such as hoods).

☐ For workers who have facial hair, beards, or facial deformities:

- Select powered air-purifying respirators or supplied-air respirators which have loose-fitting respirator coverings (e.g., helmets, hoods, blouses); or
- Prohibit workers with facial hair or beards from working in areas where respirators must be worn.

☐ For workers who must wear corrective (prescription) lenses:

- Select half- or quarter-mask respirators which accommodate corrective lenses, protective spectacle or goggle, faceshield, welding helmet or other eye and face protective device;
- Select full facepiece or other types of respirators that incorporate corrective lenses;
- Select air-purifying respirators or supplied-air respirators which have loose-fitting respirator coverings (e.g., helmets, hoods, blouses); or
- Permit wearing of corrective lenses only if it does not interfere with the seal of the respirator

☐ For workers who wear contact lenses, permit contact lenses use only when the worker has previously demonstrated that he or she had successful experience in wearing contact lenses while wearing a respirator.

☐ For workers who must wear other protective clothing and equipment, ensure that:

- Head coverings do not pass between the sealing surface of the facepiece and the wearer's face.
- Head harness straps of the respirator are not positioned over the helmets or similar headwear
- Wearing of other items does not interfere with the seal of the selected respirator.

43.3.6 Respirator Selection for Special Environments

☐ For fire fighting, select a full facepiece pressure-demand self-contained breathing apparatus (SCBA) which meets the requirements of NFPA 1981.

☐ For chemical emergency response or hazardous waste site clean up requiring Level A or B protection, select:

- Full facepiece pressure-demand self-contained breathing apparatus (SCBA)
- Combination full facepiece pressure-demand supplied-air respirator with auxiliary self-contained air supply (SCBA/SAR)

☐ For airline or air-supplied suits (without internal respiratory inlet covering), select suits which have been approved by the requirements specified by the U.S. Department of Energy.

- In Europe, air-supplied suits must meet EN 943, Part 1 for Type 1c suits.

☐ For operations involving abrasive blasting, select respirators approved for abrasive blasting:

- Powered air purifying respirators
- Type AE, BE, or CE supplied-air respirators

☐ For protection against biological airborne pathogens (Mycobacterium tuberculosis):

- Choose particulate filter facepiece air-purifying respirators equipped with a high efficiency particulate air (HEPA) filters
- Wear specified respirators when:
  -- Employees enter rooms housing individuals with suspected or confirmed infectious TB diseases
  -- Employees perform high hazard procedures on individuals who have suspected or confirmed TB diseases

-- When emergency medical response personnel or others must transport, in a closed vehicle, an individual with suspected or confirmed TB diseases

43.3.7 Comparison of Advantages and Disadvantages for Specific Respirator Types

☐ Non-powered air-purifying respirators

- Advantages
  - -- Small and compact
  - -- Lightweight
  - -- Simple construction for most
  - -- No restriction of wearer's mobility
  - -- Low initial cost
- Disadvantages
  - -- Cannot be used in oxygen-deficient atmosphere
  - -- Cannot be used in IDLH atmosphere
  - -- Gas-/vapor-removing type cannot be used for gases and vapors with poor warning properties unless for emergency escape, or unless equipped with end-of-service-life indicator, or permitted by U.S. Department of Labor standard for particular hazardous substances
  - -- Negative air pressure inside facepiece during inhalation by wearer may result in leakage of hazardous atmosphere into facepiece and inhalation of that atmosphere
  - -- Cannot be used by person having a beard or mustache because facial hair would pass between facepiece sealing surface and wearer's face
  - -- Great care must be taken to select proper type for specific air contaminants(s)
  - -- Must use fitting test to select particular make and model for each wearer
  - -- Fit of all fabric type disposable respirators cannot be checked adequately by wearer prior to entry of wearer into hazardous atmosphere; improper fit may result in excessive leakage of air contaminant(s) into respirator and its inhalation by respirator wearer

-- Wearing discomfort may limit use period

-- Resistance to breathing may cause discomfort and fatigue

-- Eye irritation caused by air contaminant(s) requires use of full facepiece

-- Many types are limited for use against only low concentrations of air contaminants

-- High cost for frequent replacement of disposable type or frequent replacement of air-purifying element in reusable type

-- High cost for maintenance of reusable type

☐ Powered air-purifying respirators

- Advantages

  -- No restriction on mobility of wearer

  -- Minimal resistance to breathing

  -- Cooling effect on wearer in warm atmospheres

  -- Less wearing discomfort than for non-powered air purifying type

  -- May be able to wear for extended time as long as air flow to respiratory-inlet covering is not reduced below required minimum value and as long as the air contaminant does not penetrate the air-purifying element to result in contaminant concentration above the permissible exposure limit in air supplied to respiratory-inlet covering

  -- Fit testing not required for devices equipped with loose-fitting respiratory-inlet coverings

  -- Many persons who are unable to get satisfactory fit with facepiece may be able to use powered air-purifying devices equipped with loose-fitting respiratory-inlet coverings

  -- Devices equipped with loose-fitting respiratory-inlet coverings can be used by wearers with beards and moustaches

- Disadvantages

  -- Cannot be used in oxygen-deficient atmosphere

  -- Cannot be used in IDLH atmosphere

  -- Gas- or vapor-removing type cannot be used for gases and vapors with poor warning properties unless for emergency escape, or unless equipped with end-of-service-life indicator, or unless permitted by

U.S. Department of Labor standard for particular hazardous substance

-- Great care must be taken to select proper type for specific air contaminant(s)

-- A fitting test of the facepiece of device equipped with facepiece type respiratory-inlet covering is required for all wearers

-- Eye irritation caused by air contaminant(s) requires use of full facepiece or loose-fitting respiratory-inlet covering that protects the entire face

-- Functional limitations of many types of air-purifying elements restrict use of respirator only to low concentrations of several air contaminants

-- Poor design of respiratory-inlet coverings of some devices, poor construction of some devices, and/or air flow rates which are too low for some devices have resulted in low values for assigned protection factors (APF) for powered air-purifying respirators

-- Failure of blower to operate while wearer is in hazardous atmosphere results in loss of protection provided to wearer and requires wearer to immediately leave the hazardous area and go to safe area

-- Use in cold atmosphere may result in severe discomfort to wearer

-- Battery for operating blower requires frequent recharging

-- Wearer generally cannot detect when inadequate quantity of respirable air is supplied to respiratory-inlet covering due to low battery power or plugging of particle-filtering element by retained particulate matter

-- High initial cost

-- High cost for frequent battery replacement

-- High speed motor of blower requires frequent replacement that is costly

-- High cost of frequent replacement of air-purifying elements

-- Complex design and construction makes maintenance difficult and costly

☐ Airline type of supplied-air form of atmosphere-supplying respirators

- Advantages

  - May be used for long periods
  - Minimal resistance to breathing
  - Low weight
  - Low bulk
  - Minimal discomfort
  - Many persons who are unable to obtain satisfactory fit with facepieces may be able to use airline devices equipped with loose-fitting respiratory-inlet coverings
  - Cooling effect on wearer in warm atmospheres
  - Fit testing not required for devices equipped with loose-fitting respiratory-inlet covering
  - Devices equipped with loose-fitting respiratory-inlet coverings can be used by wearers with facial hair such as beards and mustaches
  - Continuous-flow devices and especially pressure-demand devices offer high levels of protection
  - Can be used for oxygen-deficient atmospheres not IDLH
  - Moderate initial cost
  - Low operating cost
  - Low maintenance cost
- Disadvantages
  - Cannot be used in IDLH atmosphere unless it is a combination pressure-demand airline respirator equipped with a full facepiece and an auxiliary self-contained breathing supply of respirable air
  - Trailing air-supply hose restricts mobility of wearer unless it is a combination airline respirator with an auxiliary self-contained breathing supply of respirable air, or a combination airline respirator with a provision for air-purifying protection
  - Air-supply hose is vulnerable to damage that may result in loss of respirable air being provided to respiratory-inlet of respirator
  - Sudden loss of source of supply of respirable air due to damage to air-supply hose or failure of air pump or air compressor to provide respirable air leaves wearer without respiratory protection unless the respirator is a combination airline respirator with an auxiliary self-contained breathing supply of respirable air, or a combination airline respirator with a provision for air-purifying protection

-- Quality of air provided to respirator must be checked to ensure it meets pertinent requirements

-- Inlet of air pump or air compressor used to provide respirable air to respirator must be located in area having respirable atmosphere

-- Oil-lubricated compressor used to provide respirable air to respirator must be equipped with both a high temperature alarm and a carbon monoxide gas alarm

-- Couplings in plant used to connect air-supply hoses of airline respirators to source of respirable air must be incompatible with couplings for non-respirable plant air and any other gases

-- Demand type device does not provide wearer with high level of protection. Negative air pressure inside facepiece of demand type device during inhalation may result in leakage of hazardous atmosphere into facepiece and inhalation of that atmosphere by the wearer

-- Complexity of airflow control valve of pressure-demand unit requires that maintenance be performed only by adequately trained person

-- A fitting test of the facepiece of device equipped with facepiece type respiratory-inlet covering is required for all wearers

☐ Hose-mask type of supplied-air form of atmosphere-supplying respirators

- Advantages

-- May be used for long periods

-- Minimal resistance to breathing by hose-mask with blower

-- Can be used in oxygen-deficient atmospheres not IDLH

-- Simple construction

-- Low bulk

-- Easy maintenance

-- Low operating cost

- Disadvantages

-- Cannot be used in IDLH atmosphere

-- Cannot be used by a person having facial hair such as a beard or mustache because facial hair would pass between facepiece sealing surface and wearer's face

-- Negative air pressure inside facepiece of hose-mask without blower during inhalation by wearer may result in leakage of hazardous atmosphere into facepiece and inhalation of that atmosphere by the wearer

-- Hose-mask without blower provides wearer with low level of protection

-- A low air flow rate by hose-mask with blower to facepiece provides the wearer with a low level of protection

-- A fitting test of the facepiece is required for any potential wearer

-- Inlet of hose of hose-mask without blower and inlet of blower for hose-mask with blower must be located in area having respirable air

-- Air-supply hose is vulnerable to damage that may result in loss of respirable air being provided to facepiece of respirator

-- Sudden loss of source of respirable air due to damage of air-supply hose or failure of blower of hose-mask with blower leaves wearer without respiratory protection

-- Resistance to breathing by hose-mask without blower causes discomfort and fatigue

☐ Self-contained breathing apparatus type atmosphere-supplying respirators

- Advantages

  -- SCBA wearer carries own source of respirable air, and thus is independent of ambient atmosphere

  -- Minimal restriction of mobility of wearer

  -- Provides protection against air contaminants of all types, particulate and gaseous, and oxygen deficiency

  -- Pressure-demand open-circuit type with full facepiece provides very high level of protection, and can be used for IDLH contaminated atmosphere, IDLH oxygen-deficient atmosphere, entry into atmosphere with unknown level of contamination or oxygen deficiency, and for fire fighting

  -- Positive pressure closed-circuit type can be used for escape in mining operations.

- Disadvantages

  -- Complex construction

  - -- Most are heavy and bulky making them unsuitable for wearers engaged in strenuous work or for wearers who must enter and work in a confined space
  - -- Limited service period makes them unsuitable for use for long periods
  - -- Short service period of open-circuit type device limits its use to situations in which the worker, after donning SCBA, can travel quickly and easily to the worksite and perform a task in a short period
  - -- A fitting test of the facepiece of a device equipped with facepiece type respiratory-inlet covering is required for any person who will use the device
  - -- Extensive training of potential wearers is required
  - -- Negative pressure inside facepiece of demand open-circuit device and non-positive pressure closed-circuit device during inhalation by wearer may result in leakage of hazardous atmosphere into facepiece and inhalation of that atmosphere by wearer

- High initial cost
  - -- High operating cost
  - -- Complexity of device makes maintenance time-consuming and costly
  - -- Complexity of the two-stage flow control regulating valve employed in open-circuit device requires maintenance of valve be performed only by adequately trained individuals
  - -- Respirable air and oxygen must meet stringent requirements

43.3.8 Respirable Air and Oxygen Requirements

☐ OSHA 29 CFR 1910.134 requires the employer to provide employees using atmosphere-supplying respirators (supplied-air and SCBA) with breathing gases of high purity.

☐ Compressed air, compressed oxygen, liquid air, and liquid oxygen used for respiration must meet the following specifications

- Compressed and liquid oxygen must meet U.S. Pharmacopoeia requirements for medical or breathing oxygen.

- Compressed breathing air must meet the requirements for Type 1-Grade D breathing air specified in Compressed Gas Association Commodity Specification G-7.1, 1989, including:
    - -- Oxygen content of 19.5 to 23.5% (v/v)
    - -- Hydrocarbon (condensed) content of 5 milligrams per cubic meter of air or less
    - -- Carbon monoxide content of 10 ppm or less
    - -- Carbon dioxide content of 1,000 ppm or less
    - -- Lack of noticeable odor
- Oxygen concentrations greater than 23.5% must be used only in equipment for oxygen service or distribution.
- Cylinders used to supply breathing air to respirators must meet the following requirements:
    - -- Cylinders must be tested and maintained as prescribed in Shipping Container Specification Regulations of the Department of Transportation (49 CFR Parts 173 and 178);
    - -- Cylinders of purchased breathing air have a certificate of analysis from the supplier that the breathing air meets the requirements for Type 1-Grade D breathing air; and
    - -- The moisture content in the cylinder does not exceed a dew point of -50°F (-45°C) at 1 atmosphere pressure
- Compressors used to supply breathing air to respirators must be constructed and situated to:
    - -- Prevent entry of contaminated air into the air-supply system
    - -- Minimize moisture content so that the dew point at 1 atmosphere pressure is 10°F (6°C) below the ambient temperature
    - -- Have suitable in-line air-purifying sorbent beds and filters to further ensure breathing air quality (sorbent bed and filters must be replaced or refurbished in accordance with manufacturer instructions)
    - -- Have a tag maintained at the compressor containing the most recent change date and the signature of the person authorized by the employer to perform the change
- Compressors that are not oil-lubricated have carbon monoxide levels that do not exceed 10 ppm.

- Oil-lubricated compressors must have a high-temperature or carbon monoxide alarm, or both, to monitor carbon monoxide levels such that the air supply is monitored at intervals sufficient to prevent carbon monoxide in the breathing air from exceeding 10 ppm.
- Breathing air couplings must be incompatible with outlets for non-respirable worksite air or other gas systems.
- Asphyxiating substances must not be introduced into breathing airlines.
- Breathing gas containers must be marked in accordance with 42 CFR Part 84.

43.4 Choose Respirator Design Features

☐ Air-purifying respirator

- Choose type:
  - Reusable (except cartridges and canisters)
  - Disposable
- Choose type of respiratory inlet cover:
  - Full facepiece
  - Half mask
  - Quarter mask
- Choose style of canister or gas mask:
  - Chin style (sorbent canister attached to full facepiece)
  - Front and back style (large sized sorbent canister with flexible breathing tube fastened with harness to either the front or back to the wearer's body)
- Choose facepiece material:
  - Neoprene rubber
  - Silicone rubber
  - Other rubber
- Choose communications enhancement options for gas mask full facepiece:
  - Speaking diaphragm
  - Voice amplifier (electronic microphone with amplifier)

- Choose options:
  - Spectacle holders
  - Lens scratch resistant coating

☐ Powered air-purifying respirator

- Choose configuration:
  - Blower belt-mounted
  - Blower mask-mounted
- Choose type of respiratory inlet cover:
  - Hood/helmet (coverage areas vary significantly)
  - Full facepiece
  - Half mask
- Choose facepiece material:
  - Neoprene rubber
  - Silicone rubber
  - Other rubber
- Choose communications enhancement options for gas mask full facepiece:
  - Speaking diaphragm
  - Voice amplifier (electronic microphone with amplifier)
- Choose options:
  - Spectacle holders
  - Lens scratch resistant coating

☐ Supplied-air respirator

- Choose mode of operation:
  - Continuous flow
  - Demand
  - Pressure demand
- Choose respiratory inlet covering:
  - Half mask
  - Full facepiece
  - Hood/helmet (coverage areas vary significantly)

-- Abrasive protective covering (Type CE)

- Choose facepiece material:
  -- Neoprene rubber
  -- Silicone rubber
  -- Other rubber
- Choose type of airline harness:
  -- Belt
  -- Belt and shoulder strap
- Choose type of airline couplers:
  -- Industrial
  -- Schrader
  -- Hansen
  -- Foster
  -- Snap-tite
  -- Bayonet
- Choose type of air supply:
  -- Air pump
  -- Compressor
  -- Cascade of air cylinders
- Choose communications enhancement options for gas mask full facepiece:
  -- Speaking diaphragm
  -- Voice amplifier (electronic microphone with amplifier)
- Choose options:
  -- Nose cup
  -- Spectacle holders
  -- Lens scratch resistant coating

☐ Self-contained breathing apparatus

- Choose type of configuration, application, and mask type:
  -- Closed circuit, entry or escape, full facepiece
  -- Open circuit, entry or escape, full facepiece

- -- Closed circuit, escape, half mask
- -- Closed circuit, escape, full facepiece
- -- Closed circuit, escape, hood or helmet (coverage areas vary significantly)
- -- Open circuit, escape, mouth bit
- -- Open circuit, escape, half mask
- -- Open circuit, escape, full facepiece
- -- Open circuit, escape, hood or helmet (coverage areas vary significantly)

- Choose rated service time:
  - -- Closed circuit, entry or escape (common: 60, 120, 240, 480 minutes)
  - -- Open circuit, entry or escape (common: 30, 45, 60 minutes)
  - -- Closed circuit, escape (5 to 60 minutes)
  - -- Open circuit, escape (3 to 60 minutes)
- Choose facepiece material:
  - -- Neoprene rubber
  - -- Silicone rubber
  - -- Other rubber
- Choose type of headgear/facepiece configuration:
  - -- Four point suspension
  - -- Five point suspension
  - -- Head harness
- For open-circuit SCBA, choose type of cylinder (significantly affects weight):
  - -- Steel (heaviest)
  - -- Hoop wrap composite
  - -- Full wrap composite
  - -- Graphite (lightest)
- For open-circuit SCBA, choose regulator location:
  - -- Belt-mounted
  - -- Mask-mounted (provides higher level of positive pressure performance)
- Choose type of harness straps:
  - -- Synthetic fiber (e.g., Nylon, polyester)

-- High temperature materials (e.g., aramids)

- Choose communications enhancement options:
  - -- Speaking diaphragm
  - -- Voice amplifier (electronic microphone with amplifier)
  - -- Radio microphone
- Choose options:
  - -- Buddy breathing option
  - -- Quick fill option
  - -- Nose cup
  - -- Spectacle holders
  - -- Lens scratch resistant coating

43.5 Consider Potential Hazards from Selected Respirator

☐ Do respirator materials (especially those in the facepiece) irritate or sensitize the wearer's skin?

☐ Is respirator likely to retain contamination even after cleaning?

- Should respirator be protected from exposure to liquids and other contaminants?

☐ Does respirator have a design with loose, bands, straps or material that can be caught in moving machinery?

☐ Does respirator (full facepiece of helmet/hood configurations) provide clear and unobstructed vision for performing required tasks?

☐ Is the respirator available in a sufficient number of sizes or can the respirator be adjusted to fit organization personnel?

- Has organization fit tested each individual who must wear a respirator?

☐ Is respirator difficult to use and reservice?

☐ Is respirator uncomfortable for the wearer under use conditions?

☐ Is respirator reliable for meeting the intended service life?

Section 44

Selecting Hearing Protection

The purpose of this section is to provide a systematic approach for selecting hearing protectors in terms of the employer selection requirements, selection factors to be considered, hearing protective types, design features, material or material systems, and consideration of hazards created by the face and eyewear.

44.1 Overview of Hearing Protector Selection Parameters

☐ The basic approach for selecting hearing protectors encompasses:

- Understanding selection requirements for hearing protectors
- Selecting the type of hearing protector
- Choosing hearing protector design features and materials having performance related to the hazards identified as priorities during the risk assessment or meeting the organization's basic requirements
- Considering potential hazards from the selected hearing protectors

☐ OSHA 29 CFR 1910.95 establishes requirements for occupational noise exposure.

☐ Background information on hearing protectors is provided in Section 23.

44.2 Selection Requirements for Hearing Protectors

☐ OSHA 29 CFR 1910.95 requires places several requirements on the selection of hearing protectors.

☐ Similar practices exist for hearing protector selection in other North American, European, or International standards.

44.2.1 Employer Responsibilities for Selection

- ☐ Employers must provide hearing protectors to employees, who have noise exposures at or above the action level of an eight-hour Time Weighted Average (TWA) of 85 decibels, at no cost to the employees.
- ☐ Employers must ensure that hearing protectors are worn by all employees who:
  - Have noise exposures that exceed the levels established in Table G-16 in 29 CFR 1910.95.
  - Have noise exposures at the action level or greater and either have not had a baseline audiogram established or have experience a standard threshold shift (a change in the hearing threshold of 10 decibels or more at 2000, 3000, or 4000 Hz in either ear).
- ☐ Employees must be given a choice of hearing protectors from a variety of suitable hearing protectors provided by the employer.
- ☐ Employers must provide training in the use and care of all hearing protectors provided to employees.
- ☐ Employers must ensure proper initial fitting and supervise the correct use of all hearing protectors.

44.2.2 Hearing Protector Attenuation

- ☐ OSHA 29 CFR 1910.95 requires employers to evaluate the hearing protector attenuation of the specific noise environments in which the protector will be used by specific methods (see Section 27.8).
- ☐ Selected hearing protectors must attenuate employee exposures at least to an eight-hour TWA of 90 dB.
- ☐ For employees who have experienced a standard threshold shift, hearing protectors must attenuate employee exposures at least to an eight-hour TWA of 85 dB.
- ☐ The adequacy of hearing protector attenuation must be reevaluated whenever employee noise exposure increases to the extent that the hearing protectors provided may not longer provide adequate attenuation; employers must then provide more effective hearing protectors when necessary.

44.3 Select the Type of Hearing Protector

- ☐ Basic types of hearing protectors include:
  - Ear plugs
  - Canal caps
  - Ear muffs
- ☐ The specific hearing protector should be chosen after considering different selection factors while ensuring that OSHA noise exposure requirements are met; for this reason any type of hearing protector might be suitable for a particular application.
- ☐ Each type of hearing protector provides different design features, advantages, and disadvantages.

44.3.1 Base Selection of Hearing Protectors on Specific Selection Factors

- ☐ Factors in selecting hearing protectors include:
  - Noise Reduction Rating (NRR) of the hearing protector
  - Worker requirements for communication
  - Type of worker activity
  - Wearer preferences (often based on comfort)
  - Intended service life
  - Accommodation of other types of PPE (e.g., helmets, communications systems)
- ☐ For high noise environments, choose a hearing protector with a suitably high Noise Reduction Rating (NRR) which still permits interpersonal communications, as necessary.
- ☐ For high noise environments with requirements for communication, choose:
  - An active hearing protector which incorporates new technology for cancelling or filtering noise
  - Hearing protectors which demonstrate equal attenuation of communication frequencies (both low [below 1800 Hertz] and high [above 1800 Hertz])

- ☐ Choose a hearing protector that is comfortable for the employee and will stay in place during all expected worker activities.
- ☐ For workers who experience discomfort or poor fit with ear plugs, choose:
  - Canal plugs
  - Ear muffs
- ☐ Hearing protectors may be reusable or disposable, usually depending on the type of hearing protector and the materials of construction.
  - Many types of ear plugs are disposable.
  - Nearly all types of ear muffs are considered reusable.
- ☐ For a specific type of hearing protector, make the choice between reusable and disposable products based on:
  - Any differences in the level of performance
  - Differences in life cycles costs
  - Ease in maintaining hearing protectors versus disposal
- ☐ For protecting the wearer's hearing while wearing a head protective device, choose:
  - Ear plugs
  - Canal caps
  - Ear muffs that do not interfere with the wearing of the head protective device
  - Ear muffs that can be attached to the selected head protection device
- ☐ When separate communications is needed in high noise environments, choose a hearing protector that incorporates a communications device.

44.3.2 Consider Advantages and Disadvantages of Hearing Protector Types

- ☐ Advantages of ear plugs include:
  - Low cost
  - Disposability (for many types)
  - Conformance to wearer's ear (for moldable styles)
  - Easily dispensed to employees

- ☐ Disadvantages of ear plugs include:
  - Require wearer to correctly fit ear (may not be worn properly)
  - Easily contaminated if reused (allow bacterial growth)
  - Potential irritation of wearer ear canal
- ☐ Canal caps have the advantage that they do not enter ear canal like ear plugs and overcome most of disadvantages of ear plugs.
- ☐ Because canal caps fit on the ear canal opening they may be less effective in attenuating high noise levels.
- ☐ Advantages of ear muffs include:
  - Ease of use
  - Potential accommodation of active noise reduction or electronic noise attenuation systems (for specifically designed hearing protectors)
  - Potential attachment to head protection devices
  - Potential incorporation of communications devices
- ☐ Disadvantages of ear muffs include
  - Weight
  - Bulk
  - Expense

44.4 Select Hearing Protector Design Features and Materials

- ☐ Choose design features and materials based on face and eyewear type.

44.4.1 Ear Plugs

- ☐ Choose type of ear plug:
  - Formable (can be molded by wearer for himself or herself; usually disposable)
    - -- Expandable
    - -- Non-expandable
  - Custom-molded (permit sizing and fit to individual wearer's ear)
    - -- By factory

  - -- By user
- Pre-molded (do not allow adjustment by wearer; user must select proper size)
  - -- Sized
  - -- Universal fit

☐ Choose material:

- Glass fiber (formable)
- Wax-impregnated cotton (formable)
- Expandable plastic (formable)
- Silicone rubber (custom-molded, pre-molded)
- Plastic (custom-molded, pre-molded)
- Vinyl (custom-molded, pre-molded)

☐ Choose accessory design features:

- With cord or band (helps prevent loss, but increase cost and bulk)
- Color-coding (allows for discriminating sizes)
- Level-dependent (selectively filters harmful energy from excessive noise)

44.4.2 Canal Caps

☐ Choose band material:

- Metal (cannot be used around electrical hazards)
- Plastic

☐ Choose type of attenuation:

- Linear (no preferential attenuation)
- Level-dependent (selectively filters harmful energy from excessive noise)

44.4.3 Ear Muffs

☐ Choose type of cushion

- Foam
- Liquid

- ☐ Choose type of wearing attachment
  - Band
  - Helmet
- ☐ Choose band location
  - Under chin
  - Over head
  - Behind head
- ☐ Choose band material
  - Metal (cannot be used around electrical hazards)
  - Plastic
- ☐ Choose type of attenuation:
  - Linear (no preferential attenuation)
  - Level-dependent (selectively filters harmful energy from excessive noise)
    - -- Electronic
    - -- Passive
  - Active noise reduction
- ☐ Choose accessory design features
  - Built-in one-way communication system
  - Built-in two-way communication system
  - Cooling pads
  - Warming pads

44.5 Consider Potential Hazards from Selected Hearing Protectors

- ☐ Does the hearing protector (especially ear plugs) irritate or sensitize the wearer's skin?
- ☐ Is the ear protector or its components likely to retain contamination even after cleaning?
- ☐ Does the hearing protector design potentially allow the cord to be caught in moving machinery?
- ☐ Is the hearing protector available in sufficient sizes for proper fit of

organization personnel?

- ☐ Is the hearing protector difficult to adjust, use and reservice?
- ☐ Is the hearing protector uncomfortable for the wearer under use conditions?
- ☐ Does the hearing protector have sufficient durability for meeting the intended service life?

Section 45

Selecting Related Items of PPE

The purpose of this section is to provide a systematic approach for selecting related items of PPE including personal fall arrest systems, personal cooling garments, personal communications equipment, and personal flotation devices. This section does not provide the same level of detail provided for other items of PPE covered in this book.

45.1 Overview of Related PPE Item Selection

☐ Related PPE items include:

- Personal fall protection systems
- Personal cooling garments and systems
- Personal communications equipment
- Personal flotation devices

☐ These items may be used by themselves or as part of a clothing and equipment ensemble designed for multiple hazards.

45.2 Personal Fall Protection Systems

☐ Fall protection systems are intended to prevent worker falls or to arrest falls when they do occur.

45.2.1 Selection Requirements and Standards

☐ OSHA 29 CFR 1926.502 defines requirements for fall protection systems that include:

- Guardrail systems

- Safety net systems
- Personal fall arrest systems
- Positioning device systems
- Controlled access zones
- Safety monitoring systems
- Covers

☐ Personal fall arrest systems and positioning device systems can be considered personal protective equipment since they are worn by or tethered to the person.

☐ A personal fall arrest system is an assembly of components and subsystems used to arrest a person in a fall from a working height; personal fall arrest systems usually consist of the following components and subsystems:

- Anchorage
- Anchorage connector
- Lanyard connecting subsystem consisting of:
  - Lanyard
  - Energy absorber
  - Self-retracting lifeline or lanyard
- Fall arrest attachment
- Full body harness

☐ A positioning device system is a body harness system rigged to allow an employee to be supported on an elevated work surface, such as a wall, and work with both hands free.

☐ OSHA 29 CFR 1926.502 sets several requirements specific to personal fall arrest systems:

- Body belts and non-locking snap hooks are no longer acceptable as part of a personal fall arrest system.
- Specific structural and strength requirements are set for:
  - Connectors
  - Dee-rings and snap hooks
  - Lanyards and vertical lifelines

  - -- Anchorage points

- Connectors must:
  - -- Be drop forged, pressed, or formed steel, or made of equivalent materials
  - -- Have a corrosion resistant finish and smooth surfaces to prevent damage from interfacing parts of the system
- Dee-rings and snap hooks must:
  - -- Have a minimum tensile strength of 5,000 lbs
  - -- Be proof-tested to a minimum load of 3,600 lbs with cracking or breaking
  - -- Be of the locking type (snap hooks only)
- Lanyards and vertical lifelines must have a minimum breaking strength of 5,000 lbs.
  - -- Self-retracting lifelines and lanyards that automatically limit free fall distance to 2 feet or less must sustain a minimum tensile load of 3,000 lbs applied to the device with the lifeline or lanyard fully extended.
  - -- Self-retracting lifelines and lanyards that do not limit free fall distance to 2 feet or less, rip-stitch lanyards, and tearing and deforming lanyards must sustain a minimum tensile load of 5,000 lbs applied to the device with lifeline or lanyard fully extended.
- Anchorages used for the attachment of the fall arrest system must be independent of any anchorage being used to support or suspend platforms, support at least 5000 lbs per employee attached, or be designed, installed and used as follows:
  - -- As part of a complete personal fall arrest system that maintains a safety factor of at least 2
  - -- Under the supervision of a qualified person
- Personal fall arrest systems must:
  - -- Limit maximum arresting force on an employee to 1,800 lbs. when used with a body harness
  - -- Be rigged so an employee can neither free fall for more than 6 feet nor contact any lower level
  - -- Bring an employee to a complete stop and limit maximum deceleration distance to 3.5 feet

-- Have strength to withstand twice the potential energy of an employee falling a distance of 6 feet, or the free fall distance permitted by the system, whichever is less

☐ OSHA 29 CFR 1926.502 requires that positioning device systems:

- Be rigged so that an employee cannot free fall more than 2 feet
- Be secured to an anchorage able to support at least twice the potential impact load of the employee's fall, or 3,000 lbs, whichever is greater
- Use connecting assemblies that have a minimum tensile strength of 5,000 lbs
- Meet other requirements for connectors, dee-rings, and snap hooks as specified for personal fall arrest systems

☐ Additional federal regulations require personal fall protection systems for:

- Scaffolding (OSHA 29 CFR 1910.28)
- Powered platforms for building maintenance (OSHA 29 CFR 1910.66)
- Respiratory protection - in immediately dangerous to life and health (IDLH) atmospheres for rescuing downed persons (OSHA 29 CFR 1910.134)
- Welding, cutting, and brazing (OSHA 29 CFR 1910.252)
- Pulp, paper and paperboard mills (OSHA 29 CFR 1910.261)
- Sawmills (OSHA 29 CFR 1910.265)
- Telecommunications (OSHA 29 CFR 1910.268)

☐ Additional requirements for personal fall arrest systems are provided in ANSI Z359.1-1992, *American National Standard on Safety Requirements for Personal Fall Arrest Systems, Subsystems and Components*; this standard specifically addresses:

- Design and performance requirements for:
  -- Full systems
  -- Components and elements
  -- Subsystems
- Qualification testing
- Marking and instructions
- User inspection, maintenance and storage of equipment

- Equipment, selection, rigging, use, and training
- Test methods

☐ Other requirements are given in ANSI A10,14-1991, American National Standard for Construction and Demolition Operations—Requirements for Safety Belts, Harnesses, Lanyards, and Lifelines for Construction and Demolition Use.

- This standard is limited to construction and demolition operations
- The standards establish different classes of systems:
- -- Class I systems are used for restraint and/or fall arrest where vertical free fall hazards exist.
- -- Class II devices are used for restraint but are not for use where any vertical free fall hazards exist.
- The standard addresses:
  - Performance requirements
  - Classification of systems for use
  - Users' responsibilities, selection, and use
  - Marking and instructions
  - Test methods

☐ ANSI Z359.1-1992 is considered more stringent than ANSI A10.14-1991 and is more closely followed by many fall protection system manufacturers.

☐ NFPA 1983 provides specific requirements for fire service life safety rope and system components, including:

- Life safety ropes
- Personal escape ropes
- Life safety harnesses
- Belts
- Auxiliary equipment

☐ Various safety factors are used throughout different industries for comparing the maximum load that will cause failure of a fall protection system to the recommended maximum load that can be used on the same system; safety factors ranging from 2 to 15 appear in different standards; higher safety factors will account for additional load and other conditions which can cause failure of a fall protection system.

45.2.2 Selection of Specific Fall Protection System Components

☐ Common fall protection systems components include:

- Positioning belts
- Full body harnesses
- Lanyards
- Lifelines
- Connectors
- Fall-arrest energy absorbers
- Rope grabs
- Confined space fall arrest/retrieval systems

☐ Positioning belts or harnesses are chest-waist harnesses that should only be used when no vertical free fall hazard is anticipated and where the intended use is restraint; position belt or harness characteristics and design features include:

- Nylon web shoulder straps attached to belts around the chest and waist
- Slide buckles for size adjustment
- Distribution of impact force between chest and waist in event of a fall

☐ Full body harnesses are a series of straps which are fastened around the wearer in a manner to contain the torso and are used whenever the danger for free fall exists; full body harness characteristics and design features include:

- Nylon web shoulder and leg straps attached to belts around the chest
- Fitted with ring in the back of the belt
- Distributes fall arrest forces over at least the upper thighs, pelvis, chest, and shoulders

☐ Lanyards are flexible lines of rope, wire rope, or straps that generally have connectors at each end for connecting the body support to an energy absorber, anchorage connector, or anchorage.

☐ Lifelines are flexible lines for:

  - Connection to an anchorage or anchorage connector at one end to hand vertically (vertical lifeline), or
  - Connection to anchorages or anchorage connectors at both ends to span horizontally (horizontal lifeline).

- ☐ Lanyards and lifelines may be retractable when built into a system that includes an energy absorber.

- ☐ Connectors are components of a fall protection system that are used to couple parts of the system; connectors are typically hardware such as:
  - Dee-rings
  - Snap hooks
    - -- Carabineer type
    - -- Lanyard type
    - -- Scaffold type
  - Swivel pulleys
  - Buckles

- ☐ Fall arrest energy absorbers are components whose primary function is to dissipate energy and limit deceleration forces which the system imposes on the body during fall arrest; there are three types of energy absorbers:
  - Personal energy absorbers, attached to a harness
  - Horizontal lifeline energy absorbers, attached to the top anchorage or anchorage connector of a vertical lifeline subsystem
  - Vertical lifeline energy absorbers, attached to one of the end anchorages or anchorage connectors of a vertical lifeline subsystem

- ☐ Rope grabs are devices that attach to the lifeline as an anchoring point to provide a means for arresting the fall; characteristics and design features of rope grabs include:
  - Deceleration device that travels on vertical lifeline which is attached to overhead anchorage
  - During the fall, rope grab applies friction to engage lifeline and absorb energy of fall, and then locks to arrest the fall
  - Work on the principle of inertial locking or cam/lever locking or both

- ☐ Confined space fall arrest/retrieval systems are fall protection systems which include many of the same components used in personal fall arrest systems; some confined space fall arrest/retrieval systems include tripod and winches for vertical confined space operations.

45.3 Personal Cooling Garments and Systems

☐ Cooling garments and systems are used to provide auxiliary cooling to individuals working in hot environments and to help prevent the onset of heat stress.

☐ Types of cooling garments and systems include:

- Portable air conditioners
- Vortex tubes
- Garments with circulating chilled water systems
- Vests with insertable cooling packets

☐ Cooling garments and systems must have an Intrinsically Safe Approval for any device that is electrically powered from any source that is going to be used in an explosive or potentially explosive environment.

45.3.1 Portable Air Conditioners

☐ Portable air conditioners are self-contained and use a battery-powered blower to flow air over a finned metallic container filled with frozen solution.

☐ Portable air conditioners may be worn or carried.

☐ The principal advantage of portable air conditioners is the cooling capability provided for hot environments.

☐ Disadvantages of portable air conditioners include:

- Weight
- Tethering for carried systems (limiting mobility)
- Limited duration (requiring frequent replenishment of frozen solution)

45.3.2 Vortex Tubes

☐ Vortex tubes cool air when compressed air passes through.

☐ Vortex tubes are typically used for encapsulating or full body suits.

- Advantages of vortex tubes include:
  - Unlimited air supply when attached to a compressor (a cascade of air cylinders may provide a limited supply of cooling air)
  - Replacement of humid air within the air space between the clothing and the wearer
- Disadvantages of vortex tubes include:
  - Limited cooling capacity
  - Restriction of movement from airline

45.3.3 Garments with Circulating Chilled Water Systems

- Garments that incorporate circulating chilled water systems use a form-fitting garment that contains several small tubes for circulating chilled water that comes from a heat exchanger.
- A small battery-powered pump is used to circulate the water through the heat exchanger that is typically filled with ice or cooling packets.
- Garments may be upper torso vests or whole body garments with or without cooling to the neck and head.
- The combination of the close-fitting garment and conductive heat transfer is effective in cooling the body surface temperature and helps to limit the rate of core temperature rise.
- Disadvantages of the circulating chilled water systems include:
  - Excessive weight
  - Bulk and restriction of movement
  - Difficulty in donning
  - Limited duration and frequency of replenishment

45.3.4 Vests with Insertable Cooling Packets

- Cooling vests have compartments that contain removable packets, such as ice packs, refreezable gel packs, or cooling materials that cool the body of the wearer. Phase change materials are also starting to be used to provide vests that regulate cooling more closely woth the wearer's body needs.

- ☐ Cooling vests are constructed from heavy-duty cotton or synthetic fabrics and have hook and loop closure tapes or adjustable buckles.
- ☐ The advantages of cooling vests with insertable packets are their portability and ease of use.
- ☐ The disadvantage of cooling vests with insertable packets include:
  - Excessive weight
  - Bulk and restriction of movement
  - Limited duration and frequency of replenishment
  - Difficulty in controlling the rate of cooling

45.4 Personal Communications Equipment

- ☐ Personal communications equipment includes radios and related devices that can be worn or carried by the worker for communicating with other workers, supervisors, or monitoring stations.
- ☐ Types of personal communications equipment include:
  - Hand-held radios
  - Headsets
  - Intercom and paging systems
  - Helmet-mounted systems
  - Respirator-mounted systems
  - Personal distress alarms
- ☐ Personal communications equipment is typically used for:
  - Operations covering large or remote areas to maintain procedures, to give clearly understood instructions, and to warn of possible hazards
  - Noisy environments
  - Worker entry or activity within hazardous areas
- ☐ Personal communications equipment must have an Intrinsic Safe Approval for any device that is electrically powered by any source that is going to be used in an explosive or potentially explosive environment.

45.4.1 Hand-Held Radios

- ☐ Hand-held radios are portable communications devices that are usually carried by the user or placed in a pocket of a garment.
- ☐ Most hand-held radios are transceivers that allow both transmission and receiving communications.
- ☐ Various options are available for:
  - Radio profile
  - Battery life
  - Speaker types
    - -- Headset
    - -- Ear phones
    - -- Speaker amplifier
  - Microphone types
    - -- Push-to-talk
    - -- Voice actuated
    - -- Noise cancelling (reduces transmission noise)
  - User controls
  - Communications system interfaces
- ☐ When selected, choose hand-held radios that can be:
  - Carried without interfering with worker activities
  - Operated for effective communications at work location
    - -- Transmission clarity
    - -- Range
  - Manipulated by wearer if protective clothing and gloves are worn

45.4.2 Headsets

- ☐ Headsets are communications systems that are built into ear muffs.
- ☐ Headsets consist of fluid-filled ear cups to provide sealing around wearer's ears and comfort combined with an adjustable headframe.
- ☐ Type of headsets include:

- One way communications systems (receive only)
- Two way communications systems (receiver with microphone)

☐ Headset may combine hearing protection and are therefore useful in noisy environments.

45.4.3 Intercom and Paging Systems

☐ Intercom systems are two-way communications systems that are worn by the individual worker with communications directly to a base station.

☐ Intercom systems may be tied to the base station by a wire or may be cordless.

☐ Intercom systems are often used in confined space operations.

☐ Paging systems provide selective one-way communications between supervisors and specific workers.

☐ Pagers may provide numeric or alphanumeric messages.

45.4.4 Helmet-Mounted and Respirator-Mounted Systems

☐ Helmets provide a convenient means for mounting radio or intercom components, and may include head sets and microphone options.

☐ Microphones may be mounted into respirator face masks for improving communications of workers wearing respirators.

☐ Respirator microphones are tied to a separate radio transceiver and can have various options as for a standard radio.

45.4.5 Personal Distress Alarms

☐ Personal distress alarms send out a signal or produce an audible alarm that is activated by either the individual in trouble or automatically become activated when a person is down for a specified period of time.

☐ Personal distress alarms alert others that a period is in trouble and can allow the person in trouble to be located.

☐ Personal distress alarms may be incorporated into radios or other personal communications equipment.

☐ NFPA 1982 provides a specification for Personal Alarm Safety Systems

(PASS) typically used by fire fighters; these systems send out an audible alarm when the fire fighter has not moved within 30 seconds.

45.5 Personal Flotation Devices

☐ Personal flotation devices (PFDs) are designed to protect personnel working in marine areas where there is a danger of falling into deep water.

☐ PFDs are intended to keep endangered persons afloat until rescue can be affected.

☐ There are five types of personal flotation devices (PFDs) which are approved by the U.S. Coast Guard.

- Type I
- Type II
- Type III
- Type IV
- Type V

☐ Some PFDs also required to include lights and retroreflective material for improved visibility of persons in the water or working under nighttime conditions in the marine area.

45.5.1 Personal Flotation Device - Type I

☐ Type I PFD characteristics and features include:

- Sleeveless jackets that surround the body are designed to turn a person, even if unconscious, from a face downward position in the water to a vertical or slightly upward position
- Have at least 20 pounds of buoyancy
- Constructed of puncture and water-resistant cotton or synthetic fabric shells and filled with kapok, fibrous glass, or unicellular foam
- Have reversible designs
- Are highly visible orange
- Most effective type of PFD
- Relatively easy to don

- ☐ Choose Type I PFDs for:
    - Rough water and conditions where immersion can occur
    - Personnel working on dredges, docks, watercraft and barges
    - Offshore rigging and drilling operations
    - Fishing industry

45.5.2 Personal Flotation Device - Type II

- ☐ Type II PFD characteristics and features include
    - Yoke-style vests designed to turn a person, even if unconscious, from a face downward position in the water to a vertical or slightly upward position (less effective than Type I PFD)
    - Have at least 15.5 pounds of buoyancy
    - Constructed of synthetic fabric shell and filled with kapok, cork, or unicellular foam
    - Use front tie tapes and polypropylene webbing straps
- ☐ Choose Type II PFDs for:
    - Boating and marine operations close to shore
    - Personnel working on construction projects near water such as bridges, dredges, docks, watercraft and barges

45.5.3 Personal Flotation Device - Type III

- ☐ Type III PFD characteristics and features include
    - Vests which allow conscious persons to place themselves in a vertical position and maintain that upward position
    - Have at least 15.5 pounds of buoyancy
    - Constructed of water resistant materials
    - Affords unimpaired body and arm movement
    - Relatively most comfortable to wear
- ☐ Choose Type III PFDs for:
    - Rescue operations when other workers are nearby
    - Personnel working on docks, loading, and off loading marine vessels

45.3.4 Personal Flotation Device - Type IV

☐ Type IV PFD characteristics and features include

- Buoyant cushions or rings designed to be thrown to someone in the water and grasped and held (not worn)
- Have at least 16.5 pounds of buoyancy
- Cushions
  - Are constructed kapok, fibrous glass, or unicellular foam
  - Covered with water-resistant nylon cloth, cotton, or vinyl coated fabric
  - Have one or more looped handles
- Rings or annular buoys
  - Contain unicellular foam, cork, or balsa wood
  - Covered with waterproof coating and canvas duck cover
  - Have grab line of polyethylene or polypropylene attached to outer edge

☐ Choose Type 1V PFDs as temporary device for throwing to workers who have accidently fallen into water.

45.5.5 Personal Flotation Device - Type V

☐ Type V PFD characteristics and features include

- Typically work vests that have received approval for specified, restricted uses
- Constructed of cotton or synthetic fabric shell and filled with unicellular foam
- Must be labeled as "Work Vest Only"

☐ Choose Type V PFDs for work operations where their use has been approved.

Section 46

Integrating PPE as an Ensemble

The purpose of this section is to provide guidelines for ensuring that all selected PPE items work together as an ensemble for providing the worker protection against anticipated hazards.

46.1 Integration Issues

- ☐ The majority of PPE represents an ensemble of individual items of protective clothing and equipment which must function together to afford the intended protection against identified workplace hazards.
- ☐ PPE ensembles must be selected such that:
  - The PPE items fit together to provide uniform protection over the areas of the body or body systems which need to be protected
  - Individual PPE items do not degrade or interfere with the performance of another item
  - The wearer can perform needed tasks without substantial impacts on required productivity
  - The overall ensemble does not cause extraneous bulk or weight that creates stress on the wearer
- ☐ Many integration problems occur at interfaces between PPE items; typical interface areas requiring attention include:
  - Upper torso garment to lower torso garment
  - Upper torso garment neck area to hood
  - Upper torso garment sleeve end to glove
  - Lower torso garment trouser end to footwear
  - Lower torso garment booties to footwear

- Outer garments to inner garments
- Hood to headwear
- Hood to face and eyewear
- Hood to respirator
- Hood to hearing protector
- Outer gloves to inner gloves
- Outer footwear to inner footwear
- Headwear to face and eyewear
- Headwear to respirator
- Headwear to hearing protector
- Face and eyewear to respirator
- Face and eyewear to hearing protector
- Respirator to hearing protector

☐ One of the primary interface problem areas is the head area since garment hoods, headwear, face and eyewear, respirators, and hearing protector elements of the ensemble may be present.

☐ While the majority of standard specifications apply to individual types of PPE, some standards define the specific items making up an ensemble:

- OSHA 29 CFR 1910.120 (protective ensembles for hazardous materials emergency response and waste site cleanup)
- NFPA 1951 (technical rescue protective ensemble)
- NFPA 1971 (structural fire fighting protective ensemble)
- NFPA 1976 (proximity fire fighting protective ensemble)
- NFPA 1991 (vapor-protective suits)
- NFPA 1992 (liquid splash-protective suits)

☐ A sample of a generalized ensemble specification for protective ensembles for hazardous materials emergency response and waste site cleanup is provided in Table 46-1.

Table 46-1. EPA Levels of Protection

LEVEL	EQUIPMENT	PROTECTION PROVIDED	SHOULD BE USED WHEN	LIMITING CRITERIA
A	RECOMMENDED: ? Pressure-demand, full-facepiece SCBA or pressure-demand supplied-air respirator with escape SCBA. ? Fully-encapsulating, chemical resistant suit. ? Inner chemical-resistant gloves. ? Chemical-resistant safety boot/shoes. ? Two-way radio communications. OPTIONAL: ? Cooling unit. ? Coveralls. ? Long cotton underwear. ? Hard hat. ? Disposable gloves and boot covers.	The highest available level of respiratory, skin, and eye protection.	? The chemical substance has been identified and requires the highest level of protection for skin, eyes, and the respiratory system based on either: - measured (or potential for) high concentration of atmo-spheric vapors, gases, or particulates; or - site operations and work functions involving a high potential for splash, immersion, or exposure to unexpected vapors, gases, or particulates of materials that are harmful to skin or capable of being absorbed through intact skin. ? Substances with a high degree of hazard to the skin are known or suspected to be present, and skin contact is possible. ? Operations must be conducted in confined, poorly ventilated areas until the absence of conditions requiring Level A protection is determined.	? Fully-encapsulating suit material must be compatible with the substances involved.
B	RECOMMENDED: ? Pressure-demand, full-facepiece SCBA or pressure-demand supplied-air respirator with escape SCBA. ? Chemical resistant clothing (overalls and long-sleeved jacket; hooded, one- or two-piece chemical splash suit; disposable chemical-resistant one-piece suit. ? Inner and outer chemical-resistant gloves. ? Chemical-resistant safety boot/shoes. ? Hard hat. ? Two-way radio communications. OPTIONAL: ? Coveralls. ? Disposable boot covers. ? Face shield. ? Long cotton underwear.	The same level of respiratory protection but less skin protection than Level A. It is the minimum level recommended for initial site entries until the hazards have been further identified.	? The type and atmospheric concentration of substances have been identified and require a high level of respiratory protection, but less skin protection. This involves atmospheres: - with IDLH concentrations of specific substances that do not represent a severe skin hazard; or - that do not meet the criteria for use of air-purifying respirators. ? Atmosphere contains less than 19.5 percent oxygen. ? Presence of incompletely identified vapors or gases is indicated by direct-reading organic vapor detection instrument, but vapors and gases are not suspected of containing high levels of chemicals harmful to skin or capable of being absorbed through intact skin.	? Use only when the vapor or gases present are not suspected of containing high concentrations of chemical that are harmful to skin or capable of being absorbed through intact skin. ? Use only when it is highly unlikely that the work being done will generate either high concentrations of vapors, gases, or particulates, or splashes material that will affect exposed skin.

Table 46-1. EPA Levels of Protection (Continued)

C	RECOMMENDED: ? Full-facepiece, air-purifying canister-equipped respirator. ? Chemical resistant clothing (overalls and long-sleeved jacket; hooded, one- or two-piece chemical splash suit; disposable chemical-resistant one-piece suit. ? Inner and outer chemical-resistant gloves. ? Chemical-resistant safety boot/shoes. ? Hard hat. ? Two-way radio communications. OPTIONAL: ? Coveralls. ? Disposable boot covers. ? Face shield. ? Escape mask ? Long cotton underwear.	The same level of skin protection as Level B, but a lower level of respiratory protection.	? The atmospheric contaminants, liquid splashes, or other direct contact will not adversely affect any exposed skin. ? The types of air contaminants have been identified, concentrations have been measured, and a canister is available that can remove the contaminant. ? All criteria for the use of air-purifying respirators are met.	? Atmospheric concentration of chemicals must not exceed IDLH levels. ? The atmosphere must contain at least 19.5 percent oxygen.
D	RECOMMENDED: ? Coveralls. ? Safety boots/shoes. ? Safety glasses or chemical splash goggles. ? Hard hat. OPTIONAL: ? Gloves. ? Escape mask. ? Face shield	No respiratory protection. Minimal skin protection.	? The atmosphere contains no known hazard. ? Work functions preclude splashes, immersion, or the potential for unexpected inhalation of or contact with hazardous levels of any chemicals.	? This level should not be worn in the Exclusion Zone. ? The atmosphere must contain at least 19.5 percent oxygen.

46.2 Ensemble Integration Guidelines

- ☐ Determine if protection must be uniform over the body or specific to certain areas or body systems as determined by risk assessment.
- ☐ Identify interface areas for which exposure to hazards may occur.
- ☐ Determine the appropriate underclothing to worn by the worker or the types of clothing that PPE must be worn over.
- ☐ Determine if multiple layers of PPE (especially garments, gloves, and footwear) will be required to achieve desired protection.
- ☐ Determine the type of protection that must be provided by each interface area.
- ☐ Decide which item of PPE should be responsible for providing protection to the interface area.
- ☐ For a given performance property, use information from the risk assessment to determine if the type of overall protection can be the same or should be different for items of PPE that interface in a particular area.
- ☐ Determine what type of integrity, if any, is needed for the overall ensemble or those parts of the ensemble which must provide protection; types of integrity include:
  - Particulate-tight integrity
  - Liquid-tight integrity
  - Vapor (gas)-tight integrity
- ☐ Apply performance criteria to interface areas as appropriate.
- ☐ Choose items of PPE that are designed to work together:
  - Match characteristics from interfacing items
  - Specify integrated systems provided by a single manufacturer
- ☐ Choose PPE that collectively integrates two or more PPE items; examples include:
  - Full body garments (combining upper and lower torso garments)
  - Garments with integrated hoods
  - Headwear that incorporates face and eyewear (or face and eyewear which directly attaches to headwear)
  - Headwear that incorporates hoods, face/eyewear, and/or respirators

- Full facepiece respirators (combining face/eyewear with respiratory protection)

☐ Ensure that combination PPE meets specifications for each separate item of PPE.

46.3 Evaluation of Selected Ensembles

☐ Conduct wearing trials of specified ensemble using simulated work tasks to determine potential effects from integrated items:

- Does the ensemble provide coverage to all areas of the wearer's body which require protection during all types of wearer movement?
- Do the ensemble materials irritate or sensitize the wearer's skin?
- Is the ensemble likely to retain contamination even after cleaning?
- Do parts of the ensemble design have the potential to be caught in moving machinery?
- Does the ensemble provide adequate fit of organization personnel?
- Can the ensemble be adjusted, used and reserviced without difficulty?
- Is the ensemble uncomfortable for the wearer under use conditions?
- Does the ensemble have sufficient durability for meeting the intended service life?

☐ Based on results from wearing trials, make adjustments in PPE items to provide the necessary performance.

Section 47

Establishing a PPE Program

The purpose of this section is to describe the principal elements of an organization's PPE program and how the program should be implemented.

47.1 Overview of the PPE Program

☐ The establishment of a written PPE program provides a method for documenting all aspects of PPE selection, use, care, and maintenance.

☐ Advantages of a PPE Program for an organization include:

- Documenting organization PPE procedures
- Establishing uniform and effective PPE usage guidelines
- Controlling PPE costs
- Creating user acceptance
- Meeting OSHA regulations for selecting and providing appropriate PPE

47.2 Elements of the PPE Program

☐ The minimum elements of a PPE program are as follows:

- A risk assessment methodology
- Procedures for the evaluation of other control options
- PPE selection criteria and procedures for determining the optimum choice
- PPE purchasing specifications
- User training procedures
- PPE usage criteria
- PPE care and maintenance procedures

- A validation plan for PPE selection coupled with medical surveillance
- An auditing plan to ensure that the PPE program is properly implemented

47.2.1 Risk Assessment Methodology

☐ The risk assessment methodology should provide a systematic approach for:

- Identifying workplace hazards
- Assessing risk associated with hazards
- Documenting the risk assessment

☐ Typically, some methods of hazard identification are outlined in a facility's Hazard Communication Program.

☐ This part of the program should detail the following elements:

- The persons who will identify the hazards and evaluate risks
    - -- Safety engineer
    - -- Industrial hygienist
    - -- Other trained individuals
- The types of equipment that should used for identification or evaluation
- Methods of evaluation
- Frequency of the evaluations

☐ A number of forms are provided in Appendix C which can used or adapted for documenting the risk assessment.

47.2.2 Evaluation of Other Control Options

☐ While the primary objective of the PPE program is the proper selection, use, care, and maintenance of PPE, part of the PPE program should address the evaluation of other control options for eliminating or reducing worker exposure.

☐ In response to the results of the hazard assessment, the individual or group within the organization responsible for making decisions should first consider engineering or administrative controls.

☐ Whenever possible, other control options should be implemented in lieu of PPE, particularly for routine and repetitive operations which required PPE.

47.2.3 PPE Selection Procedures

☐ The PPE program should contain detailed procedures for selecting PPE once the need for PPE has been identified through the risk assessment.

☐ PPE selection procedures should include:

- Developing PPE design and performance criteria
- Preparing PPE specifications
- Identifying candidate PPE
- Evaluating and choosing PPE
- Purchasing PPE
- Inspection of PPE upon receipt

☐ Specific guidance for selection of PPE is offered throughout this book.

47.2.4 PPE Purchase Specifications

☐ Purchase specifications for each item of PPE used by the organization, no matter how simple, should be included in the PPE program documentation.

47.2.5 User Training Specifications

☐ Adequate education and training for users of PPE is essential.

☐ OSHA 29 CFR Subpart I requires that employers provide training to employees in the use and care of PPE, and ensure that employees understand the training.

☐ Training and education should include:

- The nature and extent of workplace hazard(s)
- When PPE should be worn
- Which PPE is necessary
- An understanding of basic principles for how the PPE provides protection
- Use limitations of the PPE assigned
- How to properly inspect, don, doff, adjust, and wear PPE
- How to select the appropriate size of PPE

- How to recognize signs of heat stress and other ailments which may be associated with wearing PPE
- Decontamination or sterilization procedures (if needed)
- Signs of PPE wear, overexposure, or failure
- How to report PPE failure if it occurs
- Use of PPE under emergency conditions
- Proper storage, service life, care, and disposal of PPE

☐ The PPE program should establish responsibilities for conducting training, or at least specify the individual or group that is responsible for ensuring that training occurs.

☐ Some standards for specific industries require training with PPE at certain levels of competency and frequencies; requirements include:

- OSHA 29 CFR 1910.120 (hazardous materials responder and waste site worker training requirements)
- NFPA 472 (hazardous material responder competencies)
- NFPA 473 (EMS personnel competencies)
- NFPA 1500 (fire department personnel training)

47.2.6 PPE Usage Criteria

☐ PPE usage criteria should include:

- The conditions under which PPE should be used
- Which PPE should be used for specific work functions
- Limitations for PPE items

☐ One of the most effective means for specifying PPE usage is to establish a matrix of organization PPE items versus specific work functions.

- Check marks or X's in the matrix indicate the items that should be used for each job or task
- Separate narratives for specifying use conditions and PPE limitations

☐ The PPE program should also establish specific worker qualifications to use different types of PPE:

- Through required training

- Prerequisites as required by particular equipment or regulations; for example:
  - Respirator fit testing for use of respirators
  - Audiometric testing for use of hearing protectors

☐ The PPE program should identify available sizes of PPE; PPE must be provided in sufficient sizes to fit personnel required to wear PPE.

☐ Certain standards exist which govern use of different types of PPE:

- OSHA 29 CFR 1910.134 provides specific use requirements for respirators.
- OSHA 29 CFR 1910.137 provides specific use requirements for electrical protective equipment.
- OSHA 29 CFR 1910.1030 provides specific use requirements for PPE used against bloodborne pathogens
- ASTM F 1461 (chemical protective clothing)

47.2.7 PPE Care and Maintenance Procedures

☐ Proper care and maintenance of PPE is important to ensure that selected PPE continues to provide the intended protection to workers over its intended service life.

☐ OSHA 29 CFR Subpart I requires employers to maintain PPE for employees even if the employer does not provide PPE.

☐ PPE care includes procedures for:

- Cleaning
- Decontamination or sterilization
- Storage

☐ Cleaning is the process for removal of non-hazardous soiling or surface contamination such as dirt, dust, grease, body oils, etc.

☐ Decontamination is the physical and/or chemical process of reducing and preventing the spread of contamination of PPE.

☐ Sterilization is the physical and/or chemical inactivation and removal of biological contamination on PPE.

☐ Storage encompasses practices and conditions for properly storing PPE.

- ☐ PPE maintenance includes:
  - Inspection and testing
  - Repair
  - Removal from service
  - Disposal
- ☐ Inspection involves practices for routinely examining PPE for signs of wear, damage, or failure.
- ☐ Some types of PPE must be periodically tested for specific performance properties to ensure adequate protection.
- ☐ Repair encompasses manufacturer-approved practices for bringing PPE back into service.

  NOTE: Some PPE cannot be repaired.
- ☐ PPE removal from service is warranted when certain retirement criteria are met or when, in the estimation of the worker or authority designated within the organization, the PPE performance might have deteriorated.
- ☐ Some PPE, particularly PPE that has been contaminated must be properly disposed of.
- ☐ Procedures for care and maintenance of PPE should be in accordance with product manufacturer instructions.
- ☐ Responsibilities for PPE care and maintenance must be established within the organization.
- ☐ A few standards are available which provide general guidelines for care and maintenance of PPE:
  - OSHA 29 CFR 1910.134 provides specific care and maintenance requirements for respirators.
  - OSHA 29 CFR 1910.137 provides specific care and maintenance requirements for electrical protective equipment.
  - OSHA 29 CFR 1910.1030 provides specific care and maintenance requirements for PPE used against bloodborne pathogens
  - ASTM F 1449 (thermal and flame resistance protective clothing)
  - ISO 2801 (heat and flame protective clothing)
  - NFPA 1851 (structural fire fighting protective clothing)

47.2.8 Validation Plan for PPE Selections

- ☐ The PPE program should establish requirements for validating PPE selection decisions and ensuring that the proper PPE is used and used correctly.
- ☐ Responsibilities should be established throughout the organization for use of PPE.
- ☐ The individual or group responsible for PPE selections should periodically evaluate how selected PPE is providing protection and allowing workers to perform their required tasks.
- ☐ Means for workers providing feedback to the selection committee or individual should also be made available.
- ☐ Injury reports should be reviewed as another means for gaining feedback on the effectiveness of selected PPE.
- ☐ Some types of PPE will require medical surveillance to determine its effectiveness.
    - Medical surveillance programs should address the frequency and type of medical examinations or testing.
    - Medical surveillance results must be handled in a confidential manner.

47.2.9 Auditing Plan for PPE Program

- ☐ Provisions for periodically auditing the PPE program must be included to ensure long term adequacy of PPE selection decisions and proper use of PPE.
- ☐ The audit process should review:
    - New or revised standards requiring protection
    - Injury statistics
    - Worker complaints
    - Changes in tasks
    - Availability of alternative controls
    - Availability of new PPE technology
- ☐ The audit should be conducted at least annually.

47.3 Implementation of the PPE Program

☐ Employers have the ultimate responsibility and duty for the execution of the PPE program for ensuring the safety of their employees and compliance with applicable regulations.

☐ Employees have the responsibility and duty for using all PPE that is provided to them in accordance with the instructions and training that they have received.

☐ The PPE program must be prepared in writing for documenting organizational procedures with respect to all elements of the PPE program.

☐ Administration of the PPE program should be made the responsibility of an individual person with the assistance of a PPE Team.

☐ One of the essential aspects to a successful program is the formation of a PPE Team for establishing responsibilities for carrying out the plan.

☐ In some smaller organizations, the administration of the PPE program can be a responsibility of the organization's Safety Committee or similar group.

☐ The team members should be individuals who represent a cross section of the facility's operational and organization units; the following types of people are necessary to provide a working core:

- Safety/Industrial Hygienist
- Operational Supervision (i.e. Manager, Superintendent)
- Purchasing (key suppliers or vendors would be helpful)
- Operational Personnel

☐ The roles and responsibilities of these individuals should be clearly spelled out prior to the initiation of the facility's PPE Program:

- Approving the results of the risk assessment
- Recommending acceptable control options as an alternative to PPE
- Establishing a formal new equipment approval process
- Reviewing PPE evaluation plans and results of any PPE evaluation process
- Documenting decisions for allowing or removing PPE from the organization
- Maintaining the list of approved PPE for the organization

☐ Significant PPE problems should be brought to the attention of the individual within the organization responsible for administrating the PPE program.

Appendix A

Sources for Information on Safety and Health Regulations

OSHA Regional Offices

Occupational Safety and Health Administration (OSHA) Offices are distributed throughout the United States and its territories. OSHA Offices are divided among ten different regions and 86 different areas or districts. These offices may be contacted regarding issues related to personnel protective equipment requirements on specific OSHA regulations.

Region	Address	Phone Number
Region I (Connecticut, Maine, Massachusetts, New Hampshire, Rhode Island, and Vermont)		
Boston Regional Office	133 Portland Street 1st Floor, Boston, MA 02114	(617) 565-7164 x150
Region II (New Jersey, New York, and Puerto Rico)		
New York Regional Office	201 Varick Street Room 670 New York, NY 10014	(212) 337-2378
Region III (Delaware, District of Columbia, Maryland, Pennsylvania, Virginia, and West Virginia)		
Philadelphia Regional Office	Gateway Building Suite 2100 3535 Market Street Philadelphia, PA 19104	(215) 596-1201
Region IV (Alabama, Florida, Georgia, Kentucky, Mississippi, North Carolina, South Carolina, and Tennessee)		
Atlanta Regional Office	1375 Peachtree Street, N.E. Suite 587 Atlanta, GA 30367	(404) 347-3573

Region	Address	Phone Number
Region V (Indiana, Illinois, Michigan, Minnesota, Ohio, and Wisconsin)		
Chicago Regional Office	230 S. Dearborn Street Room 3244 Chicago, IL 60604	(312) 353-2220
Region VI (Arkansas, Louisiana, New Mexico, Oklahoma, and Texas)		
Dallas Regional Office	525 Griffin Street Room 602 Dallas, Texas 75202	(214) 767-4731
Region VII (Iowa, Kansas, Missouri, and Nebraska)		
Kansas City Regional Office	911 Walnut Street Room 406 Kansas City, MO 64106	(816) 426-5861
Region VIII (Colorado, Montana, North Dakota, South Dakota, Utah, and Wyoming)		
Denver Regional Office	Federal Building Room 1576 1961 Stout Street Denver, CO 80204	(303) 844-3061
Region IX (American Samoa, Arizona, California, Guam, Hawaii, Nevada, and Trust Territory of the Pacific Islands)		
San Francisco Regional Office	71 Stevenson Street Suite 420 San Francisco, CA 94105	(415) 744-6670
Region X (Alaska, Idaho, Oregon, and Washington)		
Seattle Regional Office	111 Third Avenue Suite 715 Seattle, WA 98101-3212	(206) 553-5930

Sources for State Safety and Health Standards

There are 25 states and territories which operate state OSHA programs provided for under Section 18 of the Occupational Safety and Health Act of 1970. The law declares that "any state which, at any time, desires to assume responsibility for development and enforcement therein of occupational safety and health standards relating to any occupational safety and health issue with respect to which a Federal standard has been promulgated under section 6 shall submit a State plan for the development of such standards and their enforcement." State standards and their enforcement must be at least as effective in providing safe and healthful working conditions as the federal program.

Presently, 21 states and two territories (Puerto Rico and the Virgin Islands) operate complete programs covering both private and public sector employers and employees. Two states, New York and Connecticut, operate state programs covering only public sector employers and employees. Federal OSHA provides coverage of private sector employers and employees in 29 states and the District of Columbia. The Act gives federal OSHA no authority to cover public employees.

The following table provides a listing of the state offices for occupational safety and health, indicating those states which have their own state plans:

State/Territory	State Organization	Phone Number
Alabama	Department of Labor 1789 Cong. W. L. Dickinson Drive Montgomery, AL 36130	(205) 242-3460
Alaska*	Alaska OSH 3301 Eagle Street Anchorage, AK 99510	(907) 264-2597
	Labor Standards & Safety Division P. O. Box 107021 Anchorage, AK 99510-7021	(907) 269-4914
American Samoa	Department of Human Resources Pago Pago, AS 96799	011-684-633-4485
Arizona*	State of Arizona Occupational Safety and Health P.O. Box 19070 Phoenix, AZ 85005-9070	(602) 542-5795
Arkansas	OSHA Consultation Section Department of Labor 10421 W, Markham Street Little Rock, AR 72205	(501) 682-4500
California*	CAL/OSHA 455 Golden Gate Avenue Room 5202 San Francisco, CA 94102	(415) 703-4341
Colorado	Labor Standards Unit Department of Labor and Employment 1120 Lincoln Street Suit 1305 Denver, CO 80203-2140	(303) 894-7551
Connecticut*	Connecticut Department of Labor Occupational Safety and Health Division 200 Folly Brook Blvd. Wethersfield, CT 06109	(203) 566-4500
Delaware	Department of Labor 820 N. French Street Corvel State Bldg., 6th Floor Wilmington, DE 19801	(302) 577-3908

State/Territory	State Organization	Phone Number
District of Columbia	Occupational Safety and Health Office 950 Upshur Street N.W. Washington, DC 20011	(202) 576-6339
Florida	Department of Labor & Employment Security 2012 Capitol Circle, S.E. Hartman Building, Room 303 Tallahassee, FL 32399-2152	(904) 488-3044
Georgia	Department of Labor 148 International Blvd. N.E. Suite 600 Atlanta, GA 30303-1751	(404) 656-3011
Guam	Department of Labor Government of Guam P. O. Box 9970 Tamuning, GU 96911	011-671-646-9241
Hawaii*	State of Hawaii Department of Labor and Industrial Relations Div. of Occupational Safety & Health 830 Punchbowl Street Honolulu, HI 96813	(808) 586-0116
Idaho	Department of Labor & Industrial Services 277 N. Sixth Street State House Mail Boise, ID 83720-6000	(208) 334-3950
Illinois	Department of Labor 160 N. LaSalle Street Suite 1300 Chicago, IL 60601	(312) 793-2800
Indiana*	Indiana Department of Labor Room 1013, State Office Building Indianapolis, IN 46204	(317) 232-2685
Iowa*	Iowa Division of Labor 100 East Grand Des Moines, IA 50319	(515) 281-3606
Kansas	Department of Human Resources 401 SW Topeka Blvd. Topeka, KS 66603-3182	(913) 296-7474
Kentucky*	Kentucky Occupational Safety and Health Program Kentucky Labor Cabinet 1047 U.S. 127 South, Suite 4 Frankfort, KY 40601	(502) 564-2300

State/Territory	State Organization	Phone Number
Louisiana	Department of Employee Training P. O. Box 94094 Baton Rouge, LA 70804-9094	(504) 342-3011
	Office of Public Health Services P. O. Box 3214 New Orleans, LA 70160	(504) 342-8094
Maryland*	State of Maryland-MOSH Division of Labor and Industry 501 St. Paul Place Baltimore, MD 21202	(410) 333-4195
Massachusetts	Department of Labor and Industries 100 Cambridge Street Boston, MA 02202	(617) 727-3463
Michigan*	Michigan Department of Public Health Division of Occupational Health P.O. Box 30035 Lansing, MI 48909	(517) 373-9600
Minnesota*	Minnesota Occupational Health and Safety 443 Lafayette Road St. Paul, MN 55155	(612) 296-2116
Mississippi	Environmental Health Bureau P. O. Box 1700 Jackson, MS 39215-1700	(601) 987-7518
Missouri	Department of Labor and Industrial Relations P. O. Box 504 Jefferson City, MO 65102	(314) 751-4091
Montana	State Bureau Labor and Industry Department P. O. Box 1728 Helena, MT 59624	(406) 444-1605
Nebraska	Department of Labor P. O. Box 94600 Lincoln, NE 68509-4600	(402) 471-2239
Nevada*	Nevada Occupational Safety and Health Enforcement Section 1370 South Curry Street Carson City, NV 89710	(702) 687-5240
	Labor Office 1445 Hot Springs Road Suite 108 Carson City, NV 89710	(702) 678-4850

State/Territory	State Organization	Phone Number
New Hampshire	Department of Labor 95 Pleasant Street Concord, NH 03301	(603)-271-3171
	Bureau of Health Risk Assessment Division of Public Health Services Health and Welfare Building 6 Hazen Drive Concord, NH 03301-6527	(603) 271-4664
New Jersey	Occupational Safety and Health Office Department of Labor Standards and Safety Enforcement John Fitch Plaza CN 054 Trenton, NJ 08625-0054	(609) 984-3507
New Mexico*	New Mexico Occupational Health and Safety Bureau P.O. Box 26110 Santa Fe, NM 87502	(505) 827-2877
New York*	New York State Department of Labor Public Employees Safety and Health Program Room 457, Building 12 State Office Building Campus Albany, NY 12240	(518) 457-1263
North Carolina*	North Carolina Department of Labor Division of Occupational Safety and Health 319 Chapanoke Road Raleigh, NC 27603-3432	(919) 662-4575
North Dakota	Labor Department P. O. Box 5523 Bismarck, ND 58502-5523	(701) 224-3650
Ohio	OSHA On-site Consultation Department of Industrial Relations P. O. Box 825 Columbus, OH 43216	(614) 644-2631
Oklahoma	Department of Labor 4001 N. Lincoln Blvd. Oklahoma City, OK 73105	(405) 528-1500
Oregon*	Occupational Safety and Health Division (OR-OSHA) Labor and Industries Bldg., Room 430 Salem, OR 97310	(503) 378-3272

State/Territory	State Organization	Phone Number
Pennsylvania	Occupational and Industrial Safety Labor and Industry Building Room 1700 7th & Forester Streets Harrisburg, PA 17120	(717) 787-3323
Puerto Rico*	Department of Labor and Human Resources Occupational Safety and Health Offices 505 Munoz Rivera Avenue Hato Rey, PR 00918	(809) 754-2171
Rhode Island	Department of Labor 610 Manton Avenue Providence, RI 02907	(401) 457-1833
South Carolina*	South Carolina Department of Labor Box 11329 Columbia, SC 29211	(803) 734-9600
South Dakota	Department of Labor 700 Governors Drive Pierre, SD 57501-2291	(605) 773-3101
Tennessee*	Tennessee Department of Labor Division of Occupation Safety and Health 501 Union Building Nashville, TN 37219	(615) 741-2793
Texas	Environmental & Consumer Health Department of Health 1100 W. 49th Street Austin, TX 78756	(512) 458-7541
Utah*	Occupational Safety and Health Division Utah Industrial Commission 160 East 300 South, Third Floor Salt Lake City, UT 84114-6650	(801) 530-6901
Vermont*	Vermont Occupational Safety and Health Administration National Life Building, Drawer 20 Montepelier, VT 05602-3401	(802) 828-2765
Virgin Islands*	Department of Labor 2131 Hospital Street Christiansted, St. Croix, VI 00828-4660	(809) 773-1994
Virginia*	Commonwealth of Virginia Department of Labor and Industry Powers-Taylor Building 13 South Thirteenth Street Richmond, VA 23219	(804) 786-5873

State/Territory	State Organization	Phone Number
Washington*	Washington State Department of Labor and Industries P.O. Box 4401 Olympia, WA 98504-4001	(360) 902-4200
West Virginia	West Virginia Department of Labor Capitol Complex Bldg. #3 Room 319 1900 Kanawha Blvd. East Charleston, WV 25303	(304) 558-7890
	Department of Health Capitol Complex Bldg. #3 Charleston, WV 25303	(304) 558-2971
Wisconsin	Department of Industry, Labor, and Human Relations P. O. 7969 Madison, WI 53707	(608) 266-1816
Wyoming*	Wyoming Occupational Health and Safety Herschler Building, Second Floor, East 122 West 25th Street Cheyenne, WY 82002	(307) 777-7786

Sources for Canadian Provincial Safety and Health Standards

The enactment of legislation pertaining to safety and health regulations is in the domain of thirteen jurisdictions in Canada: the federal government, the ten provinces, and the two territories. This delineation is constitutional, and there is no overlap between the federal regulations and those of the provinces or territories. While each jurisdiction has promulgated safety and health regulations that address issues similar to those in United States Title 29, these regulations may vary, both by name and content, between jurisdictions. The appropriate authority for each jurisdiction must be consulted for current legislation addressing safety and health standards. The name by which this authority is identified may also vary by jurisdiction.

The federal regulations cover workers who, by the nature of their activities, fall under federal jurisdiction. One might consider these activities to be "interprovincial" rather than "intraprovincial" in nature. These activities include, for example, certain ones related to maritime navigation, railway, telecommunications, airline, banking, and, under special provision, atomic and nuclear facilities. Federal public service employees are also covered under the federal regulations. The federal regulations are developed on a consensus basis by working committees comprised of representatives of both employer and employee groups. Thus, the style of the federal regulations may vary, with some committees favoring prescriptive and others performance-based requirements.

The provincial and territorial jurisdictions address health and safety issues for activities within that jurisdiction, such as industry or construction. The regulations may differ greatly between provinces or territories, with some jurisdictions focusing on different activities or industries. The western provinces, for example, emphasize the petroleum industry, while the eastern provinces emphasize mining. The style and even the name of safety and health regulations also can vary between provinces and territories. Ontario, for example, has adopted an Occupational Health and Safety Act, while British Colombia has a Workers' Compensation Act. Some jurisdictions have enacted generic regulations (e.g., ones that address *sound* or *fall protection*), while others have enacted industry or activity-specific regulations (e.g., ones that address *construction*). The language of these regulations may be prescriptive or performance-based.

Efforts towards enacting harmonized regulations between the jurisdictions are presently in the early stages. Political, historical, and economic considerations complicate this undertaking.

Sources for information on policies and legislation pertaining to health and safety for each jurisdiction are listed in the first table The Canadian Centre of Occupational Safety and Health, a public organization funded primarily by the federal government, makes available the *Canada Labour Code, Part II* (containing all of the provincial and federal regulations, as well as the federal policy manuals), on CD-ROM. The Canada Communication Group provides the *Canada Labour Code, Part II* in a loose-leaf format. The addresses and telephone numbers for these organizations are included in second table.

Sources for Information on Health and Safety Regulations in Canada

Province	Organization	Phone/ Fax Numbers
Alberta	Workplace Health, Safety & Strategic Services Alberta Labour	(403) 427-6724 (403) 427-5698
British Columbia	Workers' Compensation Board of British Columbia	(250) 387-1755 (250) 387-0878
Manitoba	Workplace Safety & Health Division Department of Labour	(204) 945-2351 (204) 945-4556
New Brunswick	Health & Safety Services Workplace, Health, Safety & Compensation Commission of New Brunswick	(506) 444-5508 (506) 453-7982
Newfoundland	Occupational Health & Safety Department of Environment and Labour	(709) 729-3619 (709) 729-6639
Nova Scotia	Occupational Health and Safety Department of Labour	(902) 424-4328 (902) 424-3239
Ontario	Occupational Health & Safety Branch Operations Division Ministry of Labour	(416) 326-7859 (416) 326-7650

Province	Organization	Phone/ Fax Numbers
Prince Edward Island	Occupational Health & Safety Division	(902) 368-5680 (902) 368-5705
Quebec	Service des relations avec le reseau Commission de la sante et de la securite du travail Gouvernement du Quebec	(514) 873-6364 (514) 864-9985
Saskatchewan	Occupational Health & Safety Division Saskatchewan Labour	(306) 787-3369 (306) 787-7729
Northwest Territories	Workers Compensation Board	(403) 920-3869 (403) 873-4596
Yukon	Occupational Health & Safety Branch Workers' Compensation Health & Safety Board	(403) 667-8616 (403) 393-6279
Federal	Labour Branch of Human Resources Development Canada	(819) 953-0229 (819) 953-1743

Sources for the *Canada Labour Code, Part II*

Source	Address	Phone No.
Canada Communication Group -Publishing	Ottawa, Canada K1A 0S9	(819) 956-4800
Canadian Centre for Occupational Health and Safety	250 Main Street East Hamilton, Ontario L8N 1H6	(800) 668-4284 *or* (905) 570-8094

Appendix B

Subpart I of OSHA 29 CFR 1910—Personal Protective Equipment

Authority: Sections 4, 6 and 8, Occupational Safety and Health Act of 1970 (29 U.S.C. 653, 655, 657); Secretary of Labor's Order No. 12-71 (36 FR 8754), 8-76 (41 FR 25059), 9-83 (48 FR 35736), or 1-90 (55 FR 9033), as applicable.

Sections 1910.132, and 1910.138 also issued under 29 CFR part 1911.

Sections 1910.133, 1910.135, and 1910.136 also issued under 29 CFR part 1911 and 5 U.S.C. 553. 406

§ 1910.132 General requirements.

(a) *Application.* Protective equipment, including personal protective equipment for eyes, face, head, and extremities, protective clothing, respiratory devices, and protective shields and barriers, shall be provided, used, and maintained in a sanitary and reliable condition wherever it is necessary by reason of hazards of processes or environment, chemical hazards, radiological hazards, or mechanical irritants encountered in a manner capable of causing injury or impairment in the function of any part of the body through absorption, inhalation or physical contact.

(b) *Employee-owned equipment.* Where employees provide their own protective equipment, the employer shall be responsible to assure its adequacy, including proper maintenance, and sanitation of such equipment.

(c) *Design.* All personal protective equipment shall be of safe design and construction for the work to be performed.

(d) *Hazard assessment and equipment selection.* (1) The employer shall assess the workplace to determine if hazards are present, or are likely to be present, which necessitate the use of personal protective equipment (PPE). If such hazards are present, or likely to be present, the employer shall:

(i) Select, and have each affected employee use, the types of PPE that will protect the affected employee from the hazards identified in the hazard assessment;

(ii) Communicate selection decisions to each affected employee; and,

(iii) Select PPE that properly fits each affected employee.

Note: Non-mandatory Appendix B contains an example of procedures that would comply with the requirement for a hazard assessment.

(2) The employer shall verify that the required workplace hazard assessment has been performed through a written certification that identifies the workplace evaluated; the person certifying that the evaluation has been performed; the date(s) of the hazard assessment; and, which identifies the document as a certification of hazard assessment.

(e) *Defective and damaged equipment.* Defective or damaged personal protective equipment shall not be used.

(f) *Training.* (1) The employer shall provide training to each employee who is required by this section to use PPE. Each such employee shall be trained to know at least the following:

(i) When PPE is necessary;

(ii) What PPE is necessary;

(iii) How to properly don, doff, adjust, and wear PPE;

(iv) The limitations of the PPE; and,

(v) The proper care, maintenance, useful life and disposal of the PPE.

(2) Each affected employee shall demonstrate an understanding of the training specified in paragraph (f)(1) of this section, and the ability to use PPE properly, before being allowed to perform work requiring the use of PPE.

(3) When the employer has reason to believe that any affected employee who has already been trained does not have the understanding and skill required by paragraph (f)(2) of this section, the employer shall retrain each such employee. Circumstances where retraining is required include, but are not limited to, situations where:

(i) Changes in the workplace render previous training obsolete; or

(ii) Changes in the types of PPE to be used render previous training obsolete; or

(iii) Inadequacies in an affected employee's knowledge or use of assigned PPE indicate that the employee has not retained the requisite understanding or skill.

(4) The employer shall verify that each affected employee has received and understood the required training through a written certification that contains the name of each employee trained, the date(s) of training, and that identifies the subject of the certification.

(g) Paragraphs (d) and (f) of this section apply only to 1910.133, 1910.135, 1919.136, and 1910.138. Parpagraphs (d) and (f) of this section do not apply to 1910.134 and 1910.137.

[39 FR 23502, June 27, 1974, as amended at 59 FR 16334, Apr. 6, 1994; 59 FR 33910, July 1, 1994]407

§ 1910.133 Eye and face protection.

(a) *General requirements.* (1) The employer shall ensure that each affected employee uses appropriate eye or face protection when exposed to eye or face hazards from flying particles, molten metal, liquid chemicals, acids or caustic liquids, chemical gases or vapors, or potentially injurious light radiation.

(2) The employer shall ensure that each affected employee uses eye protection that provides side protection when there is a hazard from flying objects. Detachable side protectors (e.g. clip-on or slide-on side shields) meeting the pertinent requirements of this section are acceptable.

(3) The employer shall ensure that each affected employee who wears prescription lenses while engaged in operations that involve eye hazards wears eye protection that incorporates the prescription in its design, or wears eye protection that can be worn over the prescription lenses without disturbing the proper position of the prescription lenses or the protective lenses.

(4) Eye and face PPE shall be distinctly marked to facilitate identification of the manufacturer.

(5) The employer shall ensure that each affected employee uses equipment with filter lenses that have a shade number appropriate for the work being performed for protection from injurious light radiation. The following is a listing of appropriate shade numbers for various operations.

(b) *Criteria for protective eye and face devices.* (1) Protective eye and face devices purchased after July 5, 1994 shall comply with ANSI Z87.1-1989, "American National Standard Practice for Occupational and Educational Eye and Face Protection," which is incorporated by reference as specified in 1910.6, or shall be demonstrated by the employer to be equally effective.

(2) Eye and face protective devices purchased before July 5, 1994 shall comply with the ANSI "USA standard for Occupational and Educational Eye and Face Protection," Z87.1-1968, which is incorporated by reference as specified in 1910.6, or shall be demonstrated by the employer to be equally effective.

[59 FR 16360, Apr. 6, 1994; 59 FR 33911, July 1, 1994, as amended at 61 FR 9238, Mar. 7, 1996; 61 FR 19548, May 2, 1996]

Filter Lenses for Protection against Radiant Energy

Operations		Electrode Size 1/32 in.	Arc Current	Minimum* Protective Shade
Shielded metal arc welding		Less than 3	Less than 60.	7
		3-5	60-160	8
		5-8	160-250	10
		More than 8	250-550	11
Gas metal arc welding and flux cored arc welding		less than 60		7
		60-160		10
		160-250		10
		250-500		10
Gas Tungsten arc welding		less than 50		8
		50-150		8
		150-500		10
Air carbon	(Light)	less than 500		10
Arc cutting	(Heavy).	500-1000		11
Plasma arc welding		less than 20		6
		20-100		8
		100-400		10
		400-800		11
Plasma arc cutting	(light)**.	less than 300		8
	(medium)**	300-400		9
	(heavy)**	400-800		10
Torch brazing				3
Torch soldering				2
Carbon arc welding				14

Filter Lenses for Protection against Radiant Energy

Operations	Plate thickness–inches	Plate thickness–mm	Minimum* Protective Shade
Gas Welding:			
Light	Under 1/8	Under 3.2	4
Medium	1/8 to ½	3.2 to 12.7	5
Heavy	Over ½	Over 12.7	6
Oxygen cutting:			
Light	Under 1	Under 25	3
Medium	1 to 6	25 to 150	4
Heavy	Over 6	Over 150	5

* As a rule of thumb, start with a shade that is too dark to see the weld zone. Then go to a lighter shade which gives sufficient view of the weld zone without going below the minimum. In oxyfuel gas welding or cutting where the torch produces a high yellow light, it is desirable to use a filter lens that absorbs the yellow or sodium line in the visible light of the (spectrum) operation.

** These values apply where the actual arc is clearly seen. Experience has shown that lighter filters may be used when the arc is hidden by the workpiece.

§ 1910.134 Respiratory protection.

(a) *Permissible practice.* (1) In the control of those occupational diseases caused by breathing air contaminated with harmful dusts, fogs, fumes, mists, gases, smokes, sprays, or vapors, the primary objective shall be to prevent atmospheric contamination. This shall be accomplished as far as feasible by accepted engineering control measures (for example, enclosure or confinement of the operation, general and local ventilation, and substitution of less toxic materials). When effective engineering controls are not feasible, or while they are being instituted, appropriate respirators shall be used pursuant to the following requirements.

(2) Respirators shall be provided by the employer when such equipment is necessary to protect the health of the employee. The employer shall provide the respirators which are applicable and suitable for the purpose intended. The employer shall be responsible for the establishment and maintenance of a respiratory protective program which shall include the requirements outlined in paragraph (b) of this section.

(3) The employee shall use the provided respiratory protection in accordance with instructions and training received.

(b) *Requirements for a minimal acceptable program.* (1) Written standard operating procedures governing the selection and use of respirators shall be established.

(2) Respirators shall be selected on the basis of hazards to which the worker is exposed.

(3) The user shall be instructed and trained in the proper use of respirators and their limitations.

(4) [Reserved]

(5) Respirators shall be regularly cleaned and disinfected. Those used by 409more than one worker shall be thoroughly cleaned and disinfected after each use.

(6) Respirators shall be stored in a convenient, clean, and sanitary location.

(7) Respirators used routinely shall be inspected during cleaning. Worn or deteriorated parts shall be replaced. Respirators for emergency use such as self-contained devices shall be thoroughly inspected at least once a month and after each use.

(8) Appropriate surveillance of work area conditions and degree of em- ployee exposure or stress shall be maintained.

(9) There shall be regular inspection and evaluation to determine the continued effectiveness of the program.

(10) Persons should not be assigned to tasks requiring use of respirators unless it has been determined that they are physically able to perform the work and use the equipment. The local physician shall determine what health and physical conditions are pertinent. The respirator user's medical status should be reviewed periodically (for instance, annually).

(11) Respirators shall be selected from among those jointly approved by the Mine Safety and Health Administration and the National Institute for Occupational Safety and Health under the provisions of 30 CFR part 11.

(c) *Selection of respirators.* Proper selection of respirators shall be made according to the guidance of American National Standard Practices for Respiratory Protection Z88.2-1969.

(d) *Air quality.* (1) Compressed air, compressed oxygen, liquid air, and liquid oxygen used for respiration shall be of high purity. Oxygen shall meet the requirements of the United States Pharmacopoeia for medical or breathing oxygen. Breathing air shall meet at least the requirements of the specification for Grade D breathing air as described in Compressed Gas Association Commodity Specification G-7.1-1966. Compressed oxygen shall not be used in supplied-air respirators or in open circuit self-contained breathing apparatus that have previously used compressed air. Oxygen must never be used with air line respirators.

(2) Breathing air may be supplied to respirators from cylinders or air compressors.

(i) Cylinders shall be tested and maintained as prescribed in the Shipping Container Specification Regulations of the Department of Transportation (49 CFR part 178).

(ii) The compressor for supplying air shall be equipped with necessary safety and standby devices. A breathing air-type compressor shall be used. Compressors shall be constructed and situated so as to avoid entry of contaminated air into the system and suitable in-line air purifying sorbent beds and filters installed to further assure breathing air quality. A receiver of sufficient capacity to enable the respirator wearer to escape from a contaminated atmosphere in event of compressor failure, and alarms to indicate compressor failure and overheating shall be installed in the system. If an oil-lubricated compressor is used, it shall have a high-temperature or carbon monoxide alarm, or both. If only a high-temperature alarm is used, the air from the compressor shall be frequently tested for carbon monoxide to insure that it meets the specifications in paragraph (d)(1) of this section.

(3) Air line couplings shall be incompatible with outlets for other gas systems to prevent inadvertent servicing of air line respirators with nonrespirable gases or oxygen.

(4) Breathing gas containers shall be marked in accordance with American National Standard Method of Marking Portable Compressed Gas Containers to Identify the Material Contained, Z48.1-1954; Federal Specification BB-A-1034a, June 21, 1968, Air, Compressed for Breathing Purposes; or Interim Federal Specification GG-B-00675b, April 27, 1965, Breathing Apparatus, Self-Contained.

(e) *Use of respirators.* (1) Standard procedures shall be developed for respirator use. These should include all information and guidance necessary for their proper selection, use, and care. Possible emergency and routine uses of respirators should be anticipated and planned for.

(2) The correct respirator shall be specified for each job. The respirator type is usually specified in the work 410procedures by a qualified individual supervising the respiratory protective program. The individual issuing them shall be adequately instructed to insure that the correct respirator is issued.

(3) Written procedures shall be prepared covering safe use of respirators in dangerous atmospheres that might be encountered in normal operations or in emergencies. Personnel shall be familiar with these procedures and the available respirators.

(i) In areas where the wearer, with failure of the respirator, could be overcome by a toxic or oxygen-deficient atmosphere, at least one additional man shall be present. Communications (visual, voice, or signal line) shall be maintained between both or all individuals present. Planning shall be such that one individual will be unaffected by any likely incident and have the proper rescue equipment to be able to assist the other(s) in case of emergency.

(ii) When self-contained breathing apparatus or hose masks with blowers are used in atmospheres immediately dangerous to life or health, standby men must be present with suitable rescue equipment.

(iii) Persons using air line respirators in atmospheres immediately hazardous to life or health shall be equipped with safety harnesses and safety lines for lifting or removing persons from hazardous atmospheres or other and equivalent provisions for the rescue of persons from hazardous atmospheres shall be used. A standby man or men with suitable self-contained breathing apparatus shall be at the nearest fresh air base for emergency rescue.

(4) Respiratory protection is no better than the respirator in use, even though it is worn conscientiously. Frequent random inspections shall be conducted by a qualified individual to assure that respirators are properly selected, used, cleaned, and maintained.

(5) For safe use of any respirator, it is essential that the user be properly instructed in its selection, use, and maintenance. Both supervisors and workers shall be so instructed by competent persons. Training shall provide the men an opportunity to handle the respirator, have it fitted properly, test its face-piece-to-face seal, wear it in normal air for a long familiarity period, and, finally, to wear it in a test atmosphere.

(i) Every respirator wearer shall receive fitting instructions including demonstrations and practice in how the respirator should be worn, how to adjust it, and how to determine if it fits properly. Respirators shall not be worn when conditions prevent a good face seal. Such conditions may be a growth of beard, sideburns, a skull cap that projects under the facepiece, or temple pieces on glasses. Also, the absence of one or both dentures can seriously affect the fit of a facepiece. The worker's diligence in observing these factors shall be evaluated by periodic check. To assure proper protection, the facepiece fit shall be checked by the wearer each time he puts on the respirator. This may be done by following the manufacturer's facepiece fitting instructions.

(ii) Providing respiratory protection for individuals wearing corrective glasses is a serious problem. A proper seal cannot be established if the temple bars of eye glasses extend through the sealing edge of the full facepiece. As a temporary measure, glasses with short temple bars or without temple bars may be taped to the wearer's head. Wearing of contact lenses in contaminated atmospheres with a respirator shall not be allowed. Systems have been developed for mounting corrective lenses inside full facepieces. When a workman must wear corrective lenses as part of the facepiece, the facepiece and lenses shall be fitted by qualified individuals to provide good vision, comfort, and a gas-tight seal.

(iii) If corrective spectacles or goggles are required, they shall be worn so as not to affect the fit of the facepiece. Proper selection of equipment will minimize or avoid this problem.

(f) *Maintenance and care of respirators.* (1) A program for maintenance and care of respirators shall be adjusted to the type of plant, working conditions, and hazards involved, and shall include the following basic services:

(i) Inspection for defects (including a leak check),

maintenance and care of respirators shall be adjusted to the type of plant, working conditions, and hazards involved, and shall include the following basic services:

(i) Inspection for defects (including a leak check),

(ii) Cleaning and disinfecting,

(iii) Repair,

(iv) Storage.

Equipment shall be properly maintained to retain its original effectiveness.

(2) (i) All respirators shall be inspected routinely before and after each use. A respirator that is not routinely used but is kept ready for emergency use shall be inspected after each use and at least monthly to assure that it is in satisfactory working condition.

(ii) Self-contained breathing apparatus shall be inspected monthly. Air and oxygen cylinders shall be fully charged according to the manufacturer's instructions. It shall be determined that the regulator and warning devices function properly.

(iii) Respirator inspection shall include a check of the tightness of connections and the condition of the facepiece, headbands, valves, connecting tube, and canisters. Rubber or elastomer parts shall be inspected for pliability and signs of deterioration. Stretching and manipulating rubber or elastomer parts with a massaging action will keep them pliable and flexible and prevent them from taking a set during storage.

(iv) A record shall be kept of inspection dates and findings for respirators maintained for emergency use.

(3) Routinely used respirators shall be collected, cleaned, and disinfected as frequently as necessary to insure that proper protection is provided for the wearer. Respirators maintained for emergency use shall be cleaned and disinfected after each use.

(4) Replacement or repairs shall be done only by experienced persons with parts designed for the respirator. No attempt shall be made to replace components or to make adjustment or repairs beyond the manufacturer's recommendations. Reducing or admission valves or regulators shall be returned to the manufacturer or to a trained technician for adjustment or repair.

(5) (i) After inspection, cleaning, and necessary repair, respirators shall be stored to protect against dust, sunlight, heat, extreme cold, excessive moisture, or damaging chemicals. Respirators placed at stations and work areas for emergency use should be quickly accessible at all times and should be stored in compartments built for the purpose. The compartments should be clearly marked. Routinely used respirators, such as dust respirators, may be placed in plastic bags. Respirators should not be stored in such places as lockers or tool boxes unless they are in carrying cases or cartons.

(ii) Respirators should be packed or stored so that the facepiece and exhalation valve will rest in a normal position and function will not be impaired by the elastomer setting in an abnormal position.

(iii) Instructions for proper storage of emergency respirators, such as gas masks and self-contained breathing apparatus, are found in "use and care" instructions usually mounted inside the carrying case lid.

(g) *Identification of gas mask canisters.* (1) The primary means of identifying a gas mask canister shall be by means of properly worded labels. The secondary means of identifying a gas mask canister shall be by a color code.

(2) All who issue or use gas masks falling within the scope of this section shall see that all gas mask canisters purchased or used by them are properly labeled and colored in accordance with these requirements before they are placed in service and that the labels and colors are properly maintained at all times thereafter until the canisters have completely served their purpose.

(3) On each canister shall appear in bold letters the following:

(i)______________________________

Canister for (Name for atmospheric contaminant)

or

Type N Gas Mask Canister

(ii) In addition, essentially the following wording shall appear beneath the appropriate phrase on the canister label: "For respiratory protection in atmospheres containing not more than________percent by volume of _________."

(Name of atmospheric contaminant)

(4) Canisters having a special high- efficiency filter for protection against radionuclides and other highly toxic particulates shall be labeled with a statement of the type and degree of protection afforded by the filter. The label shall be affixed to the neck end 412of, or to the gray stripe which is around and near the top of, the canister. The degree of protection shall be marked as the percent of penetration of the canister by a 0.3-micron-diameter dioctyl phthalate (DOP) smoke at a flow rate of 85 liters per minute.

(5) Each canister shall have a label warning that gas masks should be used only in atmospheres containing sufficient oxygen to support life (at least 16 percent by volume), since gas mask canisters are only designed to neutralize or remove contaminants from the air.

(6) Each gas mask canister shall be painted a distinctive color or combination of colors indicated in Table I-1. All colors used shall be such that they are clearly identifiable by the user and clearly distinguishable from one another. The color coating used shall offer a high degree of resistance to chipping, scaling, peeling, blistering, fading, and the effects of the ordinary atmospheres to which they may be exposed under normal conditions of storage and use. Appropriately colored pressure sensitive tape may be used for the stripes.

[39 FR 23502, June 27, 1974, as amended at 43 FR 49748, Oct. 24, 1978; 49 FR 5322, Feb. 10, 1984; 49 FR 18295, Apr. 30, 1984; 58 FR 35309, June 30, 1993]

Table I-1

Atmospheric contaminants to be protected against	Colors assigned[1]
Acid gases	White.
Hydrocyanic acid gas	White with 1/2-inch green stripe completely around the canister near the bottom.
Chlorine gas	White with 1/2-inch yellow stripe completely around the canister near the bottom.
Organic vapors	Black.
Ammonia gas	Green.
Acid gases and ammonia gas	Green with 1/2-inch white stripe completely around the canister near the bottom.
Carbon monoxide	Blue.
Acid gases and organic vapors	Yellow.
Hydrocyanic acid gas and chloropicrin vapor.	Yellow with 1/2-inch blue strip completely around the canister near the bottom.
Acid gases, organic vapors, and ammonia gases.	Brown.
Radioactive materials, excepting tritium and noble gases.	Purple (Magenta).
Particulates (dusts, fumes, mists, fogs, or smokes) in combination with any of the above gases or vapors.	Canister color for contaminant, as designated above, with 1/2-inch gray stripe completely around the canister near the top.
All of the above atmospheric contaminants.	Red with 1/2-inch gray stripe completely around the canister near the top.

[1]Gray shall not be assigned as the main color for a canister designed to remove acids or vapors.

Note: Orange shall be used as a complete body, or stripe color to represent gases not included in this table. The user will need to refer to the canister label to determine the degree of protection the canister will afford.

§ 1910.135 Head protection.

(a) *General requirements.* (1) The employer shall ensure that each affected employee wears a protective helmet when working in areas where there is a potential for injury to the head from falling objects.

(2) The employer shall ensure that a protective helmet designed to reduce electrical shock hazard is worn by each such affected employee when near exposed electrical conductors which could contact the head.

(b) *Criteria for protective helmets.* (1) Protective helmets purchased after July 5, 1994 shall comply with ANSI Z89.1-1986, "American National Standard for Personnel Protection-Protective Headwear for Industrial Workers-Requirements," which is incorporated by reference as specified in 1910.6, or shall be demonstrated to be equally effective.

(2) Protective helmets purchased before July 5, 1994 shall comply with the ANSI standard "American National Standard Safety Requirements for Industrial Head Protection," ANSI Z89.1-1969, which is incorporated by reference as specified in 1910.6, or shall be demonstrated by the employer to be equally effective.

[59 FR 16362, Apr. 6, 1994, as amended at 61 FR 9238, Mar. 7, 1996; 61 FR 19548, May 2, 1996]413

§ 1910.136 Foot protection.

(a) *General requirements.* The employer shall ensure that each affected employee uses protective footwear when working in areas where there is a danger of foot injuries due to falling or rolling objects, or objects piercing the sole, and where such employee's feet are exposed to electrical hazards.

(b) *Criteria for protective footwear.* (1) Protective footwear purchased after July 5, 1994 shall comply with ANSI Z41-1991, "American National Standard for Personal Protection-Protective Footwear," which is incorporated by reference as specified in 1910.6, or shall be demonstrated by the employer to be equally effective.

(2) Protective footwear purchased before July 5, 1994 shall comply with the ANSI standard "USA Standard for Men's Safety-Toe Footwear," Z41.1-1967, which is incorporated by reference as specified in 1910.6, or shall be demonstrated by the employer to be equally effective.

[59 FR 16362, Apr. 6, 1994; 59 FR 33911, July 1, 1994, as amended at 61 FR 9238, Mar. 7, 1996; 61 FR 19548, May 2, 1996; 61 FR 21228, May 9, 1996]

§ 1910.137 Electrical protective equipment.

(a) *Design requirements.* Insulating blankets, matting, covers, line hose, gloves, and sleeves made of rubber shall meet the following requirements:

(1) *Manufacture and marking.* (i) Blankets, gloves, and sleeves shall be produced by a seamless process.

(ii) Each item shall be clearly marked as follows:

(A) Class 0 equipment shall be marked Class 0.

(B) Class 1 equipment shall be marked Class 1.

(C) Class 2 equipment shall be marked Class 2.

(D) Class 3 equipment shall be marked Class 3.

(E) Class 4 equipment shall be marked Class 4.

(F) Non-ozone-resistant equipment other than matting shall be marked Type I.

(G) Ozone-resistant equipment other than matting shall be marked Type II.

(H) Other relevant markings, such as the manufacturer's identification and the size of the equipment, may also be provided.

(iii) Markings shall be nonconducting and shall be applied in such a manner as not to impair the insulating qualities of the equipment.

(iv) Markings on gloves shall be confined to the cuff portion of the glove.

(2) *Electrical requirements.* (i) Equipment shall be capable of withstanding the a-c proof-test voltage specified in Table I-2 or the d-c proof-test voltage specified in Table I-3.

(A) The proof test shall reliably indicate that the equipment can withstand the voltage involved.

(B) The test voltage shall be applied continuously for 3 minutes for equipment other than matting and shall be applied continuously for 1 minute for matting.

(C) Gloves shall also be capable of withstanding the a-c proof-test voltage specified in Table I-2 after a 16-hour water soak. (See the note following paragraph (a)(3)(ii)(B) of this section.)

(ii) When the a-c proof test is used on gloves, the 60-hertz proof-test current may not exceed the values specified in Table I-2 at any time during the test period.

(A) If the a-c proof test is made at a frequency other than 60 hertz, the permissible proof-test current shall be computed from the direct ratio of the frequencies.

(B) For the test, gloves (right side out) shall be filled with tap water and immersed in water to a depth that is in accordance with Table I-4. Water shall be added to or removed from the glove, as necessary, so that the water level is the same inside and outside the glove.

(C) After the 16-hour water soak specified in paragraph (a)(2)(i)(C) of this section, the 60-hertz proof-test current may exceed the values given in Table I-2 by not more than 2 milliamperes.

(iii) Equipment that has been subjected to a minimum breakdown voltage test may not be used for electrical protection. (See the note following paragraph (a)(3)(ii)(B) of this section.)

(iv) Material used for Type II insulating equipment shall be capable of withstanding an ozone test, with no visible effects. The ozone test shall reliably indicate that the material will resist 414ozone exposure in actual use. Any visible signs of ozone deterioration of the material, such as checking, cracking, breaks, or pitting, is evidence of failure to meet the requirements for ozone-resistant material. (See the note following paragraph (a)(3)(ii)(B) of this section.)

(3) *Workmanship and finish.* (i) Equipment shall be free of harmful physical irregularities that can be detected by the tests or inspections required under this section.

(ii) Surface irregularities that may be present on all rubber goods because of imperfections on forms or molds or because of inherent difficulties in the manufacturing process and that may appear as indentations, protuberances, or imbedded foreign material are acceptable under the following conditions:

(A) The indentation or protuberance blends into a smooth slope when the material is stretched.

(B) Foreign material remains in place when the insulating material is folded and stretches with the insulating material surrounding it.

Note: Rubber insulating equipment meeting the following national consensus standards is deemed to be in compliance with paragraph (a) of this section:

American Society for Testing and Materials (ASTM) D 120-87, Specification for Rubber Insulating Gloves.

ASTM D 178-93 (or D 178-88) Specification for Rubber Insulating Matting.

ASTM D 1048-93 (or D 1048-88a) Specification for Rubber Insulating Blankets.

ASTM D 1049-93 (or D 1049-88) Specification for Rubber Insulating Covers.

ASTM D 1050-90, Specification for Rubber Insulating Line Hose.

ASTM D 1051-87, Specification for Rubber Insulating Sleeves.

These standards contain specifications for conducting the various tests required in paragraph (a) of this section. For example, the a-c and d-c proof tests, the breakdown test, the water soak procedure, and the ozone test mentioned in this paragraph are described in detail in the ASTM standards.

(b) *In-service care and use.* (1) Electrical protective equipment shall be maintained in a safe, reliable condition.

(2) The following specific requirements apply to insulating blankets, covers, line hose, gloves, and sleeves made of rubber:

(i) Maximum use voltages shall conform to those listed in Table I-5.

(ii) Insulating equipment shall be inspected for damage before each day's use and immediately following any incident that can reasonably be suspected of having caused damage. Insulating gloves shall be given an air test, along with the inspection.

(iii) Insulating equipment with any of the following defects may not be used:

(A) A hole, tear, puncture, or cut;

(B) Ozone cutting or ozone checking (the cutting action produced by ozone on rubber under mechanical stress into a series of interlacing cracks);

(C) An embedded foreign object;

(D) Any of the following texture changes: swelling, softening, hardening, or becoming sticky or inelastic.

(E) Any other defect that damages the insulating properties.

(iv) Insulating equipment found to have other defects that might affect its insulating properties shall be removed from service and returned for testing under paragraphs (b)(2)(viii) and (b)(2)(ix) of this section.

(v) Insulating equipment shall be cleaned as needed to remove foreign substances.

(vi) Insulating equipment shall be stored in such a location and in such a manner as to protect it from light, temperature extremes, excessive humidity, ozone, and other injurious substances and conditions.

(vii) Protector gloves shall be worn over insulating gloves, except as follows:

(A) Protector gloves need not be used with Class 0 gloves, under limited-use conditions, where small equipment and parts manipulation necessitate unusually high finger dexterity.

Note: Extra care is needed in the visual examination of the glove and in the avoidance of handling sharp objects.

(B) Any other class of glove may be used for similar work without protector gloves if the employer can demonstrate that the possibility of physical damage to the gloves is small and if the class of glove is one class higher than that required for the voltage involved. Insulating gloves that have been used without protector gloves may not be used at a higher voltage 415until they have been tested under the provisions of paragraphs (b) (2) (viii) and (b) (2) (ix) of this section.

(viii) Electrical protective equipment shall be subjected to periodic electrical tests. Test voltages and the maximum intervals between tests shall be in accordance with Table I-5 and Table I-6.

(ix) The test method used under paragraphs (b)(2)(viii) and (b)(2)(ix) of this section shall reliably indicate whether the insulating equipment can withstand the voltages involved.

Note: Standard electrical test methods considered as meeting this requirement are given in the following national consensus standards:

American Society for Testing and Materials (ASTM) D 120-87, Specification for Rubber Insulating Gloves.

ASTM D 1048-93, Specification for Rubber Insulating Blankets.

ASTM D 1049-93, Specification for Rubber Insulating Covers.

ASTM D 1050-90, Specification for Rubber Insulating Line Hose.

ASTM D 1051-87, Specification for Rubber Insulating Sleeves.

ASTM F 478-92, Specification for In-Service Care of Insulating Line Hose and Covers.

ASTM F 479-93, Specification for In-Service Care of Insulating Blankets.

ASTM F 496-93b Specification for In-Service Care of Insulating Gloves and Sleeves.

(x) Insulating equipment failing to pass inspections or electrical tests may not be used by employees, except as follows:

(A) Rubber insulating line hose may be used in shorter lengths with the defective portion cut off.

(B) Rubber insulating blankets may be repaired using a compatible patch that results in physical and electrical properties equal to those of the blanket.

(C) Rubber insulating blankets may be salvaged by severing the defective area from the undamaged portion of the blanket. The resulting undamaged area may not be smaller than 22 inches by 22 inches (560 mm by 560 mm) for Class 1, 2, 3, and 4 blankets.

(D) Rubber insulating gloves and sleeves with minor physical defects, such as small cuts, tears, or punctures, may be repaired by the application of a compatible patch. Also, rubber insulating gloves and sleeves with minor surface blemishes may be repaired with a compatible liquid compound. The patched area shall have electrical and physical properties equal to those of the surrounding material. Repairs to gloves are permitted only in the area between the wrist and the reinforced edge of the opening.

(xi) Repaired insulating equipment shall be retested before it may be used by employees.

(xii) The employer shall certify that equipment has been tested in accordance with the requirements of paragraphs (b)(2)(viii), (b)(2)(ix), and (b)(2)(xi) of this section. The certification shall identify the equipment that passed the test and the date it was tested.

Note: Marking of equipment and entering the results of the tests and the dates of testing onto logs are two acceptable means of meeting this requirement.

Table I-2. A - C Proof-Test Requirements

Class of equipment	Proof-test voltage rms V	Maximum proof-test current, mA (gloves only)			
		267-mm (10.5-in) glove	356-mm (14-in) glove	406-mm (16-in) glove	457-mm (18-in) glove
0	5,000	8	12	14	16
1	10,000	....	14	16	18
2	20,000	....	16	18	20
3	30,000	....	18	20	22
4	40,000	....	....	22	24

Table I-3. D - C Proof-Test Requirements

Class of equipment	Proof-test voltage
0	20,000
1	40,000
2	50,000
3	60,000
4	70,000

Note: The d-c voltages listed in this table are not appropriate for proof testing rubber insulating line hose or covers. For this equipment, d-c proof tests shall use a voltage high enough to indicate that the equipment can be safely used at the voltages listed in Table I-4. See ASTM D 1050-90 and ASTM D 1049-88 for further information on proof tests for rubber insulating line hose and covers.

Table I-4.–Glove Tests–Water Level[1 2]

Class of glove	AC proof test		DC proof test	
	mm.	in.	mm.	in.
0	38	1.5	38	1.5
1	38	1.5	51	2.0
2	64	2.5	76	3.0
3	89	3.5	102	4.0
4	127	5.0	153	6.0

[1]The water level is given as the clearance from the cuff of the glove to the water line, with a tolerance of 13 mm. (0.5 in.).
[2]If atmospheric conditions make the specified clearances impractical, the clearances may be increased by a maximum of 25 mm. (1 in.).

Table I-5. Rubber Insulating Equipment Voltage Requirements

Class of equipment	Maximum use voltage[1] a-c–rms	Retest voltage[2] a-c–rms	Retest voltage[2] d-c–avg
0	1,000	5,000	20,000
1	7,500	10,000	40,000
2	17,000	20,000	50,000
3	26,500	30,000	60,000
4	36,000	40,000	70,000

[1]The maximum use voltage is the a - c voltage (rms) classification of the protective equipment that designates the maximum nominal design voltage of the energized system that may be safely worked. The nominal design voltage is equal to the phase-to-phase voltage on multiphase circuits. However, the phase-to-ground potential is considered to be the nominal design voltage:

(1) If there is no multiphase exposure in a system area and if the voltage exposure is limited to the phase-to-ground potential, or

(2) If the electrical equipment and devices are insulated or isolated or both so that the multiphase exposure on a grounded wye circuit is removed.

[2]The proof-test voltage shall be applied continuously for at least 1 minute, but no more than 3 minutes.

Table I-6. Rubber Insulating Equipment Test Intervals

Type of equipment	When to test
Rubber insulating line hose	Upon indication that insulating value is suspect.
Rubber insulating covers	Upon indication that insulating value is suspect.
Rubber insulating blankets	Before first issue and every 12 months thereafter.[1]
Rubber insulating gloves	Before first issue and every 6 months thereafter.[1]
Rubber insulating sleeves	Before first issue and every 12 months thereafter.[1]

[1]If the insulating equipment has been electrically tested but not issued for service, it may not be placed into service unless it has been electrically tested within the previous 12 months.

[59 FR 4435, Jan. 31, 1994; 59 FR 33662, June 30, 1994]

§ 1910.138 Hand protection.

(a) *General requirements.* Employers shall select and require employees to use appropriate hand protection when employees' hands are exposed to hazards such as those from skin absorption of harmful substances; severe cuts or lacerations; severe abrasions; punctures; chemical burns; thermal burns; and harmful temperature extremes.

(b) *Selection.* Employers shall base the selection of the appropriate hand protection on an evaluation of the performance characteristics of the hand protection relative to the task(s) to be performed, conditions present, duration of use, and the hazards and potential hazards identified.

[59 FR 16362, Apr. 6, 1994; 59 FR 33911, July 1, 1994]

Pt. 1910, Subpt. I, App. A

Appendix A to Subpart I-References for Further Information (Non-mandatory)

The documents in Appendix A provide information which may be helpful in understanding and implementing the standards in Subpart I.

1. Bureau of Labor Statistics (BLS). "Accidents Involving Eye Injuries." Report 597, Washington, D.C.: BLS, 1980.

2. Bureau of Labor Statistics (BLS). "Accidents Involving Face Injuries." Report 604, Washington, D.C.: BLS, 1980.

3. Bureau of Labor Statistics (BLS). "Accidents Involving Head Injuries." Report 605, Washington, D.C.: BLS, 1980. 417

4. Bureau of Labor Statistics (BLS). "Accidents Involving Foot Injuries." Report 626, Washington, D.C.: BLS, 1981.

5. National Safety Council. "Accident Facts", Annual edition, Chicago, IL: 1981.

6. Bureau of Labor Statistics (BLS). "Occupational Injuries and Illnesses in the United States by Industry," Annual edition, Washington, D.C.: BLS.

7. National Society to Prevent Blindness. "A Guide for Controlling Eye Injuries in Industry," Chicago, Il: 1982.

[59 FR 16362, Apr. 6, 1994]

Pt. 1910, Subpt. I, App. B

Appendix B to Subpart I-Non-mandatory Compliance Guidelines for Hazard Assessment and Personal Protective Equipment Selection

This Appendix is intended to provide compliance assistance for employers and employees in implementing requirements for a hazard assessment and the selection of personal protective equipment.

1. *Controlling hazards.* PPE devices alone should not be relied on to provide protection against hazards, but should be used in conjunction with guards, engineering controls, and sound manufacturing practices.

2. *Assessment and selection.* It is necessary to consider certain general guidelines for assessing the foot, head, eye and face, and hand hazard situations that exist in an occupational or educational operation or process, and to match the protective devices to the particular hazard. It should be the responsibility of the safety officer to exercise common sense and appropriate expertise to accomplish these tasks.

3. *Assessment guidelines.* In order to assess the need for PPE the following steps should be taken:

a. *Survey.* Conduct a walk-through survey of the areas in question. The purpose of the survey is to identify sources of hazards to workers and co-workers. Consideration should be given to the basic hazard categories:

(a) Impact

(b) Penetration

(c) Compression (roll-over)

(d) Chemical

(e) Heat

(f) Harmful dust

(g) Light (optical) radiation

b. *Sources.* During the walk-through survey the safety officer should observe: (a) sources of motion; i.e., machinery or processes where any movement of tools, machine elements or particles could exist, or movement of personnel that could result in collision with stationary objects; (b) sources of high temperatures that could result in burns, eye injury or ignition of protective equipment, etc.; (c) types of chemical exposures; (d) sources of harmful dust; (e) sources of light radiation, i.e., welding, brazing, cutting, furnaces, heat treating, high intensity lights, etc.; (f) sources of falling objects or potential for dropping objects; (g) sources of sharp objects which might pierce the feet or cut the hands; (h) sources of rolling or pinching objects which could crush the feet; (i) layout of workplace and location of co-workers; and (j) any electrical hazards. In addition, injury/accident data should be reviewed to help identify problem areas.

c. *Organize data.* Following the walk-through survey, it is necessary to organize the data and information for use in the assessment of hazards. The objective is to prepare for an analysis of the hazards in the environment to enable proper selection of protective equipment.

d. *Analyze data.* Having gathered and organized data on a workplace, an estimate of the potential for injuries should be made. Each of the basic hazards (paragraph 3.a.) should be reviewed and a determination made as to the type, level of risk, and seriousness of potential injury from each of the hazards found in the area. The possibility of exposure to several hazards simultaneously should be considered.

4. *Selection guidelines.* After completion of the procedures in paragraph 3, the general procedure for selection of protective equipment is to: a) Become familiar with the potential hazards and the type of protective equipment that is available, and what it can do; i.e., splash protection, impact protection, etc.; b) compare the hazards associated with the environment; i.e., impact velocities, masses, projectile shape, radiation intensities, with the capabilities of the available protective equipment; c) select the protective equipment which ensures a level of protection greater than the minimum required to protect employees from the hazards; and d) fit the user with the protective

device and give instructions on care and use of the PPE. It is very important that end users be made aware of all warning labels for and limitations of their PPE.

5. *Fitting the device.* Careful consideration must be given to comfort and fit. PPE that fits poorly will not afford the necessary protection. Continued wearing of the device is more likely if it fits the wearer comfortably. Protective devices are generally available in a variety of sizes. Care should be taken to ensure that the right size is selected.
6. *Devices with adjustable features.* Adjustments should be made on an individual basis for a comfortable fit that will maintain the protective device in the proper position. Particular care should be taken in fitting devices for eye protection against dust and chemical splash to ensure that the devices are sealed to the face. In addition, proper fitting of helmets is important to ensure that it will not fall off during work operations. In some cases a chin strap may be necessary to 418keep the helmet on an employee's head. (Chin straps should break at a reasonably low force, however, so as to prevent a strangulation hazard). Where manufacturer's instructions are available, they should be followed carefully.
7. *Reassessment of hazards.* It is the responsibility of the safety officer to reassess the workplace hazard situation as necessary, by identifying and evaluating new equipment and processes, reviewing accident records, and reevaluating the suitability of previously selected PPE.
8. *Selection chart guidelines for eye and face protection.* Some occupations (not a complete list) for which eye protection should be routinely considered are: carpenters, electricians, machinists, mechanics and repairers, millwrights, plumbers and pipe fitters, sheet metal workers and tinsmiths, assemblers, sanders, grinding machine operators, lathe and milling machine operators, sawyers, welders, laborers, chemical process operators and handlers, and timber cutting and logging workers. The following chart provides general guidance for the proper selection of eye and face protection to protect against hazards associated with the listed hazard "source" operations.

Eye and Face Protection Selection Chart

Source	Assessment of Hazard	Protection
IMPACT--Chipping, grinding machining, masonry work, woodworking, sawing, drilling, chiseling, powered fastening, riveting, and sanding.	Flying fragments, objects, large chips, particles sand, dirt, etc.	Spectacles with side protection, goggles, face shields. See notes (1), (3), (5), (6), (10). For severe exposure, use faceshield.
HEAT--Furnace operations, pouring, casting, hot dipping, and welding.	Hot sparks	Faceshields, goggles, spectacles with side protection. For severe exposure use faceshield. See notes (1), (2), (3).
	Splash from molten metals	Faceshields worn over goggles. See notes (1), (2), (3).
	High temperature exposure	Screen face shields, reflective face shields. See notes (1), (2), (3).
CHEMICALS--Acid and chemicals handling, degreasing plating.	Splash	Goggles, eyecup and cover types. For severe exposure, use face shield. See notes (3), (11).
	Irritating mists	Special-purpose goggles.
DUST-- Woodworking, buffing, general dusty conditions.	Nuisance dust	Goggles, eyecup and cover types. See note (8).
LIGHT and/or RADIATION--..		
Welding: Electric arc	Optical radiation	Welding helmets or welding shields. Typical shades: 10-14. See notes (9), (12)
Welding: Gas	Optical radiation	Welding goggles or welding face shield. Typical shades: gas welding 4-8, cutting 3-6, brazing 3-4. See note (9)
Cutting, Torch brazing, Torch soldering	Optical radiation	Spectacles or welding face-shield. Typical shades, 1.5-3. See notes (3), (9)
Glare	Poor vision	Spectacles with shaded or special-purpose lenses, as suitable. See notes (9), (10).

Notes to Eye and Face Protection Selection Chart:

(1) Care should be taken to recognize the possibility of multiple and simultaneous exposure to a variety of hazards. Adequate protection against the highest level of each of the hazards should be provided. Protective devices do not provide unlimited protection.

(2) Operations involving heat may also involve light radiation. As required by the standard, protection from both hazards must be provided.

(3) Faceshields should only be worn over primary eye protection (spectacles or goggles).

(4) As required by the standard, filter lenses must meet the requirements for shade designations in 1910.133(a)(5). Tinted and shaded lenses are not filter lenses unless they are marked or identified as such.

(5) As required by the standard, persons whose vision requires the use of prescription (Rx) lenses must wear either protective devices fitted with prescription (Rx) lenses or protective devices designed to be worn over regular prescription (Rx) eyewear.

(6) Wearers of contact lenses must also wear appropriate eye and face protection devices in a hazardous environment. It should be recognized that dusty and/or chemical environments may represent an additional hazard to contact lens wearers.

(7) Caution should be exercised in the use of metal frame protective devices in electrical hazard areas.
(8) Atmospheric conditions and the restricted ventilation of the protector can cause lenses to fog. Frequent cleansing may be necessary.
(9) Welding helmets or faceshields should be used only over primary eye protection (spectacles or goggles).
(10) Non-sideshield spectacles are available for frontal protection only, but are not acceptable eye protection for the sources and operations listed for "impact."
(11) Ventilation should be adequate, but well protected from splash entry. Eye and face protection should be designed and used so that it provides both adequate ventilation and protects the wearer from splash entry.
(12) Protection from light radiation is directly related to filter lens density. See note (4). Select the darkest shade that allows task performance.

9. *Selection guidelines for head protection.* All head protection (helmets) is designed to provide protection from impact and penetration hazards caused by falling objects. Head protection is also available which provides protection from electric shock and burn. When selecting head protection, knowledge of potential electrical hazards is important. Class A helmets, in addition to impact and penetration resistance, provide electrical protection from low-voltage conductors (they are proof tested to 2,200 volts). Class B helmets, in addition to impact and penetration resistance, provide electrical protection from high-voltage conductors (they are proof tested to 20,000 volts). Class C helmets provide impact and penetration resistance (they are usually made of aluminum which conducts electricity), and should not be used around electrical hazards.

Where falling object hazards are present, helmets must be worn. Some examples include: working below other workers who are using tools and materials which could fall; working around or under conveyor belts which are carrying parts or materials; working below machinery or processes which might cause material or objects to fall; and working on exposed energized conductors.

Some examples of occupations for which head protection should be routinely considered are: carpenters, electricians, linemen, mechanics and repairers, plumbers and pipe fitters, assemblers, packers, wrappers, sawyers, welders, laborers, freight handlers, timber cutting and logging, stock handlers, and warehouse laborers.

10. *Selection guidelines for foot protection.* Safety shoes and boots which meet the ANSI Z41-1991 Standard provide both impact and compression protection. Where necessary, safety shoes can be obtained which provide puncture protection. In some work situations, metatarsal protection should be provided, and in other special situations electrical conductive or insulating safety shoes would be appropriate.

Safety shoes or boots with impact protection would be required for carrying or handling materials such as packages, objects, parts or heavy tools, which could be dropped; and, for other activities where objects might fall onto the feet. Safety shoes or boots with compression protection would be required for work activities involving skid trucks (manual material handling carts) around bulk rolls (such as paper rolls) and around heavy pipes, all of which could potentially roll over an employee's feet. Safety shoes or boots with puncture protection would be required where sharp objects such as nails, wire, tacks, screws, large staples, scrap metal etc., could be stepped on by employees causing a foot injury.

Some occupations (not a complete list) for which foot protection should be routinely considered are: shipping and receiving clerks, stock clerks, carpenters, electricians, machinists, mechanics and repairers, plumbers and pipe fitters, structural metal workers, assemblers, drywall installers and lathers, packers, wrappers, craters, punch and stamping press operators, sawyers, welders, laborers, freight handlers, gardeners and grounds-keepers, timber cutting and logging workers, stock handlers and warehouse laborers.

11. *Selection guidelines for hand protection.* Gloves are often relied upon to prevent cuts, abrasions, burns, and skin contact with chemicals that are capable of causing local or systemic effects following dermal exposure. OSHA is unaware of any gloves that provide protection against *all* potential hand hazards, and commonly available glove materials provide only limited protection against many chemicals. Therefore, it is important to select the most appropriate glove for a particular application and to determine how long it can be worn, and whether it can be reused.

It is also important to know the performance characteristics of gloves relative to the specific hazard anticipated; e.g., chemical hazards, cut hazards, flame hazards, etc. These performance characteristics should be assessed by using standard test procedures. Before purchasing gloves, the employer should request documentation from the manufacturer that the gloves meet the appropriate test standard(s) for the hazard(s) anticipated.

Other factors to be considered for glove selection in general include:

(A) As long as the performance characteristics are acceptable, in certain circumstances, it may be more cost effective to regularly change cheaper gloves than to reuse more expensive types; and,

(B) The work activities of the employee should be studied to determine the degree of dexterity required, the duration, frequency, and degree of exposure of the hazard, and the physical stresses that will be applied.

With respect to selection of gloves for protection against chemical hazards:

(A) The toxic properties of the chemical(s) must be determined; in particular, the ability of the chemical to cause local effects on the skin and /or to pass through the skin and cause systemic effects;

(B) Generally, any "chemical resistant" glove can be used for dry powders;

(C) For mixtures and formulated products (unless specific test data are available), a glove should be selected on the basis of the chemical component with the shortest breakthrough time, since it is possible for solvents to carry active ingredients through polymeric materials; and,

(D) Employees must be able to remove the gloves in such a manner as to prevent skin contamination.

12. *Cleaning and maintenance.* It is important that all PPE be kept clean and properly 420maintained. Cleaning is particularly important for eye and face protection where dirty or fogged lenses could impair vision.

For the purposes of compliance with 1910.132(a) and (b), PPE should be inspected, cleaned, and maintained at regular intervals so that the PPE provides the requisite protection.

It is also important to ensure that contaminated PPE which cannot be decontaminated is disposed of in a manner that protects employees from exposure to hazards.

[59 FR 16362, Apr. 6, 1994]

Appendix C

Sample Forms and Instructions for Conducting a Risk Assessment

Protection Strategy Form

Hazard Category/ Specific Hazards	Principal Body Areas or Body Systems Affected	Risk Rating	Priority Area?	Engineering or Administrative Control?	Recommended Protection Strategy (Type of PPE, Design, and Performance)

Specified PPE by Type of Hazard

Hazard Category/ Specific Hazards	Specific Personal Protective Equipment (PPE)															

Risk Assessment Form

Hazard Category/ Specific Hazards	Primary Body Area(s) or Body System(s) Affected										Probability Rating (P)	Severity Rating (S)	Risk (P x S)
	Whole Body	Torso	Head	Face or Eyes†	Arms	Hands	Legs	Feet	Resp. System	Hearing			

Appendix D

Referenced Publications and Standards

American Association of Textile Chemists and Colorists (AATCC). P. O. Box 12215, Research Triangle Park, NC 27709, Phone 919-549-8141, Fax 919-549-8933.

AATCC 16, Colorfastness to Light, 1993
AATCC 22, Water Repellency: Spray Test, 1996
AATCC 42, Water Resistance: Impact Penetration Test, 1994
AATCC 76, Electrical Resistivity of Textiles, 1995
AATCC 96, Dimensional Changes in Commercial Laundering of Woven and Knitted Fabrics Except Wool, 1995
AATCC 100, Antibacterial Finishes on Textile Materials: Assessment of, 1993
AATCC 127, Water Resistance: Hydrostatic Pressure Test, 1995
AATCC 132, Colorfastness to Drycleaning, 1993
AATCC 135, Dimensional Changes in Automatic Home Laundering of Woven or Knit Fabrics, 1995

American Conference of Governmental Industrial Hygienists (ACGIH). 1330 Kemper Meadow Drive, Cincinnati, OH 45240-1634, Phone 513-742-2020, Fax 513-742-3355.

Threshold Limit Values for Chemical Substances and Physical Agents and Biological Exposure Indices, 1996.

American National Standards Institute (ANSI). 11 West 42nd Street, New York, NY 10036, Phone 212-642-4900, Fax 212-398-0023.

ANSI A10,14, American National Standard for Construction and Demolition Operations - Requirements for Safety Belts, Harnesses, Lanyards, and Lifelines for Construction and Demolition Use, 1991
ANSI K13.1, American National Standard for Identification of Air-Purifying Respirator Canisters and Cartridges, 1973
ANSI S3.2, Method for Measuring the Intelligibility of Speech over Communication Systems, 1989
ANSI Z41, American National Standard for Personal Protection-Protective Footwear, 1991
ANSI Z80.1, American National Standard Recommendations for Prescription Ophthaalmic Lenses, 1987
ANSI Z87.1, American National Standard Practice for Occupational and Educational Eye and Face Protection, 1989
ANSI Z88.2, American National Standard for Respiratory Protection, 1992
ANSI Z89.1, American National Standard for Personnel Protection-Protective Headwear for Industrial Workers-Requirements, 1997
ANSI Z359.1, American National Standard on Safety Requirements for Personal Fall Arrest Systems, Subsystems and Components, 1992
ANSI Z535.4, American National Standard for Product Safety Signs and Labels, 1991

American Society for Testing and Materials (ASTM). 100 Barr Harbor Drive, West Conshohocken, PA 19428-2959, Phone 610-832-9585, Fax 610-832-9555.

ASTM B 117, Test Method of Salt Spray (Fog) Testing, 1994
ASTM D 120, Specification for Rubber Insulating Gloves, 1995
ASTM D 412, Test Methods for Vulcanized Rubber and Thermoplastic Rubbers and Thermoplastic Elastomers-Tension, 1992
ASTM D 471, Test Method for Rubber Property-Effect of Liquids, 1996
ASTM D 518, Test Method for Rubber Deterioration-Surface Cracking, 1991
ASTM D 543, Test Method for Resistance of Plastics to Chemical Reagents, 1995
ASTM D 573, Test Method for Rubber-Deterioration in an Air Oven, 1994
ASTM D 575, Test Methods for Rubber Properties in Compression, 1991
ASTM D 635, Test Method for Rate of Burning and/or Extent and Time of Burning of Self-Supporting Plastics in a Horizontal Position, 1991
ASTM D 638, Test Method for Tensile Properties of Plastics, 1995
ASTM D 737, Test Method for Air Permeability of Textile Fabrics, 1996
ASTM D 746, Test Method for Brittleness Temperature of Plastics and Elastomers by Impact, 1995
ASTM D 747, Test Method for Apparent Bending Modulus of Plastics by Means of a Cantilever Beam, 1993
ASTM D 751, Method of Testing Coated Fabrics, 1998
ASTM D 882, Test Methods for Tensile Properties of Thin Plastic Sheeting, 1995
ASTM D 1004, Test Method for Initial Tear Resistance of Plastic Film and Sheeting, 1993
ASTM D 1043, Test Method for Stiffness Properties of Plastics as a Function of Temperature by Means of a Torsion Test, 1992
ASTM D 1044, Test Method for Resistance of Transparent Plastics to Surface Abrasion, 1993
ASTM D 1051, Specification for Rubber Insulating Sleeves, 1995
ASTM D 1053, Test Method for Rubber Property-Stiffening at Low Temperatures; Flexible Polymers and Coated Fabrics, 1992
ASTM D 1117, Methods of Testing Nonwoven Fabrics, 1995
ASTM D 1148, Test Method for Rubber-Deterioration-Heat and Ultraviolet Light Discoloration of Light-Colored Surfaces, 1995
ASTM D 1149, Test Method for Rubber Deterioration-Surface Ozone Cracking in a Chamber, 1991
ASTM D 1171, Test Method for Rubber Deterioration-Surface Ozone Cracking Outdoors or Chamber (Triangular Specimens), 1994
ASTM D 1230, Test Method for Flammability of Clothing Textiles, 1994
ASTM D 1242, Test Methods for Resistance of Plastic Materials to Abrasion, 1993
ASTM D 1388, Test Methods for Stiffness of Fabrics, 1975
ASTM D 1424, Test Method for Tear Resistance of Woven Fabrics by Falling-Pendulum (Elmendorf) Apparatus, 1996
ASTM D 1434, Test Method for Determining Gas Permeability Characteristics of Plastic Film and Sheeting, 1992
ASTM D 1518, Test Method for Thermal Transmittance of Textile Material, 1990
ASTM D 1565, Specification for Flexible Cellular Materials-Vinyl Chloride Polymers and Copolymers (open-Cell Foam), 1990
ASTM D 1630, Test Method for Rubber Property-Abrasion Resistance (Footwear Abrader), 1994
ASTM D 1683, Test Method for Failure in Sewn Seams of Woven Fabrics, 1990
ASTM D 1777, Method for Measuring Thickness of Textile Materials, 1996
ASTM D 1790, Test Method for Brittleness Temperature of Plastic Film by Impact, 1994
ASTM D 1813, Method for Measuring Thickness of Leather Test Specimens, 1994
ASTM D 1912, Test Method for Cold-Crack Resistance of Upholstery Leather, 1993
ASTM D 1922, Test Method for Propagation Tear Resistance of Plastic Film and Thin Sheeting by Pendulum Method, 1993
ASTM D 1938, Test Method for Tear Propagation Resistance of Plastic Film and Thin Sheeting by a Single-Tear Method, 1993

ASTM D 2020, Test Method for Mildew (Fungus) Resistance of Paper and Paperboard, 1992
ASTM D 2136, Test Method for Coated Fabrics-Low-Temperature Bend Test, 1994
ASTM D 2137, Test Methods for Rubber Property-Brittleness Point of Flexible Polymers and Coated Fabrics, 1994
ASTM D 2207, Test Method for Bursting Strength of Leather by the Ball Method, 1995
ASTM D 2208, Test Method for Breaking Strength of Leather by the Grab Method, 1995
ASTM D 2209, Test Method for Tensile Strength of Leather, 1995
ASTM D 2228, Test Method for Rubber Property-Abrasion Resistance (Pico Abrader), 1994
ASTM D 2240, Test Method for Rubber Property-Durometer Hardness, 1991
ASTM D 2261, Test Method for Tearing Strength of Woven Fabrics by the Tongue (Single Rip) Method (Constant-Rate-of-Extension Tensile Testing Machine), 1996
ASTM D 2582, Test Method for Puncture-Propagation Tear Resistance of Plastic Film and Thin Sheeting, 1993
ASTM D 2821, Test Method for Measuring the Relative Stiffness of Leather by Means of a Torsional Wire Apparatus, 1993
ASTM D 2863, Test Method for Measuring the Minimum Oxygen Concentration to Support Candle-Like Combustion of Plastics (Oxygen Index), 1995
ASTM D 2986, Practice for Evaluation of Air Assay Media by the MonodisperseDOP (Dioctyl Phthalate) Smoke Test, 1991
ASTM D 3045, Practice for Heat Aging of Plastics Without Load, 1992
ASTM D 3354, Test Method for Blocking Load of Plastic Film by the Parallel Plate Method, 1989
ASTM D 3389, Method of Testing Coated Fabrics-Abrasion Resistance (Rotary Platform, Double-Head Abrader), 1994
ASTM D 3577, Specification for Rubber Surgical Gloves, 1991
ASTM D 3578, Specification for Rubber Examination Gloves, 1991
ASTM D 3659, Test Method for Flammability of Apparel Fabrics by Semi-Restraint Method, 1993
ASTM D 3772, Specification for Rubber Finger Cots, 1991
ASTM D 3884, Test Method for Abrasion Resistance of Textile Fabrics (Rotary Platform, Double-Head Method), 1992
ASTM D 3885, Test Method for Abrasion Resistance of Textile Fabrics (Inflated Diaphragm Method, 1992
ASTM D 3629, Test Method for Rubber Property-Cut Growth Resistance, 1994
ASTM D 3767, Practice for Rubber-Measurement of Dimensions, 1988
ASTM D 3776, Test Methods for Mass per Unit Area (Weight) of Woven Fabric, 1996
ASTM D 3786, Test Method for Hydraulic Bursting Strength of Knitted Goods and Nonwoven Fabrics: Diaphragm Bursting Strength Tester Method, 1987
ASTM D 3787, Test Method for Bursting Strength of Knitted Goods: Constant-Rate-of Traverse (CRT), Ball Burst Test, 1989
ASTM D 3886, Test Method for Abrasion Resistance of Textile Fabrics (Inflated Diaphragm Method, 1992
ASTM D 3939, Test Method for Snagging Resistance of Fabrics (Mace Test Method), 1993
ASTM D 3940, Test Method for Bursting Strength (Load) and Elongation of Sewn Seams of Knit or Woven Stretch Textile Fabrics, 1983
ASTM D 3985, Test Method for Oxygen Gas Transmission Rate Through Plastic Film and Sheeting Using a Coulometric Sensor, 1995
ASTM D 4032, Test Method for Stiffness of Fabrics by the Circular Bend Procedure, 1994
ASTM D 4100, Method for Gravimetric Determination of Smoke Particulates from Combustion of Plastic Materials, 1989
ASTM D 4108, Test Method for Thermal Protective Performance of Materials for Clothing by Open-Flame Method, 1987
ASTM D 4157, Test Method for Abrasion Resistance of Textile Fabrics (Oscillatory Cylinder Method), 1992
ASTM D 4158, Test Method for Abrasion Resistance of Textile Fabrics (Uniform Abrasion Method), 1992
ASTM D 4168, Test Method for Transmitted Shock Characteristics of Foam-in-Place Cushioning Materials, 1995
ASTM D 4679, Specification for Rubber Household or Beautician Gloves, 1993

ASTM D 4966, Test Method for Abrasion Resistance of Textile Fabrics (Martindale Abrasion Tester Method), 1989
ASTM D 5034, Test Method for Breaking Force and Elongation of Textile Fabrics (Grab Test), 1995
ASTM D 5035, Test Method for Breaking Force and Elongation of Textile Fabrics (Strip Test), 1995
ASTM D 5151, Test Method for Detection of Holes in Medical Gloves, 1992
ASTM D 5250, Specification for Poly(vinyl chloride) Gloves for Medical Application, 1992
ASTM D 5362, Test Method for Snagging Resistance of Fabrics (Bean Bag Test Method), 1993
ASTM D 5489, Guide for Care Symbols for Permanent Care Labels On Consumer Textile Produces, 1993
ASTM D 5585, Standard Table of Body Measurements for Adult Female Misses Figure Type Sizes 2-20, 1994
ASTM D 5586, Standard Tables of Body Measurements for Women Aged 55 and Older (All Figure Types), 1994
ASTM D 5712, Test Method for Analysis of Proteins in Natural Rubber and Its Products, 1995
ASTM D 5733, Test Method for Tearing Strength of Nonwoven Fabrics by the Trapezoid Procedure, 1995
ASTM D 5734, Test Method for Tearing Strength of Nonwoven Fabrics by Fall-Pendulum (Elmendorf) Apparatus, 1995
ASTM D 5735, Test Method for Tearing Strength of Nonwoven Fabrics by the Tongue (Single Rip) Procedure (Constant-Rate-of-Extension Tensile Testing Machine), 1995
ASTM D 5736, Test Method for Thickness of Highloft Nonwoven Fabrics, 1995
ASTM D 5802, Test Method for Sorption of Bibulous Paper Produce (Sorptive Rate), 1995
ASTM E 96, Test Methods for Water Vapor Transmission of Materials, 1994
ASTM E 810, Test Method for Coefficient of Retroreflection or Retroreflective Sheeting, 1993
ASTM E 906, Test Method for Heat and Visible Smoke Release Rates for Materials and Products, 1993
ASTM E 991, Practice for Color Measurement of Fluorescent Specimens, 1990
ASTM E 1053, Test Method for Efficacy of Virucidal Agents Intended for Inanimate Environmental Surfaces, 1991
ASTM E 1247, Test Method for Identifying Fluorescences in Object-Color Specimens by Spectrophotometry, 1992
ASTM E 1501, Specification for Nighttime Photometric Performance of Retroreflective Pedestrian Markings for Visibility Enhancement, 1992
ASTM E 1549, Specification for ESD Controlled Garments Required in Cleanrooms and Controlled Environments for Spacecraft for Non-Hazardous and Hazardous Conditions, 1995
ASTM E 1809, Test Method for Measurement of High-Visibility Retroreflective-Clothing, 1996
ASTM F 392, Test Method for Flex Durability of Flexible Barrier Materials, 1993
ASTM F 429, Test Method for Shock-Attenuation Characteristics of Protective Headgear for Football, 1992
ASTM F 489, Test Method for Static Coefficient of Friction of Shoe Sole and Heel Materials as Measured by the James Machine, 1996
ASTM F 609, Test Method for Static Slip Resistance of Footwear, Soke, Heel, or Related Materials by Horizontal Pull Slipmeter (HPS), 1989
ASTM F 696, Specification for Leather Protectors for Rubber Insulating Gloves and Mittens, 1991
ASTM F 739, Test Method for Resistance of Protective Clothing Materials to Permeation by Liquids or Gases under Conditions of Continuous Contact, 1996
ASTM F 903, Test Method for Resistance of Protective Clothing Materials to Penetration by Liquids, 1996
ASTM F 955, Test Method for Evaluating Heat Transfer Through Materials for Protective Clothing Upon Contact with Molten Substances, 1989
ASTM F 1001, Guide for Selection of Chemicals to Evaluate Protective Clothing Materials, 1993
ASTM F 1002, Performance Specification for Protective Clothing for Use by Workers Exposed to Specific Molten Substances and Related Thermal Hazards, 1996
ASTM F 1052, Practice for Pressure Testing of Gas Tight Totally Encapsulating Chemical Protective Suits, 1997
ASTM F 1060, Test Method for Thermal Protective Performance of Materials for Protective Clothing for Hot Surface Contact, 1993
ASTM F 1116, Test Method for Determining Dielectric Strength of Overshoe Footwear, 1988

ASTM F 1117, Specification for Dielectric Overshoe Footwear, 1993
ASTM F 1154, Practices for Qualitatively Evaluating the Comfort, Fit, Function, and Integrity of Chemical Protective Suit Ensembles, 1996
ASTM F 1186, Classification System for Chemicals According to Functional Groups, 1993
ASTM F 1194, Guide for Documenting the Results of Chemical Permeation Testing on Materials Used in Protective Clothing, 1994
ASTM F 1215, Test Method for Determining the Initial Efficiency of a Flatsheet Filter Medium in an Airflow Using Latex Spheres, 1993
ASTM F 1291, Test Method for Measuring the Thermal Insulation of Clothing Using a heated Manikin, 1996
ASTM F 1301, Practice for Labeling Chemical Protective Clothing, 1990
ASTM F 1342, Test Method for Protective Clothing Material Resistance to Puncture, 1996
ASTM F 1358, Test Method for Effects of Flame Impingement on Materials Used in Protective Clothing Not Designated Primarily for Flame Resistance, 1995
ASTM F 1359, Practice for Determining the Liquid-Tight Integrity of Chemical Protective Suits or Ensembles Under Static Conditions, 1997
ASTM F 1383, Test Method for Resistance of Protective Clothing Materials to Permeation by Liquids of Gases under Conditions of Intermittent Contact, 1996
ASTM F 1407, Test Method for Resistance of Chemical Protective Clothing Materials to Liquid Permeation-Permeation Cup Method, 1996
ASTM F 1414, Test Method for Measurement of Cut Resistance to Chain Saw in Lower Body (Legs) Protective Clothing, 1992
ASTM F 1458, Test Method for Measurement of Cut Resistance to Chain Saw of Foot Protective Devices, 1994
ASTM F 1506, Performance Specification for Textile Materials for Wearing Apparel for Use by Electrical Workers Exposed to Momentary Electrical Arc and Related Thermal Hazards, 1996
ASTM F 1608, Test Method for Microbial Ranking of Porous Packaing Materials, 1995
ASTM F 1670, Test Method for Resistance of Materials Used in Protective Clothing to Penetration by Synthetic Blood, 1998
ASTM F 1671, Test Method for Resistance of Materials Used in Protective Clothing to Penetration by Bloodborne Pathogens Using Phi-X174 Bacteriophage Penetration as a Test System, 1998
ASTM F 1731, Practice for Body Measurements and Sizing of Fire and Rescue Services Unifroms and Other Thermal Hazard Protective Clothing, 1996
ASTM F 1790, Test Method for Measuring Cut Resistance of Materials Used in Protective Clothing, 1997
ASTM F 1818, Specification for Foot Protection for Chain Saw Users, 1997
ASTM F 1819, Test Method for Resistance of Materials Used in Protective Clothing to Penetration by Synthetic Blood Using a Mechanical Pressure Technique, 1997
ASTM G 21, Practice for Determining Resistance of Synthetic Polymeric Materials to Fungi, 1990
ASTM G 22, Practice for Determining Resistance of Plastics to Bacteria, 1990
ASTM G 23, Practice for Operating Light-Exposure Apparatus (Carbon-Arc Type) With and Without Water for Exposure of Nonmetallic Materials, 1993
ASTM G 26, Practice for Operating Light-Exposure Apparatus (Xenon-Arc Type) With and Without Water for Exposure of Nonmetallic Materials, 1993
ASTM G 85, Practice for Modified Salt Spray (Fog) Testing, 1994
G 120, Practice for Determination of Soluble Residual Contaminants in Materials and Components by Soxhlet Extraction, 1995
G 136, Practice for Determination of Soluble Residual Contaminants in Materials by Ultrasonic Extraction, 1995
ASTM PS 57, Provisional Standard Test Method for Determining Ignitability of Clothing by Electrical Arc Exposure Method Using a Mannequin, 1997
ASTM PS 58, Provisional Standard Test Method for Determining Arc Thermal Performance (Value) of Textile Materials for Clothing by Electrical Arc Exposure Method Using a Instrumented Sensor Panels, 1997

Association for the Advancement of Medical Instrumentation (AAMI). 3330 Washington Blvd., Suite 400, Arlington, VA 22201-4598, Phone 800-332-2264, Fax 703-276-0793.

ANSI/AAMI ST41, Ethylene Oxide Sterilization and Sterility Assurance, 1992
AAMI ST46, Steam Sterilization and Sterility Assurance, 1993
AAMI TIR 11, Selection of Surgical Gowns and Drapes in Health Care Facilities, 1994

Association of the Nonwoven Fabrics Industry (INDA). 1001 Winstead Drive, Suite 460, Cary, NC 27513, Phone 919-677-0060, Fax 919-677-0211.

INDA IST 10.1, Absorbency Time, Absorbency Capacity, and Wicking Rate, 1995
INDA IST 40.1, Surface Resistivity of Nonwoven Fabrics, 1995
INDA IST 40.2, Electrostatic Decay of Nonwoven Fabrics, 1995
INDA IST 80.6, Alcohol Repellency of Nonwoven Fabrics, 1995
INDA IST 160.1, Resistance to Linting of Nonwoven Fabrics, 1995

Candian General Standards Board (CGSB). CGSB Sales Unit, Ottawa, Canada K1A 1G6, Phone 819-956-0425, Fax 819-956-0426.

CAN/CGSB-4.2 No. 5.1, Unit Mass of Fabrics, 1990
CAN/CGSB-4.2 No. 9.1, Breaking Strength of Fabrics - Strip Method - Constant-time-to-break Principle, 1990
CAN/CGSB-4.2 No. 9.2, Breaking Strength of Fabrics - Grab Method - Constant-time-to-break Principle, 1990
CAN/CGSB-4.2 No. 11.1, Bursting Strength - Diaphragm Pressure Test, 1994
CAN/CGSB-4.2 No. 11.2, Bursting Strength - Ball Burst Test, 1989
CAN/CGSB-4.2 No. 12.1, Tearing Strength - Single-Rp Method, 1990
CAN/CGSB-4.2 No. 12.2, Tearing Strength - Trapezoid Method, 1995
CAN/CGSB-4.2 No. 12.3, Tearing Strength - Elmendorf Ballistic Method, 1994
CAN/CGSB-4.2 No. 18.3, Textiles - Tests for Colorfastness - Part B02: Colorfastness to Artificial Light: Xenon Arc Fading Lamp Test, 1990
CAN/CGSB-4.2 No. 24, Colourfastness and Dimensional Change in Commercial Laundering, 1991
CAN/CGSB-4.2 No. 26.2, Textile Fabrics - Determination of Resistance to Surface Wetting (Spraty Test), 1994
CAN/CGSB-4.2 No. 26.3, Textile Fabrics - Determination of Resistance to Water Penetration -Hydrostatic Pressure Test, 1995
CAN/CGSB-4.2 No. 26.5, Water Resistance - High Pressure Penetration Test, 1989
CAN/CGSB-4.2 No. 27.1, Flame Resistance - Vertical Burning Test, 1994
CAN/CGSB-4.2 No. 27.2, Flame Resistance - Surface Burning Test, 1994
CAN/CGSB-4.2 No. 27.5, Flame Resistance - 45o Angle Test - One Second Exposure Impingement, 1994
CAN/CGSB-4.2 No. 28.1, Resistance to Microorganisms - Fungus Damage Test - Pure Culture - Quantitative, 1991
CAN/CGSB-4.2 No. 28.2, Resistance to Microorganisms - Surface-Growing Fungus Test - Pure Culture, 1991
CAN/CGSB-4.2 No. 28.3, Resistance to Microorganisms - Soil Burial Test, 1991
CAN/CGSB-4.2 No. 28.4, Resistance to Microorganisms - Fungus Damage Test - Pure Culture - Qualitative, 1991
CAN/CGSB-4.2 No. 30, Dimensional Change in Dry Cleaning, 1990
CAN/CGSB-4.2 No. 32.2, Breaking Strength of Seams in Woven Fabrics, 1989
CAN/CGSB-4.2 No. 36, Air Permeability, 1989
CAN/CGSB-4.2 No. 37, Fabric Thickness, 1992
CAN/CGSB-4.2 No. 38, Resistance to Insect Pests, 1989

CAN/CGSB-4.2 No. 49, Resistance of Materials to Water Vapour Diffusion, 1991
CAN/CGSB-4.2 No. 51.1, Resistance to Piling, Rotating Box Method, 1995
CAN/CGSB-4.2 No. 51.1, Resistance to Piling - Random Tmbling Piling Tester, 1987
CAN/CGSB-4.2 No. 58, Colourfastness and Dimensional Change in Domestic Laundering of Textiles, 1990
CAN/CGSB-4.2 No. 60, Resistance to Snagging - Mace Test, 1989
CAN/CGSB-4.2 No. 70.1, Thermal Insulation Performance of Textile Materials, 1994
CAN/CGSB 20.25, Sterile Surgical Gloves, Disposable, 1991
CAN/CGSB 20.27, Sterile or Non-sterile Medical Examination Gloves for Single Use, 1991
CAN/CGSB 38.16, Operating Surgical Gown, 1992
CAN/CGSB-49, Canadian standards on garment sizing (several separate standards)
CAN/CGSB 65.16, Marine Abandonment Immersion Suit Systems, 1989
CAN/CGSB 65.21, Marine Anti-exposure Work Suit Systems, 1995
CAN/CGSB-86.1, Care Labelling of Textiles, 1987
CAN/CGSB 155.1, Firefighter's Protective Clothing for Protection Against Heat and Flame, 1988

Compressed Gas Association. 1725 Jefferson Davis Highway, Arlington, VA 22202-4102, Phone 703-412-0900, Fax 703-412-0128.

Compressed Gas Association Commodity Specificaiton for Air, G-7.1, 1996.

European Committee for Standardization (CEN). Central Secretariat: rue de Stassart 36, B-1050 Brussels, Belgium (Copies of standards may be obtained by standards selling services in the United States).

EEC/89/656, Minimum Health and Safety Requirements for the Use by Workers of PPE at the Workplace, 1989
EEC/89/686, Council Directive on the Approximation of the Laws of the Member States Related to Personal Protective Equipment, 1989
EN 340, Protective clothing - General requirements, 1993
EN 342, Protective clothing against cold (proposed)
EN 343, Protective clothing against foul weather (proposed)
EN 344, Requirements and test methods for safety, protective, and occupational footwear for professional use, 1992
EN 344-2, Additional requirements and test methods for safety, protective, and occupational footwear for professional use, 1996
EN 345, Specification for safety footwear for professional use, 1992
EN 345-2, Additional specifications for safety footwear for professional use, 1994
EN 346, Specification for protective footwear for professional use, 1992
EN 347, Specification for occupational footwear for professional use, 1992
EN 348, Protective clothing - Test method: Determination of behaviour of material on impact of small splashes of molten metal, 1993
EN 366, Protective clothing - Protection against heat and fire - Method of test: Evaluation of materials and material assemblies when exposed to a source of radiant heat, 1993
EN 367, Clothing for protection against heat and fire - Method of determining heat transmission on exposure to flame, 1992
EN 368, Protective clothing for use against liquid chemicals - Test method: Resistance of materials to penetration by liquids, 1992
EN 369, Protective clothing - Protection against liquid chemicals - Test method: Resistance of material to permeation by liquids, 1993
EN 373, Protective clothing - Assessment of resistance of materials to molten metal splash, 1993
EN 374-1, Protective gloves against chemicals and microorganisms - Part 1: Terminology and performance requirements, 1994

EN 374-2, Protective gloves against chemicals and microorganisms - Part 2: Determination of resistance to penetration, 1994
EN 374-3, Protective gloves against chemicals and microorganisms - Part 3: Determination of resistance to permeation by chemicals, 1994
EN 381-1, Protective clothing for users of hand held chain saws - Part 1: Test rig for testing resistance to cutting by a chainsaw, 1993
EN 381-2, Protective clothing for users of hand held chain saws - Part 2: Test rig for leg protection, 1995
EN 381-3, Protective clothing for users of hand held chain saws - Part 3: Test methods for footwear, 1996
EN 381-4, Protective clothing for users of hand held chain saws - Part 4: Test method for chain saw protective gloves, 1997
EN 381-5, Protective clothing for users of hand held chain saws - Part 5: Requirements for leg protection, 1995
EN 381-7, Protective clothing for users of hand held chain saws - Part 7: Requirements for chain saw protective gloves (proposed)
EN 381-8, Protective clothing for users of hand held chain saws - Part 8: Test method for chain saw protective gaiters (proposed)
EN 381-9, Protective clothing for users of hand held chain saws - Part 9: Requirements for chain saw protective gaiters (proposed)
EN 381-10, Protective clothing for users of hand held chain saws - Part 10: Requirements for jackets with protection against cuts by hand held chain saws (proposed)
EN 381-11, Protective clothing for users of hand held chain saws - Part 11: Test methods for jackets with protection against cuts by hand held chain saws (proposed)
EN 388, Protective gloves against mechanical risks, 1994
EN 407, Protective gloves against thermal risks, 1994
EN 412, Protective aprons for use with hand knives, 1993
EN 412-2, Protective clothing - Part 2: Aprons, trousers, and vests protecting against cuts and stabs by hand knives (proposed)
EN 420, General requirements for protective gloves, 1994
EN 421, Protective gloves against ionizing radiation and radioactive contamination, 1994
EN 443, Helmets for firefighters, 1997
EN 463, Protective clothing - Protection against liquid chemicals - Test method: Determination of resistance to penetration by a jet of liquid (Jet Test), 1994
EN 464, Protective clothing - Protection against liquid and gaseous chemicals - Test method: Determination of leak-tightness of gas-tight suits (Internal Pressure Test), 1994
EN 465, Protective clothing - Protection against liquid chemicals - Performance requirements for chemical protective clothing with spray-tight connections between different parts of the clothing (Type 4 Equipment), 1995
EN 466, Protective clothing - Protection against liquid chemicals - Performance requirements for chemical protective clothing with liquid-tight connections between different parts of the clothing (Type 3 Equipment), 1995
EN 466-2 Protective clothing - Protection against liquid chemicals - Performance requirements for chemical protective clothing with liquid-tight connections between different parts of the clothing for emergency teams (Type 3 ET) (proposed)
EN 467, Protective clothing - Protection against liquid chemicals - Performance requirements for garments providing chemical protection to parts of the body, 1995
EN 468, Protective clothing - Protection against liquid chemicals - Test method: Determination of resistance to penetration by spray (Spray Test), 1994
EN 469, Protective clothing for firefighters - Requirements and test methods for protective clothing for firefighting, 1995
EN 470-1, Protective clothing for use in welding and allied processes - Part 1: General requirements, 1995
EN 470-2, Protective clothing for use in welding and allied processes - Part 2: Additional requirements (proposed)
EN 471, High-visibility warning clothing, 1994

EN 511, Protective gloves against cold, 1994
EN 530, Abrasion resistance of protective clothing material - Test method, 1994
EN 531, Protective clothing for industrial workers exposed to heat (excluding firefighters' and welders' clothing), 1995
EN 532, Protective clothing - Protection against heat and flame - Method of test for limited flame spread, 1994
EN 659, Protective gloves for firefighters, 1995
EN 702, Protective clothing - Protection against heat and flame - Test method: Determination of the contact heat transmission through protective clothing or its materials, 1994
EN 863, Protective clothing - Mechanical properties - Test method: Puncture resistance, 1995
EN 943-1, Protective clothing against liquid and gaseous chemicals, including aerosols and solid particles - Part 1: Performance requirements for ventilated and non-ventilated "gas tight" (Type 1) protective clothing and "non-gas-tight" (Type 2) protective clothing, 1995
EN 943-2, Protective clothing against liquid and gaseous chemicals, including aerosols and solid particles - Part 2: Performance requirements for "gas tight" (Type 1) protective clothing for emergency teams (ET) (proposed)
EN 1082-1, Protective clothing - Gloves and arm guards protecting against cuts and stabs by hand knives - Part 1: Chain mail gloves and arm guards, 1996
EN 1082-2, Protective clothing - Gloves and arm guards protecting against cuts and stabs by hand knives - Part 2: Requirements for gloves made of other material than chain mail (proposed)
EN 1082-3, Protective clothing - Gloves and arm guards protecting against cuts and stabs by hand knives - Part 3: Impact cut test for fabric, leather and other material (proposed)
EN 1486, Protective clothing for firefighters - Test methods and requirements for reflective clothing for specialized firefighting, 1996
EN 1511, Protective clothing against liquid chemicals - Performance requirements for limited use chemical protective clothing with liquid-tight connections between different parts of the clothing (Type 3 limited use clothing) (proposed)
EN 1512, Protective clothing against liquid chemicals - Performance requirements for limited use chemical protective clothing with spray-tight connections between different parts of the clothing (Type 4 limited use clothing) (proposed)
EN 1513, Protective clothing against liquid and solid chemicals - Performance requirements for limited use chemical protective garments providing chemical protection to parts of the body (proposed)
EN 1913-1, Survival suits - Part 1: Constant wear suits, requirements (proposed)
EN 1913-2, Survival suits - Part 2: Abandonment suits, requirements (proposed)
EN 1913-3, Survival suits - Part 3: Test methods (proposed)
EN 10819, Mechanical vibration and shock - Hand arm vibration - Method of measurement and evaluation of the vibration transmissibility of gloves at the palm of the hand, 1996
EN 12477, Protective gloves for welders (proposed)
EN 13982, Protective clothing against solid particulate chemicals - Part 1: Performance requirements for limited use and re-usable chemical protective clothing offering limited particle protection (Type 5 equipment) (proposed)
EN 20811, Textile Fabrics - Determination of resistance to water penetration - Hydrostatic pressure test, 1996
EN 13934-1, Textiles-Woven fabrics-Determination of breaking strength and elongation (Strip method), 1996
EN 13934-2, Textiles-Woven fabrics-Determination of breaking strength-Grab method, 1996

U.S. Federal Regulations, Specifications and Test Methods. General Services Administration, Specifications Activity, Printed Materials Supply Division, Building 197, Naval Weapons Plan, Washington, DC 20407.

Code of Federal Regulations (CFR)
Title 16, Part 1610, Standard for the Flammability of Clothing Textiles
Title 29, Part 1910, General Industry Occupational Safety and Health Standards

Title 29, Subpart I, Sections 1910.132 through 1910.140, Personal Protective Equipment
Title 29, Part 1915, Maritime Occupational Safety and Health Standards
Title 29, Part 1917, Maritime Occupational Safety and Health Standards
Title 29, Part 1918, Maritime Occupational Safety and Health Standards
Title 29, Part 1926, Construction Occupational Safety and Health Standards
Title 42, Part 84, Respiratory Protective Devises, Test for Permissibility

Federal Test Method Standard (FTMS) 101C, Plastics: Methods of Testing
Method 4046, Electrostatic Decay Method

Federal Test Method Standard (FTMS 191A)
Method 1534, Melting Point of Synthetic Fibers
Method 5030, Thickness of Textile Materials; Determination of
Method 5041, Weight of Textile Materials, Small Specimen Method; Determination of
Method 5100, Strength and Elongation, Breaking of Woven Cloth: Grab Method
Method 5102, Strength and Elongation, Breaking of Woven Cloth: Cut Strip Method
Method 5120, Strength of Cloth; Ball Bursting Method
Method 5122, Strength of Cloth; Diaphragm Bursting Method
Method 5132, Strength of Cloth, Tearing: Fall-Pendulum Mehod
Method 5134, Strength of Cloth, Tearing: Tongue Method
Method 5300, Abrasion Resistance of Cloth; Flexing, Folding Bar (Stoll) Method
Method 5302, Abrasion Resistance of Cloth; Inflated Diaphragm (Stoll) Method
Method 5304, Abrasion Resistance of Cloth; Oscillatory Cylinder (Wyznebeek) Method
Method 5306, Abrasion Resistance of Cloth; Rotary Platform, Double-Head (Taber)
Method 5504, Water Resistance of Coated Cloth; Spray Absorption Method
Method 5512, Water Resistance of Coated Cloth; High Range, Hydrostatic Pressure
Method 5514, Water Resistance of Coated Cloth; Low Range, Hydrostatic Pressure
Method 5516, Water Resistance of Coated Cloth; Water Permeability, Hydrostatic Pressure
Method 5522, Water Resistance of Cloth; Water Impact Penetration Method
Method 5524, Water Resistance of Cloth; Rain Penetration Method
Method 5671, Colorfastness of Textile Materials to Weather; Accelerated Weathering
Method 5750, Mildew Resistance of Textile Materials; Single Culture Method
Method 5760, Mildew Resistance of Textile Materials; Mixed Culture Method
Method 5764, Insect Resistance of Textiles
Method 5850, Accelerated Aging of Cloth: Oven Method
Method 5870, Temperature, High; Effect on Cloth Flexibility
Method 5872, Temperature, High; Effect on Cloth Blocking
Method 5903.1, Flame Resistance of Cloth; Vertical
Method 5905.1, Flame Resistance of Material; High Heat Flux Flame Contact
Method 5908, Flammability, Burning Rate of Cloth, 45o
Method 5930 (for textile materials)

Federal Trade Commision Trade Regulation Rule, Care Labeling of Textile Wearing Apparel and Certain Piece Goods, January 2, 1974.

Military Specification, GGG-M-125d, Mask, Air Line and Respirator, Air Filtering, Industrial, July 20, 1969 (available from NIOSH, Certification and Quality Assurance Branch, 1095 Willowdale Road, Morgantown, WV 26505-2888)

Military Specification, MIL-M-36954C, Masks, Surgical, Disposable

Military Standard, MIL-STD-662, Ballistic Test for Armor

Military Standard, MIL-STD-810E, Environmental Test Methods and Engineering Guidelines

Industrial Safety Equipment Association (ISEA). 1901 N. Moore Street, Arlington, VA 22209, Phone 703-525-1695, Fax 703-528-2148.

ANSI/ISEA 101, American National Standard for Limited-Use and Disposable Coveralls - Size and Labeling Requirements, 1996

ANSI/ISEA 103, American National Standard for Chemical Protective Clothing Classification and Performance (proposed)
ANSI/ISEA 105, American National Standard for Hand Protection Selection Criteria (proposed)
ANSI/ISEA 107, American National Standard for High-Visibility Safety Apparel (proposed)
Use and Selection Guide for Non-Powered Air Purifying Particulate Respirators, 1996

Institute of Environmental Sciences (IES). 940 E. Northwest Highway, Mount Prospect, IL 60056, Phone 847-255-1561, Fax 847-255-1699.

IES-RP-CC003.2, Garments required in cleanrooms and controlled environments, 1993
IES-RP-CC005.2, Cleanroom gloves and finger cots, 1996

International Standards Organization (ISO). 1, rue de Varembé, Case postale 56, CH-1211 Genèva 20, Switzerland, Phone +41 22 749 0111, Fax +41 22 733 3430 (Copies of standards may be obtained by standards selling services in the United States).

ISO 105, Tests for colour fastness (methods A01 through Z09)
ISO 675, Textiles-Woven fabrics-Determination of dimensional change on commercial laundering near the boiling point, 1979
ISO 811, Textile fabrics-Determinatio of resistance to water penetration-Hydrostatic pressure test, 1981
ISO 1421, Fabrics coated with rubber or plastics-Determination of breaking strength and elongation at break, 1977
ISO 1511, Protective helmets for road users, 1970
ISO 2023, Rubber footwear-Lined industrial vulcanized-rubber boots-Specification, 1994
ISO 2024, Rubber footwear, lined conducting-Specification, 1981
ISO 2025, Lined industrial rubber boots with general purpose oil resistance, 1972
ISO 2251, Lined antistatic rubber footwear-Specification, 1991
ISO 2252, Rubber footwear, lined industrial, for use at low temperature, 1983
ISO 2286, Rubber-or plastics-coated fabrics-Determination of roll characteristics, 1986
ISO 2801, Clothing for protection against heat and flame - General recommendations for users and for those in charge of such users, 1973
ISO 2960, Textiles-Determination of bursting strength and bursting distension-Diaphragm method, 1974
ISO 3005, Textiles-Determination of dimensional change of fabrics induced by free-steam, 1978
ISO 3011, Rubber or plastics coated fabrics-Determination of resistance to ozone cracking under static conditions, 1981
ISO 3175, Textiles-Evaluation of stability to machine dry-cleaning, 1995
ISO 3303, Rubber-or plastics coated fabrics-Determination of bursting strength, 1990
ISO 3635, Size designation of clothes-Definitions and body measurement procedure, 1981
ISO 3758, Textiles-Care labelling code using symbols, 1991
ISO 3801, Textiles-Woven fabrics-Determination of mass per unit length and mass per unit area, 1977
ISO 3873, Industrial safety helmets, 1997
ISO 3998, Textiles-Determination of resistance to certain insect pests, 1977
ISO 4007, Personal eye-protectors - Vocabulary, 1977
ISO 4045, Leather-Determination of pH, 1977
ISO 4417, Size designation of clothes - Headwear, 1997
ISO 4418, Size designation of clothes - Gloves, 1978
ISO 4536, Metallic and non-organic coatings on metallic substrates-Saline droplets corrosion test (SD test), 1985
ISO 4643, Moulded plastics footwear-Lined or unlined poly (vinyl chloride) boots for general industrial use-Specification, 1992
ISO 4646, Rubber- or plastics-coated fabrics-Low temperature impact test, 1989
ISO 4674, Fabrics coated with rubber or plastics - Determination of tear resistance, 1977

ISO 4675, Rubber- or plastics-coated fabrics-Low-temperature bend test, 1990
ISO 4849, Personal eye-protectors - Specifications, 1981
ISO 4850, Personal eye-protectors for welding and related techniques - Filters - Utilization and transmission requirements, 1979
ISO 4851, Personal eye-protectors - Ultraviolet filters - Utilization and transmittance requirements, 1979
ISO 4852, Personal eye-protectors - Infrared filters - Utilization and transmittance requirements, 1978
ISO 4854, Personal eye-protectors - Optical test methods, 1981
ISO 4855, Personal eye-protectors - Non-optical test methods, 1981
ISO 4856, Personal eye-protectors - Synoptic tables of requirements for oculars and eye-protectors, 1982
ISO 4869-1, Acoustics - Hearing protectors - Part 1: Subjective method for measurement of sound attenuation, 1990
ISO 4869-2, Acoustics - Hearing protectors - Part 2: Estimation of effective A-weighted sound pressure levels when hearing protectors are worn, 1994
ISO/TR 4869-3, Acoustics - Hearing protectors - Part 3: Simplified method for the measurement of insertion loss of earmuff type protectors for quality inspection purposes, 1990
ISO 4920, Textiles-Determination of resistance to surface wetting (spray test) of fabrics, 1981
ISO 5077, Textiles-Determination of dimensional change in washing and drying, 1984
ISO 5081, Textiles-Woven fabrics-Determination of breaking strength and elongation (Strip method), 1977
ISO 5082, Textiles-Woven fabrics-Determination of breaking strength-Grab method, 1982
ISO 5084, Textiles-Determination of thickness of textiles and textile products, 1996
ISO 5085, Textiles-Determination of thermal resistance-Part 2; High thermal resistance, 1990
ISO 5423, Moulded plastics footwear-Lined or unlined polyurethane boots for general industrial use-Specification, 1992
ISO 5978, Rubber-or plastics-coated fabrics-Determination of blocking resistance
ISO 5979, Rubber or plastics coated fabrics-Determination of flexibility-Flat loop method, 1982
ISO 5981, Rubber or plastics coated fabrics-Determination of flex abrasion, 1982
ISO 6110, Moulded plastic footwear-Lined or unlined poly (vinyl chloride) industrial boots with chemical resistance-Specification, 1992
ISO 6111, Rubber footwear-Lined or unlined rubber industrial boots with chemical resistance, 1982
ISO 6112, Moulded plstics footwear-Lined or unlined poly(vinyl chloride) industrial boots with general-purpose resistance to animal fats and vegetable oils-Specification, 1992
ISO 6161, Personal eye protectors - Filters and eye-protectors agains laser radiation, 1981
ISO 6179, Vulcanized rubber sheet, and fabrics coated with vulcanized rubber-Determination of transmission rate of volatile liquids (Gravimetric technique), 1989
ISO 6330, Textiles-Domestic washing and drying procedures for textile testing, 1984
ISO 6529, Protective Clothing-Protection against liquid chemicals-Determination of resistance of air-impermeable materials to permeation by liquids, 1990
ISO 6530, Protective clothing-Protection against liquid chemicals-Determination of resistance of materials to penetration by liquids, 1990
ISO 6940, Textile fabrics-Burning behaviour-Determination of ease of ignition of vertically oriented specimens, 1984
ISO 6941, Textile fabrics-Burning behaviour-Measurement of flame spread properties of vertically oriented specimens, 1984
ISO 6942, Clothing for protection against heat and fire - Evaluation of thermal behavior of materials and material assemblies when exposed to a source of radiant heat, 1993
ISO 7232, Rubber or plastics footwear-Antistatic sandals, sabots and clogs, 1986
ISO 7771, Textiles-Determination of dimensional changes of fabrics induced by cold-water immersion, 1985
ISO 7854, Rubber- or plastics-coated fabrics-Determination of resistance to damage by flexing, 1995
ISO 8096, Rubber- or plastic-coated fabrics for water-resistant clothing - Specification, 1989
ISO 8782-1, Safety, protective and occupational footwear for professional use - Part 1: Requirements and test methods (proposed)
ISO 8782-2, Safety, protective and occupational footwear for professional use - Part 2: Specification for safety footwear (proposed)

ISO 8782-3, Safety, protective and occupational footwear for professional use - Part 3: Specification for protective footwear (proposed)
ISO 8782-4, Safety, protective and occupational footwear for professional use - Part 4: Specification for occupational footwear (proposed)
ISO 8194, Radiation protection-Clothing for protection against radioactive contamination-Design, selection, testing and use, 1987
ISO 9073-1, Textiles-Test methods for nonwovens-Part 1: Determination of mass per unit area, 1989
ISO 9073-2, Textiles-Test methods for nonwovens-Part 2: Determination of thickness, 1995
ISO 9073-3, Textiles-Test methods for nonwovens-Part 3: Determination of tensile strength and elongation, 1989
ISO 9073-4, Textiles-Test methods for nonwoevens-Part 4: Determination of tear resistance, 1989
ISO 9150, Protective clothing - Determination of behavior of materials on impact of small splashes of molten metal, 1988
ISO 9151, Protective clothing against heat and flame-Determiantion of heat transmission on exposure to flame, 1995
ISO 9185, Protective clothing-Assessment of resistance of amterials to molten metal splash, 1990
ISO 9865, Textiles-Determination of water repellency of fabrics by the Bundesmann rain-shower test, 1991
ISO 10078, Specification for materials intended for protective clothing designed to protect personnel against contact with liquid or solid chemical products (proposed)
ISO 10282, Single-use sterile surgical rubber gloves - Specification, 1994
ISO 10449.2, Hearing protectors - Safety requirements and testing - Ear muffs (proposed)
ISO 10452, Hearing protectors - Recommendations for selection, use, care, and maintenance - Guidance document (proposed)
ISO 10453, Hearing protectors - Safety requirements and testing - Ear plugs (proposed)
ISO 10923, Helmet-mounted earmuffs (proposed)
ISO 10993-1, Biological evaluation of medical devices-Part 1: Guidance on Selection of tests, 1992
ISO 10993-2, Biological evaluation of medical devices-Part 2: Animal welfare requirement, 1992
ISO 10993-3, Biological evaluation of medical devices-Part 3: Tests for genotoxicity, carcinogenicity and reproductive toxicity, 1992
ISO 10993-4, Biological evaluation of medical devices-Part 4: Selection of tests for interactions with blood, 1992
ISO 10993-5, Biological evaluation of medical devices-Part 5: Tests for cytotoxicity: in vitro methods, 1992
ISO 11092, Textiles-Phsiological effects-Measurement of thermal and water-vapour resistance under stteady-state conditions (sweating guarded-hotplate test), 1993
ISO 11138 (International steam sterilization and sterility assurance)
ISO 11193, Single-use rubber examination gloves - Specification, 1994
ISO 11393-1, Protective clothing for users of hand-held chain saws - Part 1: Test rig driven by a flywheel for testing resistance to cutting by a chain saw, 1998
ISO 11393-2, Protective clothing for users of hand-held chain saws - Part 2: Test methods and performance requirements for leg protection against cutting by a chain saw (proposed)
ISO 11393-3, Protective clothing for users of hand-held chain saws - Part 3: Test methods for footwear (proposed)
ISO 11393-4, Protective clothing for users of hand-held chain saws - Part 4: Test methods and performance requirements for chainsaw protective gloves (proposed)
ISO 11393-5, Protective clothing for users of hand-held chain saws - Part 5: Test methods and performance requirements for chainsaw protective gaiters (proposed)
ISO 11393-6, Protective clothing for users of hand-held chain saws - Part 6: Test methods and performance requirements for upper body protectors (proposed)
ISO 11610, Protective clothing - Vocabulary, 1998
ISO 11611, Clothing for protection against heat and flame - Test methods and performance requirements for protective clothing for use in welding and allied processes (proposed)
ISO 11612, Clothing for protection against heat and flame - Test methods and performance requirements for industrial protective clothing, 1998

ISO 11613, Protective clothing for firefighters - Laboratory test methods and performance requirements, 1998
ISO 11645, Leather - Heat stability of industrial-glove leather, 1993
ISO 12127, Clothing for protection against heat and flame - Determination of contact heat transmission through protective clothing or constituent materials, 1995
ISO 12084, Noise level dependent ear-muff - Specification (proposed)
ISO 12974-1, Textiles - Determination of abrasion resistance of fabrics by the Martindale method - Part 1: Martindale abrasion testing apparatus (proposed)
ISO 12974-2, Textiles - Determination of abrasion resistance of fabrics by the Martindale method - Part 2: Determination of specimen breakdown (proposed)
ISO 12974-3, Textiles - Determination of abrasion resistance of fabrics by the Martindale method - Part 3: Determination of mass loss (proposed)
ISO 12974-4, Textiles - Determination of abrasion resistance of fabrics by the Martindale method - Part 4: Assessment of appearance change (proposed)
ISO 13506, Protective clothing against heat and flame - Test method for complete garments - Prediction of burn injury using an instrumented manikin (proposed)
ISO 13688, Protective clothing - General requirements, 1998
ISO 13937, Textiles - Determination of Resistance to Tear for Woven Fabrics, 1997
ISO 13982-1, Chemical protective clothing - Protection against dust - Part 1: Requirements (proposed)
ISO 13982-2, Chemical protective clothing - Protection against dust - Part 2: Method of test (proposed)
ISO 13994, Clothing for protection against liquid chemicals - Determination of resistance of protective clothing materials to penetration by liquids under pressure, 1998
ISO 13995, Protective clothing - Mechanical properties - Determination of dynamic puncture-tear propagation (proposed)
ISO 13996, Protective clothing - Mechanical properties - Determination of resistance to puncture (proposed)
ISO 13997, Protective clothing - Mechanical properties - Determination of resistance to cutting by sharp objects (proposed)
ISO 13998, Protective clothing - Aprons, trousers, and vests for protection against cuts and stabs by hand knives (proposed)
ISO 13999-1, Protective clothing -Gloves and arm guards for protection against cuts and stabs by hand knives - Chain mail gloves and arm guards (proposed)
ISO 13999-2, Protective clothing -Gloves and arm guards for protection against cuts and stabs by hand knives - Gloves made of other material than mail (proposed)
ISO 13999-3, Protective clothing -Gloves and arm guards for protection against cuts and stabs by hand knives - Impact cut test for fabric, leather and other materials (proposed)
ISO 14160 (International liquid sterilization and sterility assurance)
ISO 14460, Protective clothing against fire for automobile racing drivers - Performance requirements and test methods (proposed)
ISO 14876-1, Body armour - Part 1: Bullet resistant vests (proposed)
ISO 14876-2, Body armour - Part 2: Stab resistant vests (proposed)
ISO 14877, Protective clothing for abrasive blasting operations using granular abrasives (proposed)
ISO 15025, Clothing for protection against heat and flame - Test methods for limited flame spread materials (proposed)
ISO 15383, Protective gloves for firefighters - Laboratory test methods and performance requirements (proposed)
ISO 15384, Protective clothing for firefighters - Requirements and test methods for wildland firefighting (proposed)
ISO 15538, Protective clothing for firefighters - Test methods and requirements for reflective clothing for specialized firefighting (proposed)

National Fire Protection Association (NFPA). 1 Batterymarch Park, Quincy, MA 02269, Phone 617-770-3000, Fax 617-7730-3500.

NFPA 99, Standard on Heath Care Facilities, 1998
NFPA 701, Standard Methods of Fire Tests for Flame-Resistant Textiles and Films, 1996
NFPA 702, Standard for Classification of the Flammability of Wearing Apparel, 1980 (discontinued)
NFPA 1951, Standard on Protective Ensemble for Technical Rescue Operations (proposed)
NFPA 1971, Standard on Protective Ensemble for Structural Fire Fighting, 1997
NFPA 1976, Standard on Protective Clothing for Proximity Fire Fighting, 1992
NFPA 1977, Standard on Protective Clothing and Equipment for Wildland Fire Fighting, 1998
NFPA 1981, Standard on Open-Circuit, Self-Contained Breathing Apparatus (SCBA), 1997
NFPA 1982, Standard on Personal Alarm Safety Systems (PASS), 1998
NFPA 1983, Standard on Fire Service Safety Rope and System Components, 1995
NFPA 1991, Standard on Vapor-Protective Suits for Hazardous Chemical Emergencies, 1994
NFPA 1992, Standard on Liquid Splash-Protective Suits for Hazardous Chemical Emergencies, 1994
NFPA 1993, Standard on Support Function Protective Garments for Hazardous Chemical Operations, 1994
NFPA 1999, Standard on Protective Clothing for Emergency Medical Operations, 1997

National Institute for Occupational Safety and Health (NIOSH). Publications Dissemination, 4676 Columbia Parkway, Cincinnati, OH 45226-1998, Phone 800-356-4674, Fax 513-533-8573.

Guide to Industrial Respiratory Protection (DHHS/NIOSH Publication No. 87-116, 1987)
Guide to the Selection and Use of Particulate Respirators (DHHS/NIOSH Publication No. 96-101, 1996)

National Institute of Justice (NIJ). Office of Science and Technology, 810th Street, NW, Washington, DC 20531, Phone 202-305-0645, Fax 202-307-9907.

National Institute of Justice Standard 0101.01, Ballistic Resistance of Soft Body Armor
National Institute of Justice Standard 0108.01, Ballistic Resistance of Protective Material

SFI Foundation Inc. 15708 Pomerado Road, Suite N208, Poway, CA 92064, Phone 619-451-8868, Fax 619-451-9268.

SFI Technical Bulletin 3.2, Fire Protection Material, 1998
SFI Specification 3.2A, Driver Suits, 1998
SFI Specification 3.3, Driver Accessories, 1997

Glossary

AAMI—Association for the Advancement of Medical Instrumentation; an organization in the United States which writes standards on medical devices such as surgical gowns and drapes.

AATCC—American Association of Textile Chemists and Colorists; an organization in the United States which writes standards on various textile properties.

Abandonment suit—an immersion suit designed to permit rapid donning in the event of an imminent and an unintended immersion in cold water.

Abrasion resistance—the ability of PPE to resist damage from rubbing on a rough surface.

Accelerated aging—the use of controlled environmental conditions to promote rapid physical or chemical changes in a PPE item or material.

Acclimization—the ability of a person to adapt to the temperature and humidity conditions of a new working environment.

ACGIH—American Conference of Governmental Industrial Hygienists; an organization which sets threshold limit values for a number of hazardous substances.

Active hearing protection device—a type of hearing protector which uses electronic filters to reduce the levels of harmful noise levels while allowing the wearer to hear normal conversation.

Administrative controls—Procedures or practices used in a workplace to reduce employee contact with hazards.

Aerosol—suspensions of solid, liquid, or solid and liquid particles in a gaseous medium having a negligible falling velocity (generally considered to be less than 0.25 m/s).

Afterflame time—the length of time for which a material continues to flame after the ignition source has been removed.

Afterglow—a glow in a material after the removal of an external ignition source or after the cessation (natural or induced) of flaming of the material.

Air permeability—the volume of air which passes through an area of material per unit time under a defined pressure difference.

Airborne pathogens—pathogens which can be transmitted through air and are harmful through inhalation (e.g., tuberculosis).

Air flow resistance—the resistance of a PPE (usually a respirator) to the flow of air through it.

Airline respirator—a type of respirator which is supplied with breathing quality air from a remote source such as a bank of air cylinders or a compressor

Airline suit—a full body suit in which the source of respirable air is provided by a hose or airline which is supplied with breathing quality air from a remote source such as a bank of air cylinders or a compressor

Air-purifying respirator (APR) —a type of respirator which relies on filtering media to remove contaminants from the ambient air.

Alpha Particles—a form of ionizing radiation; doubly ionized nuclei of helium.

Anchorage—a part of a fall arrest system which is used for supporting the other fall arrest system components.

Anesthesia—contamination which causes loss of feeling or sensation.

Anisotropic—having different values for a specific property in different directions of a flat material.

ANSI—American National Standards Institute; the official standards body for the United States. Many U.S. standards from other standards organization are endorsed by ANSI.

Antifragmentation clothing—protective clothing design to provide protection to the wearer from high energy flying debris resulting from explosions.

Antimicrobial resistance—the ability of a material to prevent the growth of microorganism or to inactivate microorganisms which contact the material.

Apron a garment covering the front of the body from the chest to the legs.

Arc thermal performance exposure value (AVTPV) —in electrical testing of materials, the measurement of insulation provided by a material when exposed to an electric arc, expressed as the amount of arc energy required to cause a second degree burn.

Arm guard—a protective device covering the forearm. It may be permanently attached to or held in place by a glove with a special cuff while both are used.

Aseptic—sterile, free from viable microbiological contamination.

Assay—analysis of a biological mixture to determine the presence or concentration of a particular microorganism.

Asphyxiation—contamination that interferes with the supply or utilization of oxygen in the body.

Assay fluid—a sterile liquid used to wash the test specimen surface to determine microbiological penetration.

Assigned protection factor—a protection factor which has been determined to represent a specific class of respirators based on extensive testing.

ASTM—American Society for Testing and Materials; an organization which writes consensus standards.

Atmosphere-supplying respirator—a respirator which provides an independent source or respirable air (e.g., a supplied-air respirator or self-contained breathing apparatus)

Back belt—a harness or belt-like device intended to provide support to the wearer's back and eliminate strain.

Bacteria—a unicellular microorganism, existing as a free-living organism or a parasite.

Bacterial filtration efficiency (BFE)—the ability of a PPE material to prevent the passage of bacteria.

Bacteriophage—a type of virus which infects bacteria; often used as a surrogate for measuring viral penetration.

Ballistic resistance—the ability of PPE or a material to absorb the energy of a projectile.

Beta particles—a form of ionizing radiation; negatively charged particles identical to electrons with a wide range of energies.

Biocompatibility—refers to the compatibility of PPE materials with skin, blood, or internal tissue such that contact causes no ill effects.

Biogenic toxin—a naturally occurring substance which can cause acute toxic disease in addition to long-term reproductive and carcinogenic effects.

Biological fluid resistance—the ability of a PPE material to prevent the penetration of blood and related body fluids.

Biological fluid penetration resistance—the ability of a PPE material to totally prevent the penetration of blood and related body fluids.

Biological hazards—hazards arising from contact with biological agents, e.g., exposure to bloodborne pathogens.

Biosafety level—the combination of laboratory practices and techniques, safety equipment, and laboratory facilities for providing adequate protection against specific type of biological hazards.

Blocking—undesired adhesion between touching layers of material, such as might occur under moderate pressure, during storage or use.

Bloodborne pathogen—a pathogen which is transmitted by blood-to-blood exposure, e.g, hepatitis virus and human immunodeficiency virus (HIV).

BNQ—Bureau of Normalization for Quebec; an organization within the Province of Quebec which writes standards.

Body fluid simulant—a liquid which is used to act as a model for human body liquids.

Body heat balance—the sum of all heat gains and losses by the body from the body's metabolism, convection, radiation, evaporation, and conduction.

Boot—an item of footwear that covers the wearer's feet and ankles, and can extend up part of the wearer's lower legs.

Boot cover—an item of footwear that fits over the wearer's boots; generally made of relatively soft textile, rubber, or plastic materials.

Bootie—a sock-like extension of the garment leg designed to protect the wearer's feet when worn with other footwear.

Bouffant—a partial body garment which covers the top of the wearer's head and is secured by elastic.

Brannock scale—a standardized scale in the United States for measuring the wearer's foot size.

Breakdown voltage—in electrical insulating testing of PPE, the voltage where current leakage occurs.

Break-open—in testing thermal protective material, a response evidenced by the formation of a hole in the material which allows the heat, not liquid or a molten substance to pass through the material.

Breaking strength—the force required to break PPE materials or components when the item is pulled in opposite directions.

Breakthrough (detection) time—the elapsed time measured from the start of a permeation test to the time that chemical is first detected.

Breathability—the ability of a PPE material or item to pass air and moisture vapor

Breathing bag—a device which is part of a respirator (usually a closed-circuit self-contained breathing apparatus) which is used to providing mixing of air.

Breathing hose or tube (low pressure—a flexible hose (for instance corrugated hose) connected to the face-piece or hood through which air or oxygen enters at atmospheric pressure or at a pressure slightly above atmospheric pressure.

Breathing resistance—the resistance experienced by an individual during inhalation and exhalation while wearing a respirator.

Bridge—the part of the safety glasses which secure the device on the nose of the wearer.

Burn distance—the measurement from the bottom edge of the specimen to the farthest point that shows evidence of damage due to combustion.

Burning behavior—all the changes that take place when materials or products are exposed to a specified ignition source.

Burst strength—the force or pressure required to rupture PPE materials or components when the force is applied perpendicularly to the item.

Calorimeter—a device usually constructed of a conductive metal used as a heat sink in measuring the insulation of PPE materials exposed to thermal energy; calorimeters are intended to simulate the response of skin to thermal energy.

Canal caps—a type of hearing protector which fits over the wearer's ears with a headband that goes over or behind the wearer's head.

Canister—a part of an air-purifying respirator (gas mask) which contains filter medium or elements.

Carcinogen—a cancer-causing agent.

Care—procedures for cleaning, sterilization, decontamination, and storage of PPE.

Cartridge—a part of an air-purifying respirator which contains filter medium or elements.

Caution—a signal word used to indicate a potentially hazardous situation which, if not avoided, may result in minor or moderate injury or to alert the user against unsafe practices.

CEN—European Committee on Standardization; an organization representing countries of the European Economic Community which write standards.

Certification organization—an organization completely independent of the PPE manufacturer which tests PPE to specific standards and provides listing of the product in its list of certified products.

CGSB—Canadian General Standards Board; an organization within Canada which writes standards.

Char length—in flame resistance testing of materials, the damaged length of material as determined by specimen tearing when a weight is applied.

Charring—in the thermal testing of materials, the formation of carbonaceous residue on a material as the result of pyrolysis or incomplete combustion.

Chaps—partial pants which are open at the sides or back and intended to provide the wearer front leg protection.

Chemical cartridge respirator—a type of air-purifying respirator which uses chemical cartridges mounted directly on the facepiece.

Chemical degradation resistance—the ability of a PPE item or material to resist reduction of properties as the result of chemical exposure.

Chemical flash fire—the sudden ignition of a chemical-containing atmosphere resulting in a fireball which produces extreme radiant heat.

Chemical hazards—hazards arising from contact from chemicals, e.g., inhalation, injection, or skin absorption

Chemical liquid penetration resistance—the ability of a PPE item or material to resist penetration by liquid chemicals..

Chemical particulate penetration resistance—the ability of a PPE item or material to resist penetration by chemical particulates.

Chemical permeation resistance—the ability of a PPE material to resist permeation by chemicals.

Chemical vapor penetration resistance—the ability of a PPE item or material to resist penetration by chemicals vapors or gases.

Chin strap—a strap used to secure the headwear on the wearer's head by passing under the chin.

Clarity (of vision) —the ability of an individual to see clearly through that part of the PPE intended to protect the individual's face and eyes.

Cleaning—the removal of non-hazardous surface contamination such as dirt, dust, grease, etc.

Cleanroom—refers to a room in which release of particulates and/or chemicals is carefully controlled to avoid contamination of the working environment.

Clo—a unit of thermal resistance (insulation) equal to 0.155 K-m^2/W.

Closed-circuit—refers to a type of self-contained breathing apparatus which removes carbon dioxide from the exhaled air and restores the oxygen content of the air.

Closure—a device to close openings for the donning of protective clothing; e.g., zipper, hook and loop fastener, etc.

Closure system—a method of fastening openings in the garment including combinations of more than one method of achieving a secure closure, e.g. a slide fastener covered by an overflap fastened down with a touch and close fastener. This term does not cover seams.

Clute cut—a style of glove design in which the front of the glove liner is made from one piece, while the back is sewn from a number of parts.

Coated fabric—a flexible material composed of a textile fabric and an adherent polymeric material applied to one or both surfaces.

Cold temperature resistance—the ability of PPE to resist effects or damage by exposure to cold temperatures (usually below freezing).

Collection medium—in chemical permeation resistance testing, a liquid or gas that does not affect the measured permeation and in which the test chemical is freely soluble or adsorbed to a saturation concentration greater than 0.5 weight or volume percent.

Colorfastness—the resistance of a material to change in any of its color characteristics, to transfer of its colorant(s) to adjacent materials, or both, as the result of exposure to a specific set of environmental conditions or specific process.

Combustion—a chemical process of oxidation that occurs at a rate fast enough to produce heat and usually light either as glow or flames.

Compression resistance—the ability of a PPE item or material to resist deformation when a force is applied or a heavy object is dropped on it.

Compression set—the ease of permanently changing a material's thickness.

Conduction—heat transfer as the result of direct contact of the individual's body with a hot surface.

Conductive heat resistance—the ability of a PPE item or material to resist heat transfer by conduction.

Contamination—undesirable substances present in the workplace on items of equipment.

Continuous-flow—refers to a type of atmosphere-supplying respirator which maintains a flow of respirable air into the respirator at all times.

Convection—heat transmission through the movement and density of surrounding gases or liquids.

Convective heat resistance—the ability of a PPE item or material to resist heat transfer by convection.

Corrosion resistance—the ability of PPE to resist corroding.

Corrosivity—a property of a substance which can cause other substances to dissolve or become corroded.

Coverall—a one-piece full-body garment.

Crown straps—the part of headwear suspension that passes over the wearer's head.

Cuff—the part of the trousers which covers the ankles or the part of a garment of glove which covers the wrist.

Cumulative permeation—in chemical permeation resistance testing, the total mass of chemical that permeates during a specified time from when the material is first contacted.

Cumulative trauma disorder (CTD)—wear and tear on the joints and surrounding tissues because of overuse; includes specific conditions such as tendinitis and bursitis.

Curie—the amount of radioactivity that decays at the same rate of 1 gram of Radium 226.

Current leakage—in electrical insulation testing of PPE, the passage of current through a PPE item or material.

Cut resistance—the ability of a PPE component or material to withstand forces from a sharp blade.

Cytotoxicity—the measurement of PPE biocompatibility by determining the effects of PPE extracts on cell tissue.

Danger—a signal word used to indicate an imminently hazardous situation which, if not avoided, may result in death or serious injury.

Decibel—a unit of measuring sound energy.

Decontamination—the removal of a contaminant or contaminants from the surface or matrix, or both, of protective clothing to the extent necessary for its next intended action (for example, reuse or disposal).

Degradation—reduction of PPE item or material properties as the result of exposure to a specific environment.

Demand-type—refers to a type of atmosphere-supplying respirator which uses an air-regulating valve to permit the flow of respirable air to the facepiece only when the wearer inhales and creates a negative pressure inside the facepiece.

Dermatitis—a localized inflammatory reaction following a single or repeated exposure to a chemical or biological agent.

Design criteria—requirements related to the physical design of PPE.

Dexterity—manipulative ability to perform a task.

Dimensional change—a generic term for changes in length and width of a fabric specimen subjected to specific conditions.

Disposable (PPE) —PPE that is intended for a single use.

Doffing—the act of taking off PPE.

Donning—the act of putting on PPE.

Dripping—in thermal protection testing, a response evidence by flowing of the fiber polymer.

Durability—the ability of PPE to retain original performance properties during and after use.

Dust, fume, or mist respirator—a type of air-purifying respirator intended to provide protection from particulates.

Dusts—a type of aerosol in which solids are dispersed into air by mechanical means.

Ear muffs—a hearing protector which consists of a cup which covers each ear and that uses a headband for holding the cups onto the wearer's head.

Ear plugs—a hearing protector which directly inserts into the wearer's ear canals.

Ear protector—a type of PPE used to protect the wearer's hearing from excessive noise.

Ease—in garment construction, the difference between garment measurement and body measurement.

Elastomer—a term often used for rubber and polymers that have properties similar to rubber.

Electric arc—electrical energy passed between two electrodes in ionized gases or vapors producing a high amount of energy in a very short period of time.

Electric shock—contact of the body with a voltage source causing sensation or damage to the body.

Electrostatic decay—the dissipation of static charge on a material surface.

Elongation—the ratio of the extension of a material to the length of the material prior to stretching.

Embrittlement—in thermal protection testing, the formation of a brittle residue as a result of pyrolysis or incomplete combustion.

Encapsulating suit—a full-body garment which completely encloses the wearer.

End-of-service life indicator (ESLI)—a device on a respirator which provides the wearer with an indication for how much service time or respirable air is left (e.g., a gauge).

Endotoxin—part of the cell envelope of Gram-negative bacteria which primarily affect the lungs causing difficulty in breathing, coughing, and fever.

Engineering controls—changes to the design of a task or process which prevents exposure of employees to workplace hazards.

Ensemble—the combination of PPE items to provide complete protection to the wearer.

Environmental hazards—hazards created by the ambient environment (e.g., high heat and humidity, cold, wind, wetness).

Escape—refers to respirators which provide a limited supply of respirable air for escape purposes only.

Evaporative heat loss—body heat loss manifested as sweating which results from the difference in water vapor pressure of a person's skin and the relative humidity and air velocity of the working environment.

Exhalation valve—in a respirator or encapsulating suit, the valve through which exhaled breathing gases pass.

Exhalation valve leakage—the amount of leakage that occurs through an exhalation valve when a vacuum is applied to its interior side.

Explosive—materials which are capable of detonation or the rapid release of destructive energy.

Face and eyewear—protective devices intended to provide protection to the wearer's eyes and face.

Facepiece—the part of a respirator which provides a tight-fitting enclosure on the wearer's face.

Faceshield—face and eyewear intended to shield the wearer's face or portion of the face in addition to the eyes.

Fall arrest energy absorber—a component of a fall arrest system whose primary function is dissipate energy and limit deceleration forces which the system imposes on the body during fall arrest.

Fall protection device—a system which may consist of lifelines, ropes, straps, body harnesses, and related hardware designed to arrest a person's fall from an elevated surface.

Field of vision—the extent of peripheral vision possible for a person while wearing PPE.

Field trial—an evaluation of PPE conducted in the field to evaluate the effectiveness or acceptability of specific PPE.

Finger cots—apparel items which fit around the finger tips of the wearer's hands.

Finger guards—apparel items which fit around the fingers of the wearer's hands.

Fingerless gloves—gloves in which the wearer's fingertips are not covered.

Finish—a chemical or mechanical modification, or both, of a protective clothing fabric for a specific performance result.

Fit—the quality, state or manner in which the length and closeness of PPE, when worn, relates to the human body.

Fit testing—in the evaluation of respirators, the measurement of overall respirator or facepiece effectiveness in preventing the inward leakage of outside contaminants; fit testing may be either qualitative or quantitative.

Five finger glove—any glove covering both the back and palm of the hand and wrist, and having separate individual fingers and thumb.

Flame impingement—direct contact between a flame and a material.

Flame resistance—the ability of a PPE item or material to resist damage or burning from flame contact.

Flame-resistant—referring to a material which resists ignition or burning when contacted by flame; may include materials which are intrinsically flame-resistant or materials treated chemically to be flame-retardant.

Flammability—those characteristics of a material that pertain to its ignition and support of combustion.

Flex fatigue resistance—the ability of PPE items or materials to resist wear or other damage when repeatedly flexed.

Fluorescent material—a material that emits optical radiation at wavelengths longer than absorbed.

Footwear—items intended to be worn on the foot such as shoes, boots, or booties, or over existing footwear such as shoe or boot covers.

Full body garment—a garment which covers the wearer's torso, arms, and legs, and may cover the wearer's head when a hood is attached, feet when footwear is attached, and hands when gloves are attached.

Full body harness—a series of straps which are fastened around the wearer in such a manner as to contain the torso; used whenever the danger for free fall exists.

Full facepiece—a facepiece which covers the wearer's entire face.

Fumes—aerosols which are solids dispersed in the air by condensation.

Fungus—either a unicellular yeast or branching filaments of cells in molds.

Gaiter—for chain saw cut resistance, a foot protective device worn outside the footwear.

Gas mask—a type of air-purifying respirator which includes a canister attached either directly to a facepiece or via a breathing hose.

Gauntlet—an extension of a glove up the wearer's arm.

Gamma rays—a form of ionizing radiation; electromagnetic radiation of extremely short wavelengths and intensely high energy.

Garment—a single item of clothing (for example, shirt).

Gas-tight—referring to PPE where no penetration of contaminants is detected or where the PPE shows no leakage when pressurized.

Gas-tight integrity—the resistance of PPE to the penetration or inward leakage by gases and vapors.

Glow—visible, flameless combustion of the solid phase of a material.

Goggles—face and eyewear which are intended to fit over the face surrounding the eyes in order to shield the eyes from certain hazards.

Gown—full-length partial body garments which provide protection to the wearer's front torso, upper legs, and arms.

Grip—the ability of an individual to grasp an object (while wearing PPE).

Gunn cut—a glove design in which the back of the glove liner is made from one piece, the second and third fingers are set in with a seam along the base of the fingers, and the seams in the finger areas extend two-thirds of the way around each finger.

Half-mask—a respirator facepiece that covers the wearer's nose and mouth which has a lower sealing surface resting under the chin.

Hand function—the ability of a person to use their hands through dexterity, tactility, and grip.

Hand pads—apparel items which are shaped like an envelope in which the wearer inserts his or her hands.

Handshield—a special type of face and eyewear held in front the wearer's face intended to provide protection to the wearer's eyes from the non-ionizing radiation and thermal hazards associated with welding.

Handwear—gloves, mittens, and other apparel items which cover the all or part of the wearer's hands.

Hazard assessment—the identification of hazards in a task or workplace setting with the determination of those areas of the body which will be affected and the likelihood of exposure.

Hazardous chemical—any solid, liquid, gas, or mixture thereof that can potentially cause harm to the human body through inhalation, ingestion, or skin absorption.

Head cover—a partial body garment which covers the top of the wearer's heat or an item of headwear.

Headband—bands in the interior of a helmet shell which allow adjustment of the headwear to fit the wearer's head; also describes part of face and eyewear which can be used to secure the item onto the wearer's head; also describes the part of a hearing protection which can be used to secure the item onto the wearer's head.

Headgear—the part of a faceshield which suspends the faceshield on the wearer's head.

Headwear—items worn on the head, such as a helmet.

Hearing protector—an item of PPE used to reduce occupational noise levels to acceptable levels.

Heat cramps—muscle spasms or pain in the hands, feet, or abdomen caused by heavy sweating with inadequate fluid replacement.

Heat exhaustion—stress on body organs caused by inadequate blood circulation from dehydration usually manifested by pale, cool, moist skin, heavy sweating, dizziness, nausea, and fainting.

Heat flux—the thermal intensity indicated by the amount of energy transmitted per unit area and per unit time (cal/cm^{2}s) (watts/cm^2).

Heat rash—a rash resulting from continuous exposure to heat or humid air.

Heat stress—disorders related to the inability of the body to dissipate heat.

Heat stroke—a serious form of heat stress where temperature regulation in the body fails as signified by red, hot, dry skin, lack of reduced perspiration, nausea, dizziness, rapid pulse, and sometimes coma.

Heat transfer index—in testing the thermal insulation of protective clothing materials, the time to achieve a specific rise in temperature as measured by a calorimeter or thermocouple.

HEPA filter—a high efficiency particulate air filter; a filter which is capable of filtering a very high proportion of submicron sized particles.

Helmet—an item of headwear which is worn on the top of the head intended to offer at least physical protection to the top of the wearer's head.

Hemocompatibility—biocompatibility of PPE with blood components.

High-visibility clothing—warning clothing intended to provide conspicuousness at all time.

Hood—a one-piece garment designed to fit closely over the entire head and to extend downwards to cover the neck.

Hose mask—a type of supplied-air respirator which supplies respirable air through a large diameter flexible hose.

Human factors—aspects of PPE related to its use by or impact on the wearer.

ICRP—International Commission on Radiation Protection; an organization which set guidelines for protection from radiation.

IDLH—immediately dangerous to life and health; refers to atmospheres which produce physical discomfort immediately, chronic poisoning after exposure, or acute physiological symptoms after prolonged exposure.

IES—Institute for Environmental Sciences; an organization which represents manufacturers and users in cleanroom environments and also writes standards.

Immersion suit—a suit intended to protect the wearer from the effects of immersion in cold water.

Impact resistance—the ability of a PPE item to withstand the transmitted energy or acceleration associated with the controlled drop of a heavy object.

INDA—Association for the Nonwoven Fabrics Industry; a trade organization which represents manufacturers of nonwoven materials and also writes standards.

Infection control—use of PPE to prevent the transmission of disease from the health care provided to the patient.

Inhalation—the introduction of a substance into the body via the lungs by respiration.

Inhalation valve—in a respirator, the valve through which inhaled gases pass.

Ingestion—the introduction of a substance into the body via the digestive system by accidental or intentional eating.

Injection—the introduction of a substance into the body (usually the blood stream) by puncturing of the skin by a sharp contaminated object.

Insect resistance—the ability of PPE to repel insects.

Insole—a fixed or removable item on the inside bottom of the footwear intended to provide cushioning to the bottom of the wearer's feet.

Integrity—a determination as to how well PPE prevents the entry of penetrating substances.

Interface—areas where PPE items joint to provide protection to the wearer (e.g., glove to clothing sleeve end).

Intrusion coefficient—in testing the gas-tight integrity of PPE, the ratio of outside (challenge) concentration to the measured inside concentration.

Ionizing radiation—X-rays and other radiation emitted from radioactive materials.

Irritation—a property of a substance which causes inflammation, roughness, or soreness of the body.

ISEA—Industrial Safety Equipment Association; a trade organization representing manufacturers of PPE which also writes standards.

ISO—International Standards Organization; an international standards writing body.

Ladder shank—a protective device installed in the sole of footwear to prevent the sole from bending when the wearer steps on ladder rungs.

Lanyard—a flexible line of rope, wire rope, or straps that usually have connectors at each end for connecting the body support to an energy absorber, anchorage connector, or anchorage.

Laundering—the process of cleaning PPE using water and detergent, usually in a machine.

Leggings—partial body garments intended to protect the wearer's lower legs.

Legibility—the ability of an ordinary person to be able to read a product label.

Life cycle—the full use, care, and maintenance history of a PPE item.

Life cycle cost—the cost associated with the life cycle of a PPE item.

Lifeline—a flexible line of rope for connecting to an anchorage or anchorage connector at one or both ends of a span.

Lift-front—a feature of face and eyewear in which the protective feature of the face and eyewear can be lifted up from the wearer's face or eyes.

Light attenuation—the ability of PPE to reduce the harmful effects of incident light.

Limited use (of PPE)—PPE which is intended to be used for a limited number of occasions.

Liner—the interior part of a PPE item which contacts the wearer's skin when the PPE item is worn.

Liquid-tight—refers to the absence of detectable liquid penetration inside protective clothing when it is sprayed with the liquid at specified pressure, direction and duration.

Liquid-tight integrity—the resistance of PPE to penetration or inward leakage by liquids.

Loose-fitting respirator inlet cover—a PPE-like structure such as a hood or a helmet which provides a barrier from the hazardous atmosphere for the wearer and that connects the wearer's respiratory system with the respirator.

Maintenance—procedures for inspection, repair, and removal from service for PPE.

Mean median aerodynamic diameter (MMAD) —the average diameter of particles which are suspended in air.

Medical device—PPE which is used in a medical setting which is subject to regulation by the U.S. Food and Drug Administration.

Metatarsal footwear—footwear that is specifically designed to provide protection to the top of the wearer's foot between the ankle and toes.

Metatarsal plate—a plate installed in the top area of the footwear between the ankle and toe for physical (impact) protection.

Microorganism filtration efficiency—the ability of a PPE material to prevent the passage of microorganisms.

Microporous films—plastic layers used in protective clothing materials which exhibit a degree of breathability while being resistant to the penetration of liquids.

Mists—aerosols which are liquids dispersed in the air by condensation.

Mitten—any glove covering both the back and palm of the hand and wrist, and having a separate thumb and a common covering for the fingers.

Mobility—ease of movement while wearing PPE.

Molten substances—metals in their liquified, elevated temperature state, as well as related non-metallic substances also handled at elevated temperatures such as slag, dross, and salt. Excluded are liquid hot substances that may be associated with metal processing such as water, oil, and caustic solutions.

Mouthpiece—a part of a respirator which is designed to fit in the wearer's mouth to provide respirable air (usually found in only escape respirators).

Nape strap—a strap on headwear which passes behind the wearer's head to secure the headwear to the wearer's head.

NCRP—National Council on Radiation Protection and Measurement; an organization which set guidelines for protection from radiation.

NFPA—National Fire Protection Association; an organization which writes standards related to fire protection, general safety, and PPE.

Noise attenuation—the ability of PPE to reduce noise levels.

Noise reduction rating (NRR) —a measure of hearing protector performance which indicates the level of sound energy reduction in decibels which is accomplished by the item.

Non-ionizing radiation—electromagnetic radiation which does not change the structure of atoms of elements that it contacts.

Normalized breakthrough detection time—the elapsed time measured from the start of a permeation resistance test to the time when the chemical permeation rate is equal to a specific permeation rate.

Open-circuit—refers to a type of self-contained breathing apparatus which utilizes a portable compressed air cylinder to provide respirable air.

OSHA—Occupational Safety and Health Administration; an administration under the U.S. Department of Labor responsible for regulating the workplace for safety and health issues.

Overall—a garment which covers the legs and lower torso of the wearer and includes straps which go over the wearer's shoulders to suspend the garment on the wearer.

Overboot—an item of footwear which is worn over a boot; overboots typically do not provide physical protection.

Overshoe—an item of footwear which is worn over a shoe; overboots typically do not provide physical protection.

Oxygen-deficient—refers to an atmosphere which contains less than 19.5% oxygen.

Ozone resistance—the ability of a PPE or materials to withstand effects or damage by exposure to ozone.

Paint spray respirator—a type of air-purifying respirator intended to provide protection from particulates.

Partial body garment—a garment which covers only a portion of the wearer's body.

Partial gloves—gloves which do not cover the entire hand, usually open in the wrist or back of hand areas.

Particulate removal efficiency—the percentage of particles that is removed by a filter medium based on the number of particles in the incoming gas stream.

Particulate-tight integrity—the resistance of PPE to the penetration or inward leakage of particles.

Penetration—the process by which a substance (gas, liquid, or particulate) moves through porous materials, seams, pinholes, or other imperfections in a material on a non-molecular level.

Performance criteria—requirements related to specific PPE performance properties.

Performance property—a characteristic of PPE which demonstrates its protective qualities against a specific hazard.

Permeability index—in the thermal (comfort) insulation testing of PPE, a dimensionless number, from 0 to 1, where 0 represents no evaporative heat transfer for the material and 1 indicates that the clothing has achieved the theoretical maximum amount of heat transport.

Permeation—the process by which a chemical moves through a material on a molecular level. Permeation involves: (1) absorption of the molecules of the chemical into the contacted (outside) surface of a material; (2) diffusion of the absorbed molecules in the material, and; (3) desorption of the molecules from the opposite (inner) surface of the material.

Permeation rate—the mass of test chemical permeating the glove per unit time per unit area.

Permissible Exposure Limit (PEL) —the levels of exposure to a specific chemical in air set by OSHA which is believed to cause harmful effects on employees; PELs are normally expressed as a time-weighted average (over an 8-hour workday), and can also be a ceiling level, the maximum level to which an employee may be exposed.

Person-equipment hazards—hazards which arise from the wearing of PPE.

Person-position hazards—hazards which arise from the location or position of the person in the workplace.

Personal fall arrest system—an assembly of components and subsystems used to arrest a person in a fall from a working height.

Personal flotation device (PFD) —a vest or ring which can be used to maintain a person's buoyancy in water for preventing drowning.

Pesticide respirator—a type of air-purifying respirator intended to provide protection from particulates.

Phosphorescent materials—materials which absorb energy and reradiate light after the energy source has been removed.

Physical hazards—hazards which arise from contact with physical items or surfaces.

Physical properties—PPE or material properties related to strength, durability, or resistance to physical hazards

Polymer—a substance consisting of molecules characterized by repetition (neglecting ends, branches, junctions, and other minor irregularities) of one or more chemically bonded types of monomeric units.

Positioning belt or harness—a chest-waist harness used when no vertical free fall hazard is anticipated.

Powered air-purifying respirator (PAPR) —an air-purifying respirator which uses a portable or fixed-location blower to blow air through the air-purifying material or element.

PPE—personal protective equipment, protective clothing and equipment worn to provide protection to the wearer from specific hazards. PPE includes, but is not limited to, protective garments, protective gloves, protective footwear, protective headwear, protective face and eyewear, respirators, and hearing protectors.

Pressure-demand type—refers to a type of atmosphere-supplying respirator which uses a regulative valve to maintain respirable air at positive pressure inside the facepiece at all times.

Product label—label placed on or affixed to PPE for identifying the product and providing warnings, care instructions, and other information to the end user.

Protective clothing—clothing or apparels used to provide protection to the wearer from specific hazards.

Protection factor—in testing the gas-tight integrity of respirators, the ratio of outside (challenge) concentration to the measured inside concentration.

Puncture resistance—the ability of a PPE component or material to withstand penetration by a pointed object.

Puncture resistant device—a plate provided in footwear above the sole to protect the bottom of the wearer's feet from puncture hazards.

Qualitative fit testing—in the evaluation of respirators, the qualitative measurement of overall respirator or facepiece effectiveness in preventing the inward leakage of outside contaminants based on the test subject's ability to detect inward leakage.

Quantitative fit testing—in the evaluation of respirators, the quantitative measurement of overall respirator or facepiece effectiveness in preventing the inward leakage of outside contaminants based on actual measurement of contaminant inside the respirator.

Quarter-mask—a facepiece which covers the wearer's nose and mouth and which has a lower sealing surface which rests between the mouth and chin.

Rad—a radiation absorbed dose; the amount of radiation that results in the absorption of 100 ergs of energy by 1 gram of a material.

Radiant heat—heat communicated by energy propagated through space and transmitted by electromagnetic waves.

Radiant heat resistance—the ability of a PPE item or material to resist heat transfer by radiant heat.

Radiant protective performance (RPP) —the protective qualities of PPE material in providing insulation against radiant heat exposures.

Rate of flame spread—in flame resistance testing of materials, the time it takes for the flame to travel up to the top of the specimen.

RBE—relative biological effectiveness; the ratio of the absorbed dose of gamma radiation to the absorbed dose of the give radiation which gives the same biological effect.

Reactivity—a property of a substance in which the substance reacts with another substance to produce harmful effects.

Rem—the relative biological effectiveness multiplied by the dose indicating the amount of biological damage from radiation.

Repellency—the ability of a material to shed liquid that is applied to its surface.

Respirator—an item of PPE intended to protection the wearer from inhalation of harmful dusts, chemicals, and other respirable substances.

Respirator inlet cover—the part of the respirator which provides a barrier from the hazardous atmosphere for the wearer and that connects the wearer's respiratory system with the respirator.

Retroreflective material—a material that reflects light beams back to their point of origin.

Reusable (of PPE) —PPE that is intended for multiple uses.

Risk assessment—the process by which hazards are identified in the workplace and the risk associated with those hazards is determined as an approach to minimum employee exposure to those hazards by using administrative controls, engineering controls, or PPE.

Roentgen—the amount of radiation that produces sufficient ion pairs in a cubic centimeter of air to carry one electrostatic unit of electrical charge.

Safety glasses—eyewear which are intended to shield the wearer's eyes from certain hazards.

Salt spray resistance—the resistance to damage or corrosion of PPE when subjected to a salt spray of specified concentration, duration, and exposure.

Seam—a permanent fastening between two or more pieces of material.

Seam strength—the force required to break or separate a seam.

Selection—the process of determining what PPE is necessary for protection of employees during a specific operation or application from anticipated hazards or activity.

Self-contained breathing apparatus (SCBA)—an atmosphere-supplying respirator which provides an independent and transportable source of respirable air.

Self-extinguishing—a term used to indicate whether a PPE item or material will continue to burn once removed from the flame.

Sensitization—repeated exposure to a chemical or biological agent such that a lower level of exposure will cause a harmful effect or allergic reaction.

Service life—the useful history for an item of PPE.

Service time—the length of time that a respirator provide respirable air.

Serviceability—the ease and success with which PPE can be maintained for its original performance.

Shell—the main exterior part of a helmet, or the exterior material layer of a clothing item.

Shock absorption—the ability of PPE items to provide cushioning against vibration or related light impact.

Shoe—an item of footwear that covers the wearer's feet but not the ankles.

Shoe cover—an item of footwear that fits over the wearer's shoes generally made of relatively soft textile, rubber, or plastic materials.

Shrinkage—a decrease in one or more dimensions of an object or material.

Sideshields—the part of safety glasses which provides side protection to the wearer's eyes.

Skin absorption—the introduction of a substance into the body by contact with the skin and subsequent reaction, penetration or permeation.

Sleeve protector—partial body garment which provides protection to the wearer's arm.

Slip resistance—the ability of PPE items to maintain grip or traction with a surface.

Smock—partial body garment which provides the wearer with front torso, arm, and upper leg protection.

Splash suit—a protective garment intended to protection against splashes of liquid chemicals.

Snag resistance—the ability of a fabric to withstand individual fibers being pulled away from the base fabric.

Sole—the bottom of the footwear which contacts the walking surface.

Spats—partial body garments intended to protect the wearer's ankles and upper feet.

Speaking diaphragm—a device used in respirators for improving an individual's ability to communicate to others while wearing a facepiece.

Specifications—a description of PPE in terms of its construction, design, and performance.

Spectacles—see safety glasses.

Sprays—aerosols which are liquids dispersed in the air by mechanical means.

Static charge accumulation resistance—the ability of PPE or material to resist the build-up of static charge from rubbing action.

Sterile—free from viable microorganisms.

Sterilization—the process by which PPE is rendered free of biological contamination.

Stiffness—resistance of a PPE material or component to flexing.

Supplied-air respirator (SAR)—an atmosphere-supplying respirator which provides a source of respirable air that is stationary and removed from the wearer.

Surface Resistivity—in electrical testing of materials, the resistance in ohms as determined by using specified electrodes placed on the surface of the material.

Suspension—the part of headwear which allows the headwear to rest on the wearer's head.

Synthetic blood—a mixture of a red dye/surfactant, thickening agent, and distilled water having a surface tension and viscosity, and a red color representative of blood and other body fluids; the red color makes visible detection easier.

Systemic Poison—a contaminant which produces injury to specific organs or specific parts of the body.

Tactility—the ability of a person to sense through touch variations in surfaces of objects.

Tear resistance—the ability of a material to withstand forces which continue a tear once initiated.

Temples—part of the safety glasses which secure the device on the wearer's ears.

Thermal insulation—the resistance to dry heat transfer via conduction, convection, and radiation.

Thermal protective performance (TPP) —the protective qualities of PPE material in providing insulation against combined radiant and convective heat exposures.

Thermal resistance (insulation) —a property of PPE material which defines the amount of the "sensible" heat flow (consisting of parts of conduction, convection and radiation) in a given field of temperature as a result of a temperature gradient between the outside environment and the interior of the PPE item.

Threshold Limit Value (TLV)—levels of exposure to specific hazards which are believed to cause harmful effects on employees; TLVs may be expressed as a time-weighted average (over an 8-hour workday), short term exposure limit for a 15 minute or similar short period exposure, or as a ceiling level, the maximum level to which an employee may be exposed.

Third party laboratory—a laboratory which is completely independent of the PPE manufacturer.

Tight-fitting respirator inlet cover—a facepiece which provides a gas- or particle-tight seal on the wearer's face.

Time-to-burn—in the thermal testing of PPE materials, the elapsed time between initial exposure and time that a sufficient amount of energy has passed through the material and been absorbed by the sensor to cause a second degree burn.

Time-to-pain—in the thermal testing of PPE materials, the elapsed time between initial exposure and time that a sufficient amount of energy has passed through the material and been absorbed by the sensor to cause the sensation of pain.

Toe caps—footwear which is worn over the wearer's toes for physical protection only.

Tolerance time—in the thermal testing of PPE materials, the elapsed time between initial exposure and time that a sufficient amount of energy has passed through the material and been absorbed by the sensor to cause the sensation of pain or damage to the skin.

Top line—the very top of the footwear item.

Total heat loss—in the thermal (comfort) insulation testing of PPE, the measurement of conductive and evaporative heat loss through a material system in a manner simulating use.

Toxicity—the propensity of a substance to produce adverse biochemical or physiological effects.

Traction—the ability of the footwear sole to grip a walking surface.

Triboelectric—referring to the generation of static charge on a material surface by rubbing.

Universal precautions—in preventing exposure to biological hazards, the practice of assuming that all blood and body fluids are contaminated with harmful pathogens.

UV—ultraviolet light.

UV-visible light resistance—the ability of PPE to withstand the effects or damage by ultraviolet or visible light.

Vest—an upper torso garment usually used for physical protection or for mounting high visibility materials.

Viral penetration resistance—the ability of a PPE material to prevent the penetration of viruses.

Virus—a submicroscopic pathogen consisting of a single nucleic acid enclosed by a protein coat, able to replicate only within a living cell.

Vortex tube—a tube used in a full body suit which cools the air as it passes through.

Warning—a signal word used to indicate a potentially hazardous situation which, if not avoided, may result in death or serious injury.

Warnings—information provided on product labels or in product instructions which indicate limitations of PPE or hazards for the use of PPE or its applications.

Warranty—an agreement between the PPE manufacturer and the purchaser usually related to workmanship, performance in use, or service life.

Water absorption resistance—the ability of a material to resist saturation by water when held in contact with water under specified conditions.

Water penetration resistance—the ability of a material to resist the passage of water through the it, as measured by the hydrostatic head supported by the material.

Water vapor permeability—the weight of water vapor transmitted through an area of material for a given period of time, under specified conditions of temperature and humidity.

Water vapor resistance—a property of clothing material which defines the amount of the flow of "latent" evaporative heat (consisting of parts of diffusion and convection) in a given partial water vapor pressure field. The evaporative heat flows through the material under the effect of a partial water vapor pressure gradient in a stationary condition at right angles to the plane. The measured water vapor resistance is in consequence a measure for the stationary ability of water vapor to permeate.

Welding helmet—a special type of face and eyewear worn on the wearer's head and intended to provide protection to the wearer's eyes from the non-ionizing radiation and thermal hazards associated with welding.

Work/rest cycles—alternating periods of work and rest for reducing the potential for heat stress.

X-rays—a form of ionizing radiation; electromagnetic radiation of extremely short wavelengths emitted as the result of electron transitions in the inner orbits of heavy atoms bombarded by cathode rays in a vacuum tube.

Index

I

Government Institutes Mini-Catalog

PC #	ENVIRONMENTAL TITLES	Pub Date	Price*
949	ABCs of Environmental Regulation, Second Edition	2002	$79
812	Achieving Environmental Excellence	2003	$79
898	Book of Lists for Regulated Hazardous Substances, Tenth Edition	2001	$99
846	Clean Water Handbook, Third Edition	2003	$125
822	Environmental Compliance Auditing & Management Systems (ECAMS)	2001	$1,395
897	EH&S CFR Training Requirements, Fifth Edition	2001	$125
4300	EH&S Daily Federal Register Notification Service (e-mail)	2003	$150
890	Environmental Biotreatment: Technologies for Air, Water, Soil, and Wastes	2002	$149
4045	Environmental, Health & Safety CFRs on CD-ROM, single issue	2003	$450
825	Environmental, Health and Safety Audits, 8th Edition	2001	$115
4900	Environmental Law Expert, CD-ROM	2003	$695
955	Environmental Law Handbook, Seventeenth Edition	2003	$99
688	Environmental Health & Safety Dictionary, Seventh Edition	2000	$95
956	Environmental Statutes, 2003 Edition	2003	$125
957	Fundamentals of Environmental Sampling	2003	$79
689	Fundamentals of Site Remediation	2000	$85
907	Hazardous Materials Transportation Training, Student's Manual	2002	$89
958	Information Technology Solutions for EH&S Professionals	2003	$115
823	Integrating Environmental, Health, and Safety Systems	2001	$495
588	International Environmental Auditing	1998	$179
819	ISO 14001: Positioning Your Organization for Environmental Success	2001	$115
936	Managing Your Hazardous Wastes	2002	$89
830	RCRA CFRs Made Easy	2002	$115
841	Risk Management Planning Handbook, Second Edition	2002	$115
816	Stormwater Discharge Management	2003	$89
946	Wastewater and Biosolids Treatment Technologies	2002	$129

PC #	SAFETY and HEALTH TITLES	Pub Date	Price*
697	Applied Statistics in Occupational Safety and Health	2000	$105
893	Beyond Safety Accountability	2001	$79
894	Building Successful Safety Teams	2001	$79
843	Emergency Preparedness for Facilities	2003	$125
904	Ergonomics: A Risk Manager's Guide	2002	$159
949	Excavation Safety	2003	$85
663	Forklift Safety, Second Edition	1999	$85
814	Fundamentals of Occupational Safety & Health, Third Edition	2003	$79
818	Job Hazard Analysis	2001	$79
888	Keys to Behavior Based Safety	2001	$85
662	Machine Guarding Handbook	1999	$75
838	Managing Chemical Safety	2002	$95
889	Managing Electrical Safety	2001	$69
4800	OSHA Compliance Master CD-ROM	2003	$375
668	Safety Made Easy, Second Edition	1999	$75
947	Safety Metrics	2003	$85
815	So, You're the Safety Director, Third Edition	2003	$89

Government Institutes
4 Research Place, Suite 200 • Rockville, MD 20850-3226
Tel. (301) 921-2323 • FAX (301) 921-0264
Email: giinfo@govinst.com • www.govinst.com

Please call our customer service department at (301) 921-2323 for a free publications catalog.

CFRs now available online. Call (301) 921-2323 for info.

*All prices are subject to change. Please call for current prices and availablity.

Government Institutes Order Form

4 Research Place, Suite 200 • Rockville, MD 20850-3226
Tel (301) 921-2323 • Fax (301) 921-0264
Internet: http://www.govinst.com • E-mail: giinfo@govinst.com

4 EASY WAYS TO ORDER

1. Tel: **(301) 921-2323**
Have your credit card ready when you call.

2. Fax: **(301) 921-0264**
Fax this completed order form with your company purchase order or credit card information.

3. Mail: **Government Institutes Division**
ABS Group Inc.
P.O. Box 846304
Dallas, TX 75284-6304 USA
Mail this completed order form with a check, company purchase order, or credit card information.

4. Online: Visit http://www.govinst.com

PAYMENT OPTIONS

❑ **Check** *(payable in US dollars to* ***ABS Group Inc. Government Institutes Division****)*

❑ **Purchase Order** *(This order form must be attached to your company P.O. Note: All International orders must be prepaid.)*

❑ **Credit Card** ☐ VISA ☐ MasterCard ☐ American Express

Exp. ___ /___

Credit Card No. ____________________

Signature ____________________

(Government Institutes' Federal I.D.# is 13-2695912)

CUSTOMER INFORMATION

Ship To: (Please attach your purchase order)

Name ____________________

GI Account # *(7 digits on mailing label)* ____________________

Company/Institution ____________________

Address ____________________
(Please supply street address for UPS shipping)
____

City ____________ State/Province ____________

Zip/Postal Code ____________ Country ____________

Tel ( ) ____________________

Fax ( ) ____________________

E-mail Address ____________________

Bill To: (if different from ship-to address)

Name ____________________

Title/Position ____________________

Company/Institution ____________________

Address ____________________
(Please supply street address for UPS shipping)
____

City ____________ State/Province ____________

Zip/Postal Code ____________ Country ____________

Tel ( ) ____________________

Fax ( ) ____________________

E-mail Address ____________________

Qty.	Product Code	Title	Price

Subtotal ________
MD Residents add 5% Sales Tax ________
Shipping and Handling (see box below) ________
Total Payment Enclosed ________

30 DAY MONEY-BACK GUARANTEE

If you're not completely satisfied with any product, return it undamaged within 30 days for a full and immediate refund on the price of the product.

SOURCE CODE: BP03

Shipping and Handling

Within U.S:
1-4 products: $6/product
5 or more: $4/product

Outside U.S:
Add $15 for each item (Global)

Sales Tax

Maryland 5%
Texas 8.25%
Virginia 4.5%